This book examines case studies of North American Quaternary mammalian evolution within the larger domain of modern evolutionary theory. The book presents previously unpublished studies of a variety of taxa (xenarthrans, rodents, carnivores, ungulates) examined over several temporal scales, from a few thousand years during the Holocene to millions of years of late Pliocene and Pleistocene time. Different organizational levels are represented, from mosaic population variation to a synopsis of Quaternary evolution of an entire order (Rodentia). In addition to specific case histories, the book includes purely theoretical and methodological contributions, for example, on the statistical recognition of stasis in the fossil record, new ways to calculate evolutionary rates, and the use of digital image analysis in the study of dental ontogeny. Perhaps the most important aspect of the studies reported in this book is that they span the time between the "ecological moment" and "deep time." Modern taxa can be traced back into the fossil record, and variation among extant taxa can be used as a control against which variation in the extinct ones can be understood.

The book will interest vertebrate paleontologists, modern ecologists concerned with the origin of biological diversity, and also evolutionists interested in the competing evolutionary models of punctuated equilibrium and phyletic gradualism.

T0215728

Morphological change in
Quaternary mammals of North America

Morphological change in Quaternary mammals of North America

Edited by

ROBERT A. MARTIN
Murray State University

ANTHONY D. BARNOSKY
University of California at Berkeley

CAMBRIDGE
UNIVERSITY PRESS

CAMBRIDGE UNIVERSITY PRESS
Cambridge, New York, Melbourne, Madrid, Cape Town, Singapore, São Paulo

Cambridge University Press
The Edinburgh Building, Cambridge CB2 2RU, UK

Published in the United States of America by Cambridge University Press, New York

www.cambridge.org
Information on this title: www.cambridge.org/9780521404501

First published 1993
This digitally printed first paperback version 2005

A catalogue record for this publication is available from the British Library

Library of Congress Cataloguing in Publication data

Morphological change in Quaternary mammals of North America / edited
    by Robert A. Martin, Anthony D. Barnosky.
        p.   cm.
    Includes index.
    ISBN 0-521-40450-9
        1. Mammals, Fossil – North America – Evolution.   2. Mammals –
North America – Evolution.   3. Paleontology – North America.
    4. Paleontology – Quaternary.   I. Martin, Robert A. (Robert Allen)
    II. Barnosky, Anthony D.
    QE881.M769   1993
    569'.097 – dc20   93-12136
                                                              CIP

ISBN-13 978-0-521-40450-1 hardback
ISBN-10 0-521-40450-9 hardback

ISBN-13 978-0-521-02081-7 paperback
ISBN-10 0-521-02081-6 paperback

Contents

vi   *Contents*

Contributors

Deborah K. Anderson
Division of Natural Science
St. Norbert College
De Pere, Wisconsin 54115

Anthony D. Barnosky
Department of Integrative
  Biology
University of California
Berkeley, California 94720

Jean Chaline
Centre de Paléontologie
  Analytique et Géologie
  Sédimentaire
URA CNRS 157
Laboratoire de Préhistoire et
  Paléoécologie du Quaternaire
  de l'EPHE
Centre des Sciences de la Terre
6 Bd. Gabriel
2100 Dijon
France

Andrew P. Czebieniak
Department of Geological
  Sciences
University of Texas at Austin
Austin, Texas 78712

Tamar Dayan
Department of Zoology
Tel Aviv University
Ramat Aviv, Tel Aviv 69978
Israel

Philip P. Gingerich
Museum of Paleontology
The University of Michigan
Ann Arbor, Michigan 48109–
  1079

H. Thomas Goodwin
Department of Natural Sciences
Loma Linda University
Loma Linda, California 92350

Richard C. Hulbert, Jr.
Department of Geology and
  Geography
Georgia Southern University
Statesboro, Georgia 30460–8149

Eric Le Boulengé
Unité de Biométrie
Université Catholique de
  Louvain
2 Place Croix du Sud
B-1348 Louvain-la-Neuve
Belgique

Adrian M. Lister
Department of Zoology
University of Cambridge
Cambridge CB2 3EJ
United Kingdom

Larry D. Martin
Museum of Natural History and
    Department of Systematics and
    Ecology
University of Kansas
Lawrence, Kansas 66045

Robert A. Martin
Department of Biological
    Sciences
Murray State University
Murray, Kentucky 42071

Gary S. Morgan
Florida Museum of Natural
    History
University of Florida
Gainesville, Florida 32611–2035

James R. Purdue
Zoology Section
Illinois State Museum
Corner Spring and Edwards
Springfield, Illinois 62706

Elizabeth J. Reitz
Zooarchaeology Laboratory
Museum of Natural History
Natural History Building
The University of Georgia
Athens, Georgia 30602

John M. Rensberger
Department of Geological
    Sciences and Burke Memorial
    Washington State Museum
DB-10
University of Washington
Seattle, Washington 98195

André Schaaf
Centre des Sciences de la Terre
URA CNRS 11
27–43 Bd du 11 November
69622 Villeurbanne cedex
France

Kevin Seymour
Department of Vertebrate
    Paleontology
Royal Ontario Museum
100 Queen's Park
Toronto, Ontario, Canada M5S
    2C6

Daniel Simberloff
Department of Biological
    Sciences
Florida State University
Tallahassee, Florida 32306–2043

Eitan Tchernov
Department of Evolution,
    Systematics and Ecology
The Hebrew University
Jerusalem 91904
Israel

Laurent Viriot
Centre de Paléontologie
    Analytique et Géologie
    Sédimentaire
URA CNRS 157
Laboratoire de Préhistoire et
    Paléoécologie du Quaternaire
    de l'EPHE
Centre des Sciences de la Terre
6 Bd. Gabriel
2100 Dijon
France

Richard J. Zakrzewski
Department of Geosciences and
    Sternberg Memorial Museum
Fort Hays State University
Hays, Kansas 67601–4099

Acknowledgments

We are most grateful to our editors at Cambridge University Press, Kathleen Zylan, Robin Smith, Eric Newman, and Camilla Palmer for agreeing to publish the results of our symposium and for their professional help in editing the manuscript.

Kathy Gann at Berry College typed the manuscripts and index on computer and incorporated changes as the editing progressed. Her astute professionalism and punctuality are greatly appreciated. Angela Vitale, Berry undergraduate biology major, helped in numerous ways with the symposium.

Our thanks also go to the administration at Berry College for sponsoring the meeting and providing accommodations for the participants.

1

Quaternary mammals and evolutionary theory: introductory remarks and historical perspective

ROBERT A. MARTIN AND ANTHONY D. BARNOSKY

The Quaternary period marks the transition from environments populated with fabled, extraordinary creatures such as saber-toothed cats, giant ground sloths, dire wolves, cave bears, woolly rhinos, and mammoths to the more familiar landscapes of modern time. Anatomically modern humans appeared during the late Quaternary. The Quaternary is composed of the Pleistocene epoch, or Ice Ages, which lasted from about 1.7 Ma to 10 ka, and the Holocene epoch, or present interglacial, which begins where the Pleistocene ends and extends to the present. The Quaternary was characterized by dramatic geographic shifts of continental glaciers, accompanied by major transgressions and regressions of sea level. During the height of the last major North American glacial period, the Wisconsinan, much of the continental shelf was exposed and probably represented a vast lowland savannah. Fishermen have recovered proboscidean teeth and bones from the shelf at depths of more than 90 m (Whitmore et al., 1967). Within this dynamic context the geographic ranges of many species of animals and plants expanded and contracted, and a concomitant pattern of originations and extinctions developed.

To evolutionary biologists the Quaternary record represents a vast laboratory, full of natural experiments, in which hypotheses of evolutionary tempo and mode may be examined. The benefits of working in this laboratory are manifold. Modern species are often represented, and their past distributions in some cases often provide important, independent climatic signals in their own right, and in other cases can be tested against independent climate data to ascertain the relative importance of physical perturbations in causing evolutionary events. Critical infor-

1

mation on population variation is available for some extant taxa, and these data can be used in an experimental control function. Quaternary mammalian remains are more abundant than those from other intervals and can be radiocarbon dated on a time scale (hundreds to thousands of years) intermediate between ecological time (tens of years) and geological time (hundreds of thousands to millions of years). Additionally, Quaternary osteological and dental samples occasionally represent transitions from ancestral to extant character complexes. Despite these benefits, few North American evolutionary biologists have entered the Quaternary laboratory until recently.

This volume represents an attempt to rectify this deficiency in the data base of evolutionary theory. The book is not meant to minimize the efforts of many investigators who, through the years, have contributed to our knowledge of Quaternary North American mammals with detailed taxonomic, biogeographic, or paleoecologic perspectives. Rather, it draws together new treatments, the focus of which is to directly or indirectly test hypotheses of evolutionary change on the Quaternary time scale, or to present analytical techniques that facilitate linking Quaternary data with data from other time scales. Some of the studies extend back into the Pliocene, because the lineages of many modern taxa can be traced effectively to that time.

We respectfully recognize the contributions of prior studies to any science, and therefore we begin this introductory chapter with a synopsis of earlier investigations that utilized North American Quaternary mammals primarily in evolutionary studies. We then consider in further detail the rationale for this volume and provide an overview of the included topics. Consistency of terminology is of particular importance in the subject matter of this volume, and we therefore close the introduction by explaining the terms and conventions we favor.

Historical perspective

Evolutionary biology is today an exciting and dynamic discipline, partly because of the increasing use of the fossil record in effectively formulating and testing evolutionary hypotheses (e.g., Eldredge and Gould, 1972; Stanley, 1979; Barnosky, 1987; Vrba, 1987; R. Martin, 1992). To a certain extent this was recognized many years ago by Simpson (1944), but few quantitative examinations of morphological change using North American fossil materials were attempted until the data base of specimens collected with stratigraphic control began to reach a critical mass in the late 1960s and 1970s (e.g., Gingerich, 1976). Nearly all of these early studies in North America concentrated on Tertiary sequences. However, in Europe, Björn Kurtén (e.g., 1960, 1964) demonstrated the value of using the special qualities of the Quaternary record to study evolutionary processes, and a few North American paleontologists began to follow suit.

During the 1960s the late C. W. Hibbard and his students began to investigate size and character changes in a variety of rodent lineages from superposed (or otherwise dated) deposits in the midwestern and western United States. Semken's (1966) initial examination and proposal of a chronocline in muskrat first lower molars eventually led to the current interest in the muskrat phyletic sequence (Barnosky, 1985; L. Martin, Chapter 10, this volume; R. Martin, Chapter 11, this volume). In an important contribution, Zakrzewski (1969) documented size and character modifications in muskrats and voles of the late Blancan Hagerman local fauna. Two other students of Hibbard published an investigation of dental evolution in grasshopper mice, genus *Onychomys* (Carleton and Eshelman, 1979), in which estimates of evolutionary rates in darwins were provided.

A scattering of other quantitative examinations of character changes appeared during the 1960s and 1970s in the North American literature. For example, Guthrie (1965) proposed an evolutionary progression in dental complexity from the extinct *Microtus paroperarius* through modern *M. pennsylvanicus*, and R. Martin (1970) examined the potential application of Coon's (1962) line-and-grade theory to Quaternary cotton rat evolution. The North American muskrat record was further examined by L. Martin (1979), and he was the first to demonstrate that changes in muskrat dentition did not conform to monotonic phyletic gradualism. In his original presentation of the Red Queen hypothesis, Van Valen (1973) used data on mammalian species turnover from the late Pleistocene of Florida.

Although other authors, such as Guilday, Martin, and McCrady (1964), commented peripherally on the evolutionary meaning of the size changes in various mammals across the Pleistocene-Holocene boundary, there were no further consequential contributions that focused on evolutionary processes in fossil North American Quaternary mammals until the 1980s, when a few investigators began to examine character and size changes in mammals using increasingly good data bases and modern statistical techniques and evolutionary models (e.g., McDonald, 1981; MacFadden, 1986; R. Martin, 1986; Repenning et al., 1987; Morlan, 1989). By the late 1980s it was clear that studies of Quaternary mammals could contribute substantively to resolving debates about the relative importance of the then-polarized models of phyletic gradualism and punctuated equilibrium (Barnosky, 1987). More recent investigations have become increasingly directed in the goal of using Quaternary mammalian data for understanding evolutionary patterns (Repenning and Grady, 1988; Barnosky, 1990; Lich, 1990; Martin and Prince, 1990).

Explanation of this volume

Despite the studies noted earlier, when one examines the theoretical literature in evolutionary biology there is comparatively sparse information from the North American Quaternary. In 1990 the editors agreed that a symposium on morphological change in Quaternary mammals would be both practical and desirable, as a number of projects either had been recently completed or were near completion. We also agreed that the meeting would present a unique opportunity for paleontologists to engage in a dialectic designed to examine case studies in the light of modern evolutionary theory and to consolidate the studies in a single volume. This book is a compendium of papers presented at that symposium, entitled "Morphological Change in Quaternary Mammals of North America," held at Berry College, April 9–10, 1991, co-organized by the editors; see a review of that meeting by Lister (1991).

Initially we intended the volume to be limited to case studies of North American taxa. However, as we communicated with prospective contributors, it became obvious that both the symposium and the resulting publication would be considerably enhanced by a relaxation of the original guideline. The matter of measurements of evolutionary rates, for example, has been somewhat in turmoil since Gingerich (1983) discovered that evolutionary rates measured in darwins (Haldane, 1949) are not independent of interval lengths. P. Gingerich summarizes here some of his recent work on this subject, using Quaternary case studies. There is still some controversy dealing with the statistical recognition of morphological stasis in the fossil record, a topic treated in this volume by D. Anderson and A. Czebieniak from two quite different perspectives.

L. Viriot, J. Chaline, A. Schaaf, and E. Le Boulengé introduce image processing techniques for analyzing ontogenetic change in muskrat mo-

lars, a technique also used by A. Barnosky in this volume that appears
to have wide application for future studies of mammalian dental evolution. L. Martin describes a functional and histological scenario for
increase in molar hypsodonty and enamel complexity in Quaternary
rodent lineages that balances studies that simply describe or quantify
changes in dental parameters. The recognition of character displacement
in modern mammals has had a lengthy, turbulent history, and therefore
the identification of this phenomenon in Quaternary African and North
American canids by T. Dayan, D. Simberloff, and E. Tchernov has
special significance.

The remainder of the contributions are either individual or collective
examinations of case studies, with varying emphasis on theory. The
organizational level of focus ranges from individuals (L. Viriot et al.,
muskrats) to populations (e.g., A. Barnosky, meadow voles), through
species and genera, and includes the following taxonomic groups: edentates (R. Hulbert and G. Morgan), large herbivores (J. Purdue and
E. Reitz; A. Lister), carnivores (K. Seymour), and rodents (A. Barnosky; T. Goodwin; R. Martin; R. Zakrzewski; J. Rensberger and A.
Barnosky).

The case studies vary from the ontogenetic time scale (L. Viriot et
al.), to the ecological time scale (T. Dayan, D. Simberloff, and E.
Tchernov; examples reported by R. Martin), to hundreds or thousands
of years (J. Purdue and E. Reitz; A. Barnosky), to hundreds of thousands through millions of years (D. Anderson; A. Czebieniak; A. Lister;
T. Goodwin; R. Hulbert and G. Morgan; K. Seymour; J. Rensberger
and A. Barnosky; R. Martin; L. Martin; R. Zakrzewski), to integration
of all time scales (P. Gingerich).

Chronological terminology

In this book we consider the Pliocene to extend from 5.2 Ma to 1.7
Ma, the Pleistocene from 1.7 Ma to 10 ka, and the Holocene from
10 ka to the present (Berggren, 1985; Lundelius et al., 1987). These
epochs can be further subdivided into several North American Land
Mammal Ages (NALMA), which are defined by faunal content, then
independently dated by radiometric and paleomagnetic techniques: the
Blancan (4.8–4.0 Ma to 1.9 Ma), Irvingtonian (1.9 Ma to 550–200 ka),
and Rancholabrean (end of Irvingtonian to ~10 ka) (Lundelius et al.,
1987).

The following abbreviations are used to denote chronological intervals: my, million years; ky, thousand years; BP, years before the present (where "present" is defined as 1950 AD); Ma and ka, 1 million and 1,000 years ago, respectively, in the radioisotopic time scale, *sensu* Woodburne (1987, p. xiv); for example, 2 Ma refers to the 2-million-year level of the radioisotopic time scale.

Evolutionary models, patterns, and processes

We recognize punctuated equilibrium, phyletic gradualism, mosaic evolution, and staircase evolution as the four dominant *models* of evolutionary change commonly discussed in the evolutionary literature and especially the paleobiological literature. These models differ in assuming that different *processes* dominate in evolution, and accordingly they predict certain statistical *patterns* among measured samples of fossil materials. Among the processes considered most important are the microevolutionary processes of mutation, recombination, selection, and random drift, as well as the macroevolutionary processes of selection or sorting at or above the species level. Microevolutionary processes have a reality at this point in the history of science that macroevolutionary forces do not. Microevolutionary processes have been observed, measured, and expressed mathematically in a vast array of studies. Macroevolutionary processes, while in theory sound, have yet to be tested and confirmed to the same extent. Providing appropriate tests for these processes remains among the greatest challenges in paleontology.

Patterns are the representations of morphological traits as superposed sequences of statistical summaries (Figure 1.1); they arise from the effects of various processes. The punctuated equilibrium model, for example, is represented by a *pattern* of measured samples interpreted as combining both speciation and stasis. This pattern is expected because of certain evolutionary *processes* presumed to operate under the model of punctuated equilibrium that are partly contained in the statement of the model and partly implied. The important assumptions regarding process are that most character change is concentrated during speciation and that further change during the lifespan of a species represents stasis, defined as the meandering of measured character or shape variables around grand means (the equilibrium). If punctuated equilibrium is presumed to dominate in the evolution of a clade of organisms, macroevolutionary processes of sorting at or above the species level are often invoked to explain long-term trends.

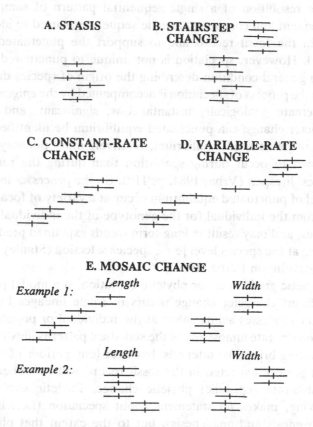

Figure 1.1. Patterns of morphological change. A horizontal bar represents a measure of variation within a sample, and a vertical bar represents the mean or median. (A) Stasis: fluctuation around a mean, with no directional change. (B) Stairstep change: sudden shift in the mean or median, but stasis below and above the shift. (C) Constant-rate change: more or less unidirectional change, and a more or less constant rate over the time interval for which data are available. (D) Variable-rate change: unidirectional change, but at a very variable rate over the time interval for which data are available. (E) Mosaic change: at least two different characters must show change by different patterns, i.e., length might show constant-rate change, but width might show stairstep change (Example 1), or length and width both might show stairstep change, but at different times (Example 2). The combination of stasis in one trait and change in another does not indicate a mosaic pattern, because most traits show stasis; only two or more traits changing in different ways will indicate a mosaic pattern.

The resolution of a single sequential pattern of samples into two subsequent and concurrent sample sequences is used to identify speciation in the fossil record and to support the punctuated equilibrium model. However, speciation is not unique to punctuated equilibrium. It is a general condition describing the origin of species diversity. Only when the process of speciation is accompanied by the *epigenetic processes* that create geologically instantaneous, significant, and recognizable character change can punctuated equilibrium be identified. Under the aegis of punctuated equilibrium, relatively more phenotypic change is assumed to occur during speciation than during the remainder of a species' lifespan (Vrba, 1984, p. 119). So the *processes* involved in the model of punctuated equilibrium occur at a variety of focal levels, ranging from the individual (or the genotype of the individual) through the species, and may result in long-term trends explained predominantly by sorting at the species level [e.g., species selection (Stanley, 1975); effect macroevolution (Vrba, 1980)].

Phyletic gradualism, or phyletic evolution, is a model proposing that significant character change occurs in single lineages by microevolutionary processes acting solely at the individual or population level. It can incorporate almost any of the statistical patterns illustrated in Figure 1.1 during brief time intervals, but over long periods of time the dominant pattern expected in the fossil record will be either constant- or variable-rate (episodic) phyletic change. Phyletic evolution, strictly speaking, makes no statement about speciation (i.e., it allows both cladogenesis and anagenesis), but to the extent that phyletic change dominates in the history of a clade, significant morphological change is not regularly expected during cladogenesis.

A stairstep pattern (Figure 1.1) can occur under any evolutionary model and at any temporal scale, and one could argue, in the absence of speciation, that staircase evolution (Stanley, 1985) simply represents an extreme case of variable-rate phyletic change.

Mosaic evolution is a model that postulates different traits evolving at different rates and times within a given character complex. It is identified by a mosaic pattern in the fossil record (Figure 1.1). Microevolutionary processes are used to explain this pattern, but the roles of macroevolutionary processes are unknown and unpostulated. To reject mosaic evolution, multiple characters in a given complex must change in concert. Mosaic evolution and the attendant pattern of mosaic change may eventually be considered as the multivariate expression of all evolutionary models, but for now we have isolated it as a distinct model to embrace the notion that related characters (e.g., in the dentition, or even on the same tooth) do not necessarily change as coordinated complexes (see A. Barnosky, Chapter 3, R. Martin, Chapter 11, this volume) and to emphasize that analyzing suites of characters provides more information about evolutionary pattern and process than do studies concentrating only on a single feature.

Although long-term patterns in the fossil record theoretically can provide support for a given evolutionary model, over brief periods of time most models can incorporate any of the patterns in Figure 1.1. As can be seen from this illustration, the pattern of stasis may also be contained in the pattern of stairstep (= staircase) change, variable-rate change, or mosaic change. Conversely, a pattern of constant-rate change may, if measured over a brief enough time interval, actually be part of a longer pattern of stasis.

Once a particular long-term pattern has been identified, the strength of the linkage among model, pattern, and process will ultimately depend on the ability of the investigator to objectively explain why one process is more or less viable than another as a cause of the pattern. Strong links demand consideration of the time scale involved, the sequence in relation to those from other geographic areas, the number and kinds of traits analyzed, and the potential influencing processes at all focal levels of causality (i.e., genetic, developmental, individual, population, species).

Although we have asked the contributors to this volume to adhere to certain general guidelines of terminology regarding pattern and process, and here have provided a brief account of our perception of this terminology, we emphatically have not sought standardized interpretations. The various investigators brought to this volume differing perspectives on research design, methodology, and philosophy. For example, the editors disagree on a very basic issue: the recognition and taxonomic identification of species in the fossil record. A. Barnosky recognizes phyletic speciation, whereas R. Martin does not. We do not wish to digress on the issue of species definition, but this difference in perspective does result in a considerably different treatment of taxa in the fossil record. However, the editors both support the extension of Mayr's (e.g., 1940, 1970) biological species concept into the fossil record. We have enjoyed the interaction of respectfully disagreeing with one another, and regardless of the differing opinions that arise from the information presented in this book, we hope that one conclusion will be unanimous: The Quaternary laboratory is worth visiting as an aid to understanding the evolution of mammals.

References

Barnosky, A. D. (1985). Late Blancan (Pliocene) microtine rodents from Jackson Hole, Wyoming: biostratigraphy and biogeography. *Journal of Vertebrate Paleontology*, 5:255–71.

(1987). Punctuated equilibria and phyletic gradualism: some facts in the Quaternary mammalian record. *Current Mammalogy*, 1:107–47.

(1990). Evolution of dental traits since latest Pleistocene in meadow voles (*Microtus pennsylvanicus*) from Virginia. *Paleobiology*, 16:370–83.

Berggren, W. A. (1985). Cenozoic geochronology. *Bulletin of the Geological Society America*, 96:1407–18.

Carleton, M. D., and R. E. Eshelman (1979). *A Synopsis of Fossil Grasshopper Mice, Genus* Onychomys, *and Their Relationships to Recent Species*. C. W. Hibbard Memorial Volume 7. Museum of Paleontology, University of Michigan, Ann Arbor.

Coon, C. S. (1962). *The Origin of Races*. New York: Knopf.

Eldredge, N., and S. J. Gould (1972). Punctuated equilibria: an alternative to phyletic gradualism. In T. J. M. Schopf (ed.), *Models in Paleobiology* (pp. 82–115). San Francisco: Freeman.

Gingerich, P. D. (1976). Paleontology and phylogeny: patterns of evolution at the species level in early Tertiary mammals. *American Journal of Science*, 276:1–28.

(1983). Rates of evolution: effects of time and temporal scaling. *Science*, 222:159–61.

Guilday, J. E., P. S. Martin, and A. D. McCrady (1964). New Paris No. 4: a Pleistocene cave deposit in Bedford County, Pennsylvania. *Bulletin of the National Speleological Society*, 26:121–94.

Guthrie, R. D. (1965). Variability in characters undergoing rapid evolution: an analysis of *Microtus* molars. *Evolution*, 19:214–33.

Haldane, J. B. S. (1949). Suggestions as to quantitative measurement of rate of evolution. *Evolution*, 3:51–6.

Kurtén, B. (1960). Rates of evolution in fossil mammals. *Cold Spring Harbor Symposium in Quantitative Biology*, 24:205–15.

(1964). The evolution of the polar bear, *Ursus maritimus* Phipps. *Acta Zoologica Fennica*, 108:1–26.

Lich, D. K. (1990). *Cosomys primus:* a case for stasis. *Paleobiology*, 16:384–95.

Lister, A. M. (1991). Evolutionary patterns in mammalian species. *Trends in Ecology and Evolution*, 6:239–40.

Lundelius, E. L., Jr., C. S. Churcher, T. Downs, C. R. Harington, E. L. Lindsay, G. E. Schultz, H. A. Semken, S. D. Webb, and R. J. Zakrzewski (1987). The North American Quaternary sequence. In M. O. Woodburne (ed.), *Cenozoic Mammals of North America: Geology and Biostratigraphy* (pp. 211–35). Berkeley: University of California Press.

McDonald, J. E. (1981). *North American Bison: Their Classification and Evolution.* Berkeley: University of California Press.

MacFadden, B. J. (1986). Fossil horses from "Eohippus" (*Hyracotherium*) to *Equus:* scaling, Cope's law, and the evolution of body size. *Paleobiology,* 12:355–69.

Martin, L. D. (1979). The biostratigraphy of arvicoline rodents in North America. *Transactions of the Nebraska Academy Sciences,* 7:91–100.

Martin, R. A. (1970). Line and grade in the extinct *medius* species group of *Sigmodon. Science,* 167:1504–6.

(1986). Energy, ecology and cotton rat evolution. *Paleobiology,* 12:370–82.

(1992). Generic species richness and body mass in North American mammals: support for the inverse relationship of body size and speciation rate. *Historical Biology,* 6:73–90.

Martin, R. A., and R. H. Prince (1990). Variation and evolutionary trends in the dentition of late Pleistocene *Microtus pennsylvanicus* from three levels in Bell Cave, Alabama. *Historical Biology,* 4:117–29.

Mayr, E. (1940). Speciation phenomena in birds. *American Naturalist,* 74:249–78.

(1970). *Populations, Species and Evolution.* Cambridge, Mass.: Belknap Press.

Morlan, R. E. (1989). Paleoecological implications of late Pleistocene and Holocene microtine rodents from the Bluefish Caves, northern Yukon Territory. *Canadian Journal of Earth Science,* 26:149–56.

Repenning, C. A., E. M. Brouwers, L. D. Carter, L. Marincovich, Jr., and T. A. Ager (1987). The Beringian ancestry of *Phenacomys* (Rodentia: Cricetidae) and the beginning of the modern Arctic Ocean borderland biota. *U.S. Geological Survey Bulletin,* 1687:1–31.

Repenning, C. A., and F. Grady (1988). The arvicolid rodents of the Cheetah Room fauna, Hamilton Cave, West Virginia, and the spontaneous origin of *Synaptomys. U.S. Geological Survey Bulletin,* 1853:1–28.

Semken, H. A., Jr. (1966). Stratigraphy and paleontology of the McPherson *Equus* beds (Sandahl Local Fauna), McPherson County, Kansas. *Contributions, Museum of Paleontology, University of Michigan,* 20:121–78.

Simpson, G. G. (1944). *Tempo and Mode in Evolution.* New York: Columbia University Press.

Stanley, S. M. (1975). A theory of evolution above the species level. *Proceedings of the National Academy of Sciences,* USA, 72:646–50.

(1979). *Macroevolution: Pattern and Process.* San Francisco: Freeman.

(1985). Rates of evolution. *Paleobiology,* 11:13–26.

Van Valen, L. (1973). A new evolutionary law. *Evolutionary Theory,* 1:1–30.

Vrba, E. S. (1980). Evolution, species and fossils: How does life evolve? *South African Journal of Science,* 76:61–84.

(1984). Patterns in the fossil record and evolutionary processes. In M. W. Ho and P. S. Saunders (eds.), *Beyond Neo-Darwinism* (pp. 115–42). London: Academic Press.

(1987). Ecology in relation to speciation rates: some case histories of Miocene–Recent mammal clades. *Evolutionary Ecology,* 1:283–300.

Whitmore, F., K. O. Emery, H. B. S. Cooke, and D. J. P. Swift (1967). Elephant teeth from the Atlantic continental shelf. *Science,* 156:1477–81.

Woodburne, M. (1987). Definitions. In M. O. Woodburne (ed.), *Cenozoic Mammals of North America: Geology and Biostratigraphy* (pp. xiii–xv). Berkeley: University of California Press.

Zakrzewski, R. J. (1969). The rodents from the Hagerman local fauna, upper Pliocene of Idaho. *Contributions, Museum of Paleontology, University of Michigan,* 23:1–36.

2

A method for recognizing morphological stasis

DEBORAH K. ANDERSON

"Stasis is a real phenomenon" (Eldredge, 1989, p. 72). Although this opinion is rarely stated so explicitly, most discussions of evolutionary rate or change in morphology over time include the assumption that stasis does occur (Futuyma, 1986; Lande, 1986; Bookstein, 1987; Stanley and Yang, 1987; Levinton, 1988; Mayr, 1988; Eldredge, 1989). Stanley and Yang (1987) present a specific example of stasis in fossil and Recent bivalve shape. Barnosky (1987) presents several examples of stasis in small fossil mammals based on an analysis of molar size. Futuyma (1986) recognizes examples of stasis in the extant genus *Plethodon*. Generally, the questions that need to be addressed are not Does stasis occur? but Why does it occur? and How can we recognize it? This chapter concentrates on the latter question.

"Stasis" has almost as many different definitions as "species" (Eldredge, 1971, 1986, 1989; Stanley, 1982; Lande, 1986; Rose and Bown, 1986; Levinton, 1988; Mayr, 1988). Because "stasis" is used to describe an absence of change in a population of a particular species, the two concepts are frequently used together. Currently, then, the interpretation of stasis may be as unique as the species definition applied to the organism(s) or referred to in discussion. Without direct evidence of reproductive isolation between populations, paleobiologists evoke the morphospecies concept for identifying species. If a morphospecies is a group of individuals that are not significantly different morphologically, then the general criterion for recognizing stasis is identical with the criterion for identifying a particular species and determining species longevity. In this case, character stasis is actually what is recognizable; species stasis is inferred. To determine whether or not the observed

13

character stasis is consistent with species stasis, one needs to know how much intraspecific and/or intrapopulational variation to expect for the organism(s).

There is a difference between "species stasis" and "character stasis." Both kinds of stasis are defined as an absence of change in morphology over time (Eldredge, 1971, 1989; Lande, 1986; Levinton, 1988; Mayr, 1988). However, "species stasis" is generally used to refer to an absence of change in several morphological characters of a species (Levinton, 1988; Eldredge, 1989), whereas "character stasis" denotes stasis in a single character of a species (Bookstein, 1978; Raup and Crick, 1981; Charlesworth, 1984; Flynn, 1986; Cheetham, 1987; Mayr, 1988). "Morphological stasis" combines species stasis and character stasis and is the kind of stasis people recognize when they infer species stasis from character stasis, that is, evolutionary stasis from the analysis of just one trait (Lich, 1990). "Evolutionary stasis" (Avers, 1989) and "species stasis" are synonymous.

Documented durations of morphological stasis vary from thousands of years (Lich, 1990) to 10 million years (Eldredge and Gould, 1972; Stanley and Yang, 1987). Although some authors (Flynn, 1986; Futuyma, 1986; Rose and Bown, 1986) believe that a time span of millions of years is necessary to recognize stasis, the definitions of Lande (1986) and Mayr (1988) require no minimum time period over which an absence of morphological change must be observed. No matter what time limits are imposed by a given definition of stasis, there is a critical topic to be addressed: How can evolutionary stasis be measured?

Bookstein (1987; Bookstein, Gingerich, and Kluge, 1978) suggests that prior to discussing any "rate" of directional evolution (including stasis), the null hypothesis of random walk must be tested to see if a directional rate exists. Those papers recognize any nonrandom fluctuation as a rate of directional evolution. The random walk procedure as a test for randomness in the pattern of morphological change or absence of change over time is useful. However, a trend that is explainable by random walk should not be dismissed as solely the result of a chance event (Mayr, 1988), because in reality it is often impossible to distinguish statistically between a random walk and a pattern caused by natural selection. Nor should trends for which the test of random walk is inapplicable (due to lack of information) be dismissed from consideration in studies of evolutionary rates. Whether or not evolutionary trends exist, it is still worthwhile to study morphological change over time.

Traditionally, the rate of character change is measured in darwins, a change of e (base of natural logarithms, 2.718) per million years (my) (Haldane, 1949). Fisher (1989) reviews several of the problems related to measuring morphological rates of evolution in darwins. Intrapopulational variability and generation time are two factors that Fisher recognizes as affecting rates of evolution. These factors would have particularly large effects on rate given the Bookstein et al. (1978) definition of the absolute rate of evolution as the amount of change between successive generations. Population size also may affect evolutionary rate. The biggest drawback to using darwins is the influence of temporal scaling, as revealed by Gingerich's (1983, 1985) recent studies. No one has yet proposed a method for recognizing stasis based on darwins.

One technique for estimating species stasis is to study the known stratigraphic range of the species (Stanley, 1979; Eldredge, 1989). A more precise definition for stasis is a "statistically lower deviation from a starting condition than expected by chance" (Levinton, 1988, p. 350). Another statistical definition for stasis is based on the time period during which changes in the mean phenotype for two successive populations are less than two standard deviations (Lande, 1986).

In order to recognize stasis statistically in a long stratigraphic sequence, Levinton's (1988) definition seems most applicable. With chance ruled out, as well as a directional rate of change, the logical interpretation of the data is stasis. One of the two previously mentioned elements is missing from other methods of recognizing morphological stasis. The following analysis of molar lengths and widths in the arvicoline *Cosomys primus* offers a practical method for recognizing morphological stasis under Levinton's (1988) definition.

Recognizing morphological stasis: a case study

Morphological stasis defined as a "statistically lower deviation from a starting condition than expected by chance" (Levinton, 1988, p. 350) can be recognized statistically by applying three types of analysis: (1) test means for a statistically significant difference using a one-way ANOVA; (2) determine the power of the ANOVA; high $= p > 0.70$ (Dowdy and Wearden, 1983), which is a low type II error; (3) test randomness of changes in means over time/elevation using a runs test (Schefler, 1980) and a random walk test (Bookstein, 1987).

These analyses were applied to 10 samples of first lower molars (some isolated, some in mandibular fragments) from a stratigraphically superposed series of *C. primus* populations collected from the Glenns Ferry

Formation near Hagerman, Idaho, and reposited in the Idaho State Museum of Natural History. Sample sizes range from 5 to 28 teeth. Each population corresponds to a known locality and elevation. The 10 populations span a time interval ranging from, at a minimum, 45,000 years to, at most, 165,000 years (Lich, 1990). Absolute time is estimated based on radiometric (K-Ar) dates of three volcanic ashes occurring at three different elevations within the 10 sample series. The interpolation of time between these dates should be fairly accurate, because there is little or no faulting throughout the section, and deposition rate appears to have been fairly constant (Bjork, 1970). The approximate numbers of years estimated between successive samples range from 3,000 to 10,000. Species identification was based primarily on qualitative taxonomic features of the first lower molar (M_1) agreed upon by most taxonomists specializing in these organisms: three or four alternating triangles, rooted teeth, a prism fold on the anterior loop, and teeth without cement. These traits were used because they are the ones that define the species. Analysis of stasis or change to study evolutionary patterns requires that species-specific characters be used. For a summary of all the descriptive statistics obtained from the measurements and the geology for the localities, see Lich (1990).

Results

A one-way ANOVA revealed no statistically significant differences between the mean molar lengths, widths, and posterior loop widths for the 10 population samples at the 0.05 level of significance (Figure 2.1). Tukey's Honestly Significant Difference (HSD) test, a multiple comparison procedure for ANOVAs that does not require a prior significant difference, was included in the analysis to control for a type I error.

If the means are not significantly different, then one cannot reject the null hypothesis that the means are all equal. On the other hand, one cannot accept the null hypothesis as stated, either. ANOVAs were designed to detect significant differences between population means. If the results of an ANOVA are used as evidence to accept the null hypothesis (i.e., evidence for stasis), then it becomes important to assess the probability of a type II error (accepting the null hypothesis when it is false) (Toft and Shea, 1983). The solution is to determine the power of the test, the degree to which the statistical procedure can distinguish a situation as different from the null hypothesis (Cohen, 1988). A robust test ($p > 0.70$) (Dowdy and Wearden, 1983) indicates

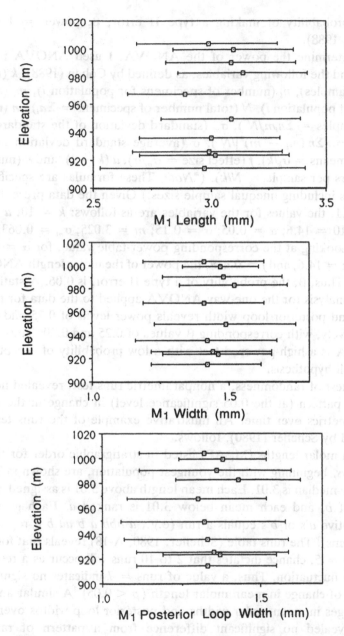

Figure 2.1. First lower molar (M_1) mean (box) length, width, and posterior loop width; two standard deviations (vertical bars) and population size (numbers) are indicated for the 10 populations of *Cosomys primus* stratigraphically arranged from oldest to youngest.

a low probability of making a type II error, β; power $= 1 - \beta$ (Cohen, 1988).

To determine the power of the ANOVA, I used ANOVA power tables and the following variables, as defined by Cohen (1988): k (number of samples), n_i (number of specimens for population i), m_i (mean metric of population i), N (total number of specimens $= \Sigma n_i$), m (mean of k samples $= \Sigma n_i m_i / N^*$), σ_m (standard deviation of the standardized means $= [\Sigma n_i (m_i - m)^2]/N^*$), σ (average standard deviation for all sample means $= \sigma_k / k$), f (effect size $= \sigma_m / \sigma$), u $(k - 1)$, and n (number of molars per sample $= N/k$). (*Note:* These formulas are specifically for cases including unequal sample sizes.) Given the data presented in Table 2.1, the values for the variables are as follows: $k = 10$; $u = 9$; $n = N/10 = 14.6$; $\alpha = 0.05$; $\sigma = 0.15$; $m = 3.025$; $\sigma_m = 0.063$; $f = 0.418$. Looking at the corresponding power-table value, for $\alpha = 0.05$, $u = 9$, $n = 14.6$, and $f = 0.418$, the power of the molar length ANOVA is 0.94. Thus, β, the probability of a type II error, is 0.06. A statistical power analysis for the one-way ANOVA applied to the data for molar width and posterior loop width reveals power levels of 0.75 and 0.80, respectively, with corresponding β values of 0.25 and 0.20. Thus, each ANOVA is a high-powered test with a low probability of accepting a false null hypothesis.

One test of randomness, a nonparametric runs test, revealed no significant pattern (at the 0.05 significance level) in change in the mean molar metrics over time. An illustrative example of the runs test, as outlined by Schefler (1980), follows.

Mean molar lengths (M_1 AP) listed in stratigraphic order for the 10 *Cosomys,* beginning with the youngest population, are shown in Table 2.2. The median is 3.01. Each mean length above 3.01 is assigned a rank value of b, and each mean below 3.01 is ranked a. Each group of consecutive a's or b's equals a run (e.g., *a bbb a b aa b a* represents seven runs). The runs table (Schefler, 1980, A-18) reveals that for $a = 5$ and $b = 5$, chance dictates that 2 to 10 runs will occur as a result of random fluctuation. Thus, a value of runs $= 7$ indicates no significant pattern of change in mean molar length ($p < 0.05$). A similar analysis of changes in mean molar widths and posterior loop widths over time also revealed no significant difference from a pattern of random fluctuation.

A second test of randomness, the random walk test, was applied to the *Cosomys* data to see if the fluctuations about the mean molar metrics fit the null hypothesis of random walk. Based on Bookstein's (1987)

Table 2.1. *Descriptive statistics and estimated time in millions of years (Ma) for the first lower molar (M_1) length (AP), width (TR), and posterior loop width (PL TR oc) of C. primus*

Time (Ma)	Elevation (m)	Molar metric	$\bar{x}$ (mm)	SD	SE	N
3.41	1,004	M_1 AP	2.99	0.15	0.041	14
		M_1 TR	1.41	0.11	0.028	14
		M_1 PL TR oc	1.32	0.10	0.028	14
3.42	1,000	M_1 AP	3.1	0.13	0.041	11
		M_1 TR	1.47	0.07	0.021	11
		M_1 PL TR oc	1.38	0.07	0.023	10
3.44	991	M_1 AP	3.09	0.14	0.061	5
		M_1 TR	1.48	0.11	0.050	5
		M_1 PL TR oc	1.36	0.12	0.051	5
3.45	983	M_1 AP	3.14	0.14	0.037	14
		M_1 TR	1.52	0.10	0.025	14
		M_1 PL TR oc	1.43	0.10	0.028	14
3.48	968	M_1 AP	2.89	0.12	0.052	5
		M_1 TR	1.41	0.11	0.048	5
		M_1 PL TR oc	1.30	0.11	0.049	5
3.51	949	M_1 AP	3.09	0.15	0.036	18
		M_1 TR	1.49	0.13	0.030	18
		M_1 PL TR oc	1.32	0.13	0.030	18
3.54	936	M_1 AP	3.00	0.15	0.045	11
		M_1 TR	1.42	0.13	0.038	11
		M_1 PL TR oc	1.33	0.13	0.039	11
3.56	926	M_1 AP	2.99	0.16	0.033	24
		M_1 TR	1.44	0.11	0.021	24
		M_1 PL TR oc	1.35	0.12	0.024	24
3.56	924	M_1 AP	3.02	0.16	0.041	16
		M_1 TR	1.46	0.10	0.026	16
		M_1 PL TR oc	1.37	0.09	0.022	16
3.58	914	M_1 AP	2.97	0.22	0.041	28
		M_1 TR	1.42	0.10	0.019	28
		M_1 PL TR oc	1.33	0.11	0.020	28

Note: $\bar{x}$, mean; SD, standard deviation; SE, standard error; N, sample size.

paper, $p < 0.05$ indicates stasis, and $p > 0.95$ indicates anagenesis. For p values falling between these two levels, the null hypothesis of random walk cannot be rejected.

Application of the random walk test required an estimate of time in millions of years for each population sample. Absolute time (Table 2.1) was estimated based on a radiometric date of $3.48 \pm 0.27 \times 10^6$ million years before the present (Ma) for an ash at an elevation of 3,171 feet and an approximate sedimentation rate of 1.8 feet per 1,000

Table 2.2. *An example runs test applied to the first lower molar (M₁) length (AP) for* Cosomys

Elevation (m)	M₁ AP (m)	Rank[a]
1,004	2.99	a
1,000	3.10	b
991	3.09	b
983	3.14	b
968	2.89	a
949	3.09	b
936	3.00	a
926	2.99	a
924	3.02	b
914	2.97	a

[a]Rank: a/b = below/above median.

years. The latter was calculated by dividing the distance separating two radiometrically dated volcanic ashes (KA 832: 994 m, 3.2×10^6; KA 831: 939 m, 3.3×10^6) by the time they spanned, 55 m/100,000 years. It is reasonable to assume a constant deposition rate for the 90 m of section based on Bjork's (1970) stratigraphic studies of the Glenns Ferry Formation.

To determine the p value for the random walk test, I computed the following variables, as defined by Bookstein (1987): mean reduced speed ($[\Sigma|m_i - m_{i+1}|/\sqrt{t_{i+1} - t_i}]/k$, where t_i is the date for the population sample, k is the number of population samples, and m_i is the mean molar metric for the i population); downspan (|maximum deviation from the most recent mean molar length| = $m_i - m_{i+?}$|); upspan (|maximum deviation from the oldest mean molar length| = |$m_k - m_{i+?}$|); mean span (average of downspan and upspan); x (mean span/{[mean reduced speed/0.798]/$t_k - t_i$), where 0.798 = a}; and p $(1 - F[x^{-2}]$, where the x values are found in the table in Appendix 2 of Bookstein's 1987 paper).

Thus, for the data presented in Table 2.1, mean reduced speed = 0.634, downspan = 0.15, upspan = 0.17, mean span = 0.16, $x = 0.489$, $p = 1 - F(4.19) = 1 - 0.99 = 0.01$. Similar computations for mean molar width and posterior loop width resulted in respective p values of 0.022 and 0.03. Because p is less than 0.05 for each molar metric, the null hypothesis of random walk is rejected, and a hypothesis of stasis in molar metrics over time is supported.

Other tests for stasis

Application of Lande's (1986) test for stasis reveals no significant difference between the 10 population sample means. The deviation between means of successive populations is less than two standard deviations for all populations and less than two standard errors, with one exception (population 70016 at 968 m for mean molar length).

Bookstein et al. (1978) and Gerrodette (1987) define a trend as a slope significantly different from zero. A Spearman rank correlation coefficient with $\log(L \times W)$ as the dependent variable and stratigraphic elevation as the independent variable revealed no significant difference ($r = 0.117$; $p > 0.05$) from a slope of zero for the *Cosomys* data.

Discussion

Levinton's (1988, p. 350) definition of morphological stasis as a "statistically lower deviation from a starting condition than expected by chance" includes valuable, practical criteria for recognizing morphological stasis in a series of closely spaced fossil population samples. The one-way ANOVA identifies any significant deviations in mean character metrics. In the event that the null hypothesis cannot be rejected, a statistical power analysis reveals the critical level at which the null hypothesis can be accepted. The runs test and random walk test evaluate the pattern of change over time to identify any trends and determine if the amount of fluctuation about the mean is less than one would expect to see based on chance alone.

A definite case for stasis in the *C. primus* molars can be made based on the results of the analysis outlined earlier. The hypothesis that *Cosomys* exhibits statistically static, nondirectional, nonrandom change over the time period represented by the 10 elevations is further supported by the results of additional tests for stasis. The mean molar length and width do not show any significant change over time. Biologically this could indicate selection for the particular molar size and morphology.

The three-step analysis presented is particularly valuable because it is easy to apply to a data set and requires a minimal amount of data (character metrics and relative dates for each population sample). The assumptions for the statistical tests used are more straightforward than when a multivariate method is employed, thus fulfilling the ideal of a parsimonious analysis. The results of applying this method of recognizing morphological stasis to many sets of data will provide valuable basic information useful for addressing the larger (more general) questions about morphological change over time.

References

Avers, C. J. (1989). *Process and Pattern in Evolution*. Oxford: Oxford University Press.

Barnosky, A. D. (1987). Punctuated equilibrium and phyletic gradualism. In H. H. Genoways (ed.), *Current Mammalogy*, Vol. 1 (pp. 109–47). New York: Plenum Press.

Bjork, P. R. (1970). The carnivora of the Hagerman local fauna (late Pliocene) of southwestern Idaho. *Transactions of the American Philosophical Society*, 60:1–51.

Bookstein, F. L. (1987). Random walk and the existence of evolutionary rates. *Paleobiology*, 13:446–64.

Bookstein, F., P. D. Gingerich, and A. G. Kluge (1978). Hierarchical linear modeling of the tempo and mode of evolution. *Paleobiology*, 4:120–34.

Charlesworth, B. (1984). Some quantitative methods for studying evolutionary patterns in single characters. *Paleobiology*, 10:310–18.

Cheetham, A. H. (1987). Tempo of evolution in a Neogene bryozoan: Are trends in single morphological characters misleading? *Paleobiology*, 13:286–96.

Cohen, J. (1988). *Statistical Power Analysis for the Behavioral Sciences*. Hillsdale, N.J.: Erlbaum.

Dowdy, S., and S. Wearden (1983). *Statistics for Research*. New York: Wiley.

Eldredge, N. (1971). The allopatric model and phylogeny in Paleozoic invertebrates. *Evolution*, 25:156–67.

  (1986). Information, economics and evolution. *Annual Review of Ecological Systematics*, 17:351–69.

  (1989). *Macroevolutionary Dynamics*. New York: McGraw-Hill.

Eldredge, N., and S. J. Gould (1972). Punctuated equilibria: an alternative to phyletic gradualism. In T. J. Schopf (ed.), *Models in Paleobiology* (pp. 82–115). San Francisco: Freeman.

Fisher, D. C. (1989). Rates of evolution. In D. E. G. Briggs and P. R. Crowther (eds.), *Paleobiology: A Synthesis* (pp. 152–9). London: Blackwell.

Flynn, L. J. (1986). Species longevity, stasis, and stairsteps in rhizomyid rodents. In K. M. Flanagan and J. A. Lillegraven (eds.), *Vertebrates, Phylogeny, and Philosophy* (pp. 273–85). University of Wyoming Contributions to Geology, Special Paper 3.

Futuyma, D. (1986). *Evolutionary Biology*. Sunderland, Mass.: Sinauer.

Gerrodette, T. (1987). A power analysis for detecting trends. *Ecology*, 68(5):1364–72.

Gingerich, P. D. (1983). Rates of evolution: effects of time and temporal scaling. *Science*, 222:159–61.

(1985). Species in the fossil record: concepts, trends, and transitions. *Paleobiology*, 11:2–41.

Haldane, J. B. S. (1949). Suggestions as to quantitative measurement of rates of evolution. *Evolution*, 3:51–6.

Lande, R. (1986). The dynamics of peak shifts and the pattern of morphological evolution. *Paleobiology*, 12:343–54.

Levinton, J. (1988). *Genetics, Paleontology, and Macroevolution*. Cambridge University Press.

Lich, D. (1990). *Cosomys primus:* a case for stasis. *Paleobiology*, 16:384–95.

Mayr, E. (1988). *Toward a New Philosophy of Biology*. Cambridge, Mass.: Harvard University Press.

Raup, D. M., and R. E. Crick (1981). Evolution of single characters in the Jurassic ammonite *Kosmoceras*. *Paleobiology*, 7:200–15.

Rose, K. D., and R. M. Bown (1986). Gradual evolution and species discrimination in the fossil record. In K. M. Flanagan and J. A. Lillegraven (eds.), *Vertebrates, Phylogeny, and Philosophy* (pp. 119–30). University of Wyoming, Contributions to Geology, Special Paper 3.

Schefler, W. (1980). *Statistics for the Biological Sciences*. Reading, Mass.: Addison-Wesley.

Stanley, S. M. (1979). *Macroevolution: Pattern and Process*. San Francisco: Freeman.

(1982). Macroevolution and the fossil record. *Evolution*, 36:460–73.

Stanley, S. M. and X. Yang (1987). Approximate evolutionary stasis for bivalve morphology over millions of years: a multivariate multilineage study. *Paleobiology*, 13:113–39.

Toft, C. A., and P. J. Shea (1983). Detecting community-wide patterns: estimating power strengthens statistical inference. *American Naturalist*, 122:618–25.

3

Mosaic evolution at the population level in *Microtus pennsylvanicus*

ANTHONY D. BARNOSKY

Microevolution and macroevolution generally are recognized as distinct phenomena that reflect two extremes of the evolutionary hierarchy. "Microevolution" denotes the evolutionary processes that operate at the genetic level (mutation, recombination, selection, random drift) among breeding pairs and their progeny (at least for mammals). These processes occur within the lifetimes and geographic ranges of individual animals: within a few months and within less than 1 km^2 in the case of rodents, up to a few decades and a few hundreds of square kilometers in the case of large herbivores. Expanded over a few generations (at most tens or hundreds of years), microevolution causes shifts in gene frequencies of populations that may or may not manifest themselves as phenotypic changes.

"Macroevolution," on the other hand, refers to the processes that cause new species to arise and interact, an enterprise that seems quite lengthy in human terms, taking place over the geological expanse of thousands of years, in some cases, though usually over hundreds of thousands to millions of years. The geographic coverage encompasses the entire range of a species during its lifespan and can vary from a few tens or hundreds of square kilometers to an entire continent or more. The motor of macroevolution still is not clear: Some argue for cumulative microevolutionary changes, whereas others claim that macroevolution is decoupled from microevolution and is driven by processes that become operative only at the species level (e.g., species selection or the "effect hypothesis"); see the literature cited by Barnosky (1987), Vrba and Eldredge (1984), and Vrba and Gould (1986).

One thing that obscures whether or not macroevolution is a natural cumulative by-product of microevolution is that the two phenomena *do* represent extreme ends of the evolutionary hierarchy. In real time and space, mammalian evolution proceeds along the following lines; steps 1 through 4 have been branded "microevolution," and step 5 has attained the status of "macroevolution."

Step 1. Organisms mate at a given location and mix their genes in their progeny.

Step 2. If the phenotypic result is viable (i.e., not for some reason strongly selected against), the progeny disperse over distances ranging from a few meters to a few kilometers, and if they're lucky they mate, thereby recombining genes again and possibly also expanding the geographic range of the genetically linked population, that is, the deme.

Step 3. At some point, the deme's geographic range becomes limited, perhaps by selective pressures, by physical barriers, or by sheer distance that prohibits gene flow (i.e., the distance a potential mate will travel in its reproductive lifetime; if an animal on the edge of the geographic range doesn't travel to the center of the geographic range in time to mate with an animal there, the two animals don't belong to the same deme).

Step 4. By dispersal of individuals over many generations, new demes are created or existing demes are connected, each of which may or may not diverge genetically from the original genotypic composition, depending on a host of factors, including selection, chance, and degree of isolation from other populations; during this phase the range of the species, of course, also is expanding.

Step 5. At this stage the species can go one of three ways:

(A) Frequent trading of individuals among demes can ensure that roughly the same genotype will prevail through all populations, or else demes can change randomly with respect to one another, but in any case they do not become reproductively isolated. No speciation will take place for as long as this goes on, even though subspecies may form if populations differentiate for a while, then reestablish reproductive contact.

(B) One or more of the populations may become reproductively isolated from the parent deme (i.e., the original genotypic composition) – one speciation event (cladogenesis) will occur if only one population becomes reproductively isolated, or several speciation events will occur if more than one population becomes reproductively isolated.

(C) As in (A), populations do not become reproductively isolated, but selective pressures across the entire geographic range of the species are similar, so that all populations change their genotypic compositions (and resultant phenotypic expression) in the same direction – in this case the original species eventually transforms into a new one through anagenesis.

Step 6.   Once a new species arises by either step 5B or 5C, the whole process starts over again at step 1 (or, in the case of 5C, anagenetic change can simply continue).

By "species," I mean essentially Mayr's (1970, p. 12) biological species concept, which is "groups of interbreeding natural populations that are reproductively isolated from other such groups." More often than not we can recognize differences between biological species of mammals by phenotypic differences that result from the lack of gene flow between species. If the phenotypic traits are hard parts that are preserved as fossils, such as the teeth analyzed in this study, the same criteria can be used for recognizing species in the paleontological record. The biological species concept offers the advantage of allowing the designation of species arising by either of the processes of morphological change that theoretically operate in nature: cladogenesis or anagenesis. Cladogenetic speciation can reflect reproductive isolation in space through time, whereas anagenetic speciation, by definition, reflects reproductive isolation in time only. I regard the endpoints of anagenetic lineages as different species if the morphological differences between them exceed the range known within a comparable living species, because in the modern species morphological boundaries enclose the individuals that interbreed and produce viable offspring under natural conditions. Operationally, this definition of a species assumes that differences in carefully chosen morphological traits correlate with reproductive isolation and that modern organisms provide the most complete sample of morphological variation within a species. It may be difficult to assign a sample to a species if it falls in the middle of a temporal or geographic cline, but this problem applies to virtually any definition of a species on some temporal or geographic scale.

A second complication in recognizing the extent to which macroevolution represents a process different from microevolution is that data bearing on step 5 of the evolutionary process commonly are examined in isolation from data that would document steps 1–4. Generally this is by necessity: The nature of the fossil record is such that it is hard to document the population-level changes that mark the transition from step 4 to step 5, because in most cases the necessary stratigraphic coverage, geographic coverage, or dating resolution is lacking.

Further complicating the picture is that within step 5 the observed patterns of morphological change through time illustrate a whole range of possibilities; see Barnosky (1987) for some examples. Stacked stratigraphic sequences of populations for different organisms have shown random fluctuation about a mean morphology, that is, stasis (scenario 5A). Other sequences have shown stratigraphically quick, significant shifts in mean morphology that were bracketed by times of stasis, which can be interpreted as support for the macroevolutionary model of punctuated equilibrium (scenario 5B) or, contrastingly, as support for a variable-rate version of phyletic change (scenario 5C) that Stanley (1985) calls "staircase evolution"; see also Flynn (1986). And still other sequences demonstrate gradual unidirectional change, supporting the prophecy of phyletic gradualism (scenario 5C) (Gingerich, 1985; Rose and Bown, 1986; Hoffman, 1989).

The background just presented throws a different light on the polarizing question: Which is more frequent in nature, phyletic gradualism or punctuated equilibrium? In fact, morphological patterns that could be used to support both models are fairly common, as the chapters in this volume and those summarized elsewhere (Barnosky, 1987) demonstrate. But there are more important questions that emerge: (1) What are the processes by which stasis, rapid change, or gradual change is maintained? (2) Is macroevolution really decoupled from microevolution?

Answering these questions will depend on understanding how and why the gene pools of populations change through thousands of years, the best paleontological proxy of which is morphological change. Accordingly, this chapter examines the population-level morphological changes that took place at the end of the Pleistocene in voles, *Microtus pennsylvanicus* (Arvicolidae) from the Appalachian region (Table 3.1), in an attempt to identify the ways in which patterns of stasis or change are maintained at the population level.

Methods and materials

The general design was to examine a series of dental characters in the third upper molar (M^3) in both modern and fossil populations. M^3 was used because it is taxonomically useful in many vole species, which

Table 3.1. *Fossil localities from which samples were obtained. All of the material is reposited in The Carnegie Museum of Natural History, Section of Vertebrate Paleontology. See Figure 3.1 for map locations.*

1. New Paris no. 4, Bedford Co., Pennsylvania: Fissure fill, excavated in 20 stratigraphic levels (top level ~2 m below ground surface, bottom level ~9.5 m down), each 0.3 m thick. Guilday et al. (1964) reported a radiocarbon date of 11,300 ± 1,000 years BP (Y-727) on charcoal from near the middle (5.7 m deep) of the stratigraphic sequence. Unidentifiable rodent-sized bone fragments lumped from approximately 6.4 to 6.7 m deep were radiocarbon-dated at 11,910 ± 350 years BP (Beta-28975, bone collagen). Two additional radiocarbon dates seem anomalously young based on the faunal composition associated with them and their low stratigraphic position: 9,540 ± 500 years BP (M-1067) from 5.2 to 6.4 m deep (Guilday et al., 1964) and 9,250 ± 120 years BP (Beta-28974, collagen) from ~7.6 m deep. Introduction of Recent bones into the bulk bone samples that were dated could be the source of contamination.

2. Strait Canyon Fissure, Highland Co., Virginia: Fissure fill, excavated in 46 stratigraphic levels, each level approximately 0.15 m thick. Samples of unidentifiable rodent-sized bone fragments were radiocarbon-dated and yielded the following ages: 18,420 ± 180 years BP (Beta-28977/ETH-4893, collagen) from the upper 0.75 m (levels +Q to +U); 29,870 (+1,800/-1,400) years BP (Krueger GX-7017-A, bone apatite) from ~2.1 m down (level +J, near the middle of the deposit); 17,880 ± 150 (Beta 28976/ETH-4892, collagen) from 4.3 to 5.3 m down (levels -B to -I at the bottom of the deposit). A. Barnosky (1990) suggested that the bottom date was anomalously young because of contamination of the sample by Recent bone fragments, but the possibility of complex stratigraphy at the base of the deposit cannot be ruled out. In any case, the upper two dates are in stratigraphic order, and conclusions based on them seem warranted. This site essentially is unpublished, but detailed notes and preliminary analysis of it by John Guilday are on file at The Carnegie Museum of Natural History, Section of Vertebrate Paleontology.

3. Baker Bluff Cave, Sullivan Co., Tennessee: Cave deposit near cave entrance. Fossiliferous units from ~0.9 to 3.0 m below modern cave floor, excavated in seven stratigraphic levels, each level 0.3 m thick. Guilday et al. (1978) reported four radiocarbon dates, the uppermost of which they considered contaminated and too young: 555 ± 185 years BP (GX-3369, bone apatite) from ~1.4 m (4-5-foot level); 10,560 ± 220 years BP (GX-3370a, bone apatite) from ~2(?) m down (listed by Guilday et al.'s text as 6-7-foot level, but plotted on their diagrams as 5-foot level); 11,640 ± 250 years BP (GX-3370b, collagen) from ~2 m down (6-7-foot level); 19,100 ± 850 years BP (GX-3495, bone apatite) from ~2.9 m down (9-10-foot level).

means that the findings might be applicable to understanding vole evolution in general. Also, the M^3 is known to be geographically variable in *M. pennsylvanicus*, which suggests that it may also have varied temporally. Only M^3s that could be associated with the second upper molar, which is species-diagnostic in having a fourth triangle (in contrast to the three triangles in other potentially sympatric *Microtus* species), were used in the analysis, to help ensure that population-level (intraspecies) differences rather than species-level (interspecific) differences were examined.

Teeth from 12 modern populations were measured to provide an estimate of morphological change across the modern geographic range (Figure 3.1 and Table 3.2). The modern samples were used to construct morphological landscapes by plotting on the appropriate map location the mean measurement for a given attribute for each population, and connecting equal values with contour lines (Figure 3.1). These morphological landscapes illustrate general trends in geographic variation today and provide a baseline for assessing the meaning of morphological changes in the past (Barnosky, in press). The kind and degree of morphological change in fossil populations were interpreted by measuring M³s stacked in stratigraphic sequences from each of three Appalachian fissure fills or cave sites (Figures 3.2–3.4). Table 3.1 lists these sites and explains their radiocarbon dates and excavation levels. The oldest of the sequences is from Strait Canyon Fissure, Virginia, and provides samples of *M. pennsylvanicus* M³s from seven stratigraphic levels that range in age from 29.9 ka in the middle of the sequence to 18.4 ka at the top. Deposits from Baker Bluff Cave, Tennessee, pick up where the Strait Canyon sequence ends, with the bottom of Baker Bluff's five stratigraphic levels dating to 19.1 ka, and the top at younger than 10.6 ka. The fissure fill at New Paris no. 4, Pennsylvania, comprises nine stratigraphic levels with *M. pennsylvanicus* M³s; these levels span at least from 12.6 to 11.3 ka and are about the same age as the upper part of the Baker Bluff record. The effects of time-averaging in comparing fossil samples with modern ones, and trapped samples with those from raptor-pellet accumulations, appear to be negligible (Barnosky, 1990, pp. 373–4).

The measurements used in this analysis are length, the width/length ratio, the closure of triangles 1 and 2 as reflected by the ratio closure/length, and how closely the shape of the posterior loop approximates a circle [calculated by the shape factor, SF, where SF $= 4\pi \times$ area/(perimeter)²]. These measurements, illustrated in Figure 3.1 and summarized in Figures 3.2–3.4, mirror some of the variables that workers in arvicolid taxonomy have used in separating species or species groups (the shape factor quantifies the complexity of the posterior loop). Measurements were captured with the BIOQUANT digitizing system, as explained by Barnosky (1990). SYSTAT was used for all statistical computations. Summary statistics for each measured population are presented in Tables 3.3 and 3.4.

The data were examined with the aim of answering the following questions: (1) Were there statistically significant morphological changes

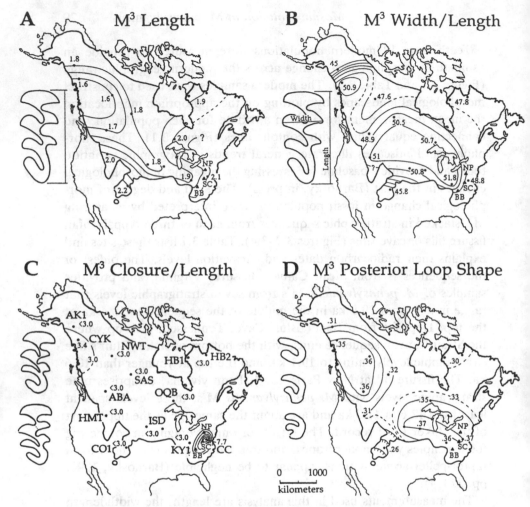

Figure 3.1. Morphological landscapes illustrating general patterns of geographic variation in molar traits of *Microtus pennsylvanicus*. The measurement in occlusal view is illustrated to the left of each map. Locations of modern measured populations are plotted as dots, the means of which were the control points for contouring. The geographic position of each fossil site is plotted on the morphoscape as a triangle (BB, Baker Bluff; NP, New Paris; SC, Strait Canyon). Fossil means were ignored when drawing the contours. (**A**) Length (mm). (**B**) Ratio width/length expressed as (width/length) × 100. (**C**) Ratio closure/length expressed as (closure/length) × 100. Upper diagram shows tooth with tight closure; lower diagram shows tooth with wide confluence, with the measurement closure illustrated by the dotted line. Contours begin at ratio 3.0, and the contour interval is 1.0. Contours below 3.0 are not plotted because it is difficult to ensure accuracy of measuring when closure approaches values small enough to result in ratios below 3.0. (**D**) Shape of the posterior loop of the upper third molar expressed as the ratio $4\pi \times \text{area}/(\text{perimeter})^2$. Diagrams show the range of variation in posterior loops; simple shapes (top) result in a high shape factor, complex shapes (bottom) in a low shape factor, and intermediate shapes (middle) in an intermediate shape factor.

Table 3.2. *Modern localities and specimens employed in the study. See Figure 3.1 for map locations.*

Sample name: AK1
Species/subspecies: *Microtus pennsylvanicus tananaensis(?)*
Location: Walker Creek, 15 miles east of Chicken, Alaska
Date of collection: July 10, 1981, by R. Peterson
Specimen numbers: CM Mammals 76635 (F), 76636 (F), 76637 (M), 76638 (F), 76639 (M), 76640 (F), 76641 (F), 76642 (F), 76643 (M), 76644 (F), 76645 (M)
Comments: 76635 has an aberrant posterior loop on the right M³

Sample name: HB1
Species/subspecies: *Microtus pennsylvanicus labradorius*
Location: East coast of Hudson Bay, Povungnituk Post, 59°40'N, 77°30'W
Date of collection: May 18, 1945, by Doutt
Specimen numbers: CM Mammals 23714 (M), 23716 (M), 23717 (?), 23718 (M), 23719 (F)
Comments: Symmetric dentition

Sample name: HB2
Species/subspecies: *Microtus pennsylvanicus labradorius*
Location: East side of Hudson Bay, mouth of Koksoak River, Ungava (CM 4175) and Fort Chino, Ungava (CM 4160-4164)
Date of collection: Koksoak, September 22, 1917; Fort Chino, August 30, 1917, by O. J. Murie
Specimen numbers: CM Mammals 4160 (M), 4162 (M), 4163 (M), 4164 (M), 4175 (M)
Comments: Symmetric dentition; very small individuals

Sample name: CO1
Species/subspecies: *Microtus pennsylvanicus modestus*
Location: Colorado, Animas Co., Weston R. Parsons Ranch, northeast of Monument Lake
Date of collection: March 31, 1965
Specimen numbers: University of Colorado 13367-13376; not sexed
Comments: Symmetric dentition

Sample name: CC
Species/subspecies: *Microtus pennylvanicus pennsylvanicus*
Location: Comers Cave, Page Co., Virginia, 38°44'59"N, 78°25'02"W
Date of collection: August 1980, by Harold Hamilton
Specimen numbers: Carnegie Museum, Guilday modern collection
Comments: From owl pellets at entrance to Comers Cave

Sample name: NWT
Species/subspecies: *Microtus pennsylvanicus drummondi*
Location: Hay River, Northwest Territory, Canada
Date of collection: June 25-26, 1958, by D. A. Simpson, M. J. Smolen, R. Peterson
Specimen numbers: CM Mammals 74711, 74712 (F), 74713 (F), 74714 (M), 74715 (F), 74716 (F), 74717 (F), 74718 (F), 76523 (F), 76524 (M)
Comments: Symmetric dentition

Table 3.2 (*cont.*)

Sample name: HMT
Species/subspecies: *Microtus pennsylvanicus*; label says *M. p. modestus*, but Hall (1981) says it should be *M. p. pullatus*
Location: Hamilton, Granite Co., Montana
Date of collection: December 25-31, 1958, by R. Peterson
Specimen numbers: CM Mammals 74755 (F), 74756 (F), 74757 (F), 74758 (M), 74759 (F), 74760 (M), 74761 (?), 74762 (F), 74763 (F), 74764 (M)
Comments: Symmetric dentition

Sample name: ISD
Species/subspecies: *Microtus pennsylvanicus pennsylvanicus*, according to Hall (1981)
Location: CM 65000, Bremer Co., Iowa, 7 miles NNW of Waverly; CM 61866, Decatur Co., Iowa, 4.5 miles NNW of Pleasanton; CM 66726-66732, Hansen Co., South Dakota, 8 miles south and 3 miles west of Alexandria; CM 70819, Cerro Gordo Co., Iowa, 0.5 mile east of Mesdervie
Date of collection: 65000 collected 5/31/79; 61866 collected 6/27/75; 66726 collected 5/23/79; 66727 collected 5/23/79; 66728 collected 5/22/79; 66729, 66730, and 66731 collected 5/27/79; 66732 collected 5/25/79; 70819 collected 7/06/80
Specimen numbers: CM Mammals 6500 (M), 61866 (F), 66726 (F), 66727 (F), 66728 (F), 66729 (F), 66730 (M), 66731 (M), 66732 (M), 70819 (M)
Comments: Collected by J. A. Groen, D. C. Lovell, N. D. Moncrief, J. R. Choate, L. L. Johnson

Sample name: KY1
Species/subspecies: *Microtus pennsylvanicus pennsylvanicus*, according to Hall (1981)
Location: Charleston Bottom, Mason Co., Kentucky; CM 60473 2.5 miles west of Maysville; CM 60474-60482 1.5 miles west of Maysville
Date of collection: 60473 collected 11/25/56; 60474 and 60475 collected 2/14/54; 60476 collected 12/25/54; 60477 collected 12/27/54; 60478 collected 12/24/54; 60479 collected 12/25/54; 60480 collected 12/27/54; 60481 and 60482 collected 12/26/54
Specimen numbers: CM Mammals 60473 (M), 60474 (F), 60475 (M), 60476 (M), 60477 (M), 60478 (M), 60479 (F), 60480 (F), 60481 (F), 60482 (M)
Comments: Collected by J. T. Wallace

Sample name: CYK
Species/subspecies: *Microtus pennsylvanicus alcorni*, or *M. p. drummondi*, according to Hall (1981); sample from near the boundary of these two subspecies; subspecies is *drummondi* on museum (CM Mammals) labels
Location: 5 miles north and 1 mile west of Carcross, Yukon Territory
Date of collection: July 1-2, 1981, by D. A. Simpson, M. J. Smolen
Specimen numbers: CM Mammals 76597 (F), 76598 (F), 76599 (M), 76600 (M), 76601 (M), 76602 (?), 76603 (F), 76604 (M), 76606 (F), 76607 (M)

Table 3.2 (*cont.*)

Sample name: ABA
Species/subspecies: Microtus pennsylvanicus drummondi
Age: Modern
Location: 61590 from Salanan Pass; 61593 and 61597 from Sheep Creek Ranger Station; 76515-76521 from 20 miles north of Hotchkiss
Date of collection: Salanan Pass, August 5, 1946, by A. Twomey; Sheep Creek, August 30 to September 1, 1946; by A. Twomey; Hotchkiss, August 5, 1981, by L. E. Carroll and D. A. Simpson
Specimen numbers: CM Mammals 61590 (M), 61593 (?), 61597 (?), 76515 (F), 76516 (M), 76517 (F), 76518 (F), 76519 (M), 76520 (M), 76521 (M)

Sample name: SAS
Species/subspecies: Microtus pennsylvanicus drummondi
Location: Emma Lake, Saskatchewan
Date of collection: July 4, 1939, by F. Banfield
Specimen numbers: CM Mammals 18216 (F), 18217 (F), 18218 (M)

Sample name: OQB
Species/subspecies: Microtus pennsylvanicus fontigenus
Location: 5983 and 5985 from Kegashka River, Gulf of St. Lawrence, Quebec, Canada; 22586-22591 from Kapusking, O'Brien Township, Ontario, Canada; 57553 from Thunder Bay District, Geraldton, Ontario, Canada
Date of collection: Kegashka, June 9, 1928, by J. K. Doutt; Kapusking, October 9 (22586) and 24 (22587), 1943, and May 14, 1944 (22588-22591); Thunder Bay, August 26, 1945
Specimen numbers: CM Mammals 5983 (M), 5985 (M), 22586 (M), 22587 (M), 22588 (M), 22589 (F), 22590 (F), 22591 (M), 57553 (?)

---

through time that exceed morphological differences across space today? (2) Did morphological changes take place episodically or at a constant rate? (3) Were changes unidirectional or fluctuating? (4) Was there congruence in the timing of changes among traits? (5) Did the timing of major changes coincide with times of late Pleistocene climate change, which would suggest selection as a driving mechanism? The answers provide a basis for interpreting the evolutionary implications of the data.

Results

Past and present morphological differences

All fossil specimens were lumped into one sample, and all modern specimens were combined in a second sample; the two samples were

Baker Bluff Cave

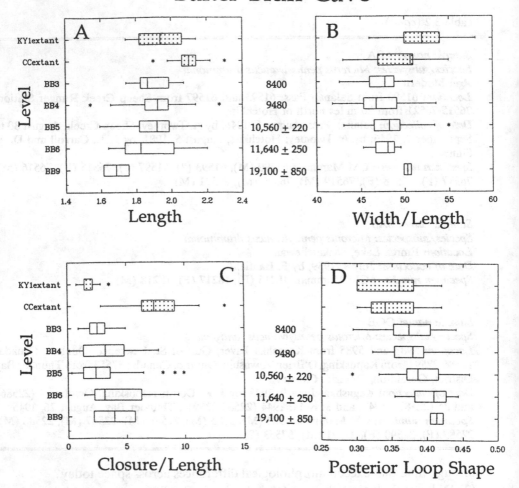

Figure 3.2. Box diagrams illustrating morphological change by stratigraphic level for dental traits of *Microtus pennsylvanicus* from Baker Bluff Cave. (A) Length (mm). (B) Ratio width/length, expressed as (width/length) × 100. (C) Ratio closure/length, expressed as (closure/length) × 100. (D) Shape factor of posterior loop, expressed as $4\pi \times \text{area}/(\text{perimeter})^2$. See Figure 3.1 for illustrations of measurements. Box diagrams for each excavation level are arranged stratigraphically, with the oldest level on bottom and the youngest on top. Stipples indicate modern samples. The median value at each stratigraphic level is shown by the vertical line within the box; edges of box (hingespread) demarcate the midrange (interquartile range) of values; horizontal lines show ranges of values that fall within 1.5 hingespreads of the box. Asterisks demarcate values that fall outside 1.5 hingespreads; circles plot values that fall outside 3.0 hingespreads. Approximate date is listed by each level; those followed by an error bar are radiocarbon dates, and those without error bars are calculated from the dated levels assuming more or less constant sedimentation rates, as explained in the text. See Table 3.3 for sample sizes, means, and standard deviations.

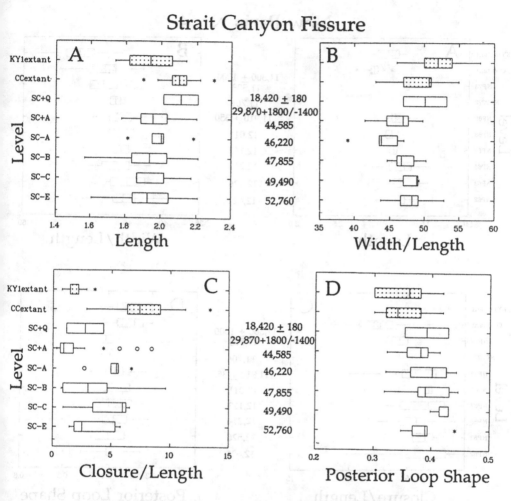

Figure 3.3. Box diagrams illustrating morphological change by stratigraphic level for dental traits of *Microtus pennsylvanicus* from Strait Canyon Fissure. See Figure 3.2 for additional explanation.

then compared by a *t* test to provide a first-order approximation of whether morphology had changed between the late Pleistocene/early Holocene and the present. The lumped fossil sample was statistically different (probability value = 0.000) from the lumped modern sample in having narrower teeth, reflected by smaller ratio of width to length, and in having a simpler posterior loop, shown by the higher shape factor (Table 3.4). This suggests that the fossil populations from the Appa-

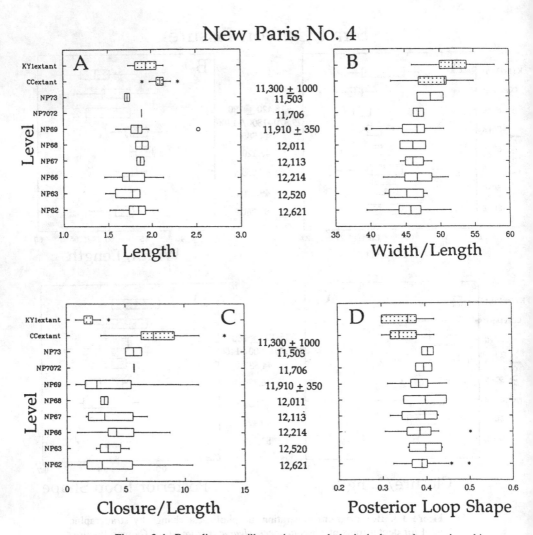

Figure 3.4. Box diagrams illustrating morphological change by stratigraphic level for dental traits of *Microtus pennsylvanicus* from New Paris no. 4. See Figure 3.2 for additional explanation.

lachians were characterized by narrower teeth and less complex posterior loops than is the modern species as a whole.

The obvious question then arises: Are the fossil Appalachian populations different from the modern Appalachian ones, which would imply temporal variation, or are both fossil and modern Appalachian populations different from modern populations in other geographic areas, which would indicate geographic variation persisting through time? To answer this question, all Appalachian fossil specimens were lumped into one sample, which was then compared with a second sample composed of only the Appalachian modern specimens: Mason, Kentucky (KY1), and Comers Cave, Virginia (CC). A *t* test was used to compare the means of the two samples (Table 3.4). Within this Appalachian subset of the data, the differences between fossil and modern specimens were statistically significant for all four traits: width/length ratios, posterior loop shape factor, closure/length ratios, and length. Thus the modern Appalachian populations are even more different from the fossil Appalachian ones than is the overall modern sample. This bears out the postulate that the differences between the overall modern sample and the Appalachian fossil sample result from temporal rather than geographic variation. Within the Appalachians on the geographic scale encompassed by the measured samples (Pennsylvania, Virginia, Kentucky), morphologies apparently have changed slightly since the late Pleistocene. The statistical differences indicate that, on average, teeth became longer, wider relative to length, and more confluent, and they attained slightly more complex posterior loops by modern times.

Were these temporal changes concentrated in one locality, thus weighting the fossil or modern averages and thereby accounting for the difference between fossil and modern means in the Appalachians? Box diagrams were constructed (Figure 3.5), and a one-way ANOVA and Tukey's HSD test comparing all Appalachian localities were performed to shed light on this question. For this part of the analysis, all levels within a given fossil locality were lumped, and then each fossil locality was compared with each modern Appalachian locality (Table 3.4 list means for each sample).

The one-way ANOVA for each measurement illustrates highly significant differences between localities ($p < 0.003$ in all cases). Tukey's HSD pairwise comparisons implicate Comers Cave as the main source of differences in length, closure, and posterior loop shape, and Comers Cave and Kentucky equally as the main sources of difference in width. Comers Cave therefore appears to be weighting the modern average and causing the statistical differences between modern and fossil means. This observation was further tested by deleting Comers Cave from the Appalachian data set. Without the Comers Cave sample, the difference between localities is not significant for posterior loop shape, and New Paris is the main source of differences between localities for length and closure. However, Kentucky remains the source of difference in width.

Table 3.3. *Summary statistics for localities and levels*

Locality	Level	Statistic	LEN	WIDLEN	CLOSLEN	PLSHP
CYK	Modern	N	10	10	10	10
		Mean	1.556	50.800	3.480	0.344
		SD	0.173	2.860	2.866	0.035
AK1	Modern	N	11	11	11	11
		Mean	1.791	43.455	2.214	0.307
		SD	0.116	3.327	1.226	0.039
CC	Modern	N	20	20	20	20
		Mean	2.084	49.341	7.785	0.356
		SD	0.104	3.397	2.614	0.044
SAS	Modern	N	3	3	3	3
		Mean	1.743	50.667	2.150	0.313
		SD	0.002	4.163	0.700	0.023
HMT	Modern	N	10	10	10	–
		Mean	2.052	48.200	2.010	–
		SD	0.116	3.327	1.226	–
NWT	Modern	N	10	10	10	–
		Mean	1.732	47.800	2.150	–
		SD	0.143	2.394	0.990	–
ABA	Modern	N	10	10	10	10
		Mean	1.668	49.400	2.570	0.360
		SD	0.162	2.503	2.169	0.034
HB1+2	Modern	N	10	10	10	10
		Mean	1.924	48.000	1.870	0.296
		SD	0.230	3.266	0.885	0.055
KY1	Modern	N	10	10	10	10
		Mean	1.932	51.800	1.800	0.356
		SD	0.147	3.584	0.889	0.047
OQB	Modern	N	9	9	9	9
		Mean	1.944	50.889	1.917	0.349
		SD	0.155	3.018	0.606	0.027
ISD	Modern	N	10	10	10	10
		Mean	1.828	50.800	2.150	0.368
		SD	0.201	3.553	1.235	0.046
CO1	Modern	N	10	10	10	10
		Mean	2.188	45.800	2.150	0.282
		SD	0.120	4.756	0.808	0.025
Baker Blf	BB4	N	47	47	47	47
		Mean	1.913	47.203	3.099	0.376
		SD	0.126	2.819	2.025	0.029
Baker Blf	BB5	N	35	35	35	35
		Mean	1.902	48.169	2.924	0.392
		SD	0.122	2.626	2.421	0.037
Baker Blf	BB6	N	3	3	3	3
		Mean	1.890	47.635	4.657	0.402
		SD	0.285	2.199	4.227	0.045

Table 3.3. (cont.)

Locality	Level	Statistic	LEN	WIDLEN	CLOSLEN	PLSHP
Baker Blf	BB9	N	2	2	2	2
		Mean	1.685	50.309	3.955	0.413
		SD	0.100	0.581	2.493	0.013
New Paris	NP73	N	2	2	2	2
		Mean	1.729	48.618	5.700	0.405
		SD	0.041	2.641	0.989	0.018
New Paris	NP7072	N	2	2	2	2
		Mean	1.891	46.905	5.725	0.396
		SD	0.001	1.049	0.044	0.026
New Paris	NP69	N	21	21	21	21
		Mean	1.855	46.102	3.617	0.386
		SD	0.179	2.942	2.898	0.033
New Paris	NP68	N	2	2	2	2
		Mean	1.892	46.046	3.120	0.398
		SD	0.099	2.607	0.450	0.071
New Paris	NP67	N	4	4	4	4
		Mean	1.874	46.274	3.668	0.385
		SD	0.046	1.883	2.446	0.052
New Paris	NP66	N	21	21	21	21
		Mean	1.779	46.757	4.512	0.380
		SD	0.162	2.434	1.974	0.047
New Paris	NP63	N	4	4	4	4
		Mean	1.724	45.084	3.621	0.398
		SD	0.190	2.911	1.229	0.043
Strait Can	SC+Q	N	2	2	2	2
		Mean	2.111	50.104	2.728	0.391
		SD	0.136	4.443	2.238	0.054
Strait Can	SC+A	N	22	22	22	22
		Mean	1.968	46.310	1.965	0.377
		SD	0.125	2.401	2.315	0.026
Strait Can	SC-A	N	5	5	5	5
		Mean	1.993	43.878	5.075	0.391
		SD	0.133	3.090	1.492	0.050
Strait Can	SC-B	N	10	10	10	10
		Mean	1.931	47.074	3.828	0.396
		SD	0.183	1.867	3.448	0.040
Strait Can	SC-C	N	3	3	3	3
		Mean	1.966	47.709	4.476	0.419
		SD	0.177	2.341	3.127	0.018
Strait Can	SC-E	N	6	6	6	6
		Mean	1.925	48.111	3.238	0.388
		SD	0.207	3.020	1.895	0.031

Note: Measurements in millimeters; *N*, sample size; SD, standard deviation; LEN, length; WIDLEN, width/length; CLOSLEN, closure/length; PLSHP, posterior loop shape.

Table 3.4. *Summary statistics for lumped fossil and lumped modern samples*

Locality	Level	Statistic	LEN	WIDLEN	CLOSLEN	PLSHP
All modern	–	N	123	123	123	103
		Mean	1.893	48.787	3.135	0.336
		SD	0.233	3.949	2.681	0.048
App. mod.[a]	–	N	30	30	30	30
		Mean	2.033	50.160	5.790	0.356
		SD	0.138	3.597	3.599	0.044
All fossil	–	N	234	234	234	234
		Mean	1.885	46.960	3.355	0.385
		SD	0.152	2.722	2.385	0.037
Baker Bluff	All	N	105	105	105	105
		Mean	1.902	47.657	3.024	0.382
		SD	0.128	2.610	2.113	0.036
New Paris	All	N	81	81	81	81
		Mean	1.818	46.218	3.979	0.388
		SD	0.154	2.666	2.455	0.041
Strait Can	All	N	48	48	48	48
		Mean	1.963	46.686	3.025	0.387
		SD	0.151	2.728	2.650	0.034

[a]Appalachian modern localities.

These data imply that since the late Pleistocene, more morphological change (in all four traits) has taken place in Virginia (Comers Cave) than in Kentucky (where only narrowing of the tooth occurred).

Episodic or constant change?

Data from the fossil localities were plotted stratigraphically to assess whether trends of morphological change were apparent over the several thousand years the samples span. Because Baker Bluff is located midway between the modern Comers Cave and Kentucky localities, these data were also added to the stratigraphic interpretations. Strait Canyon is very near Comers Cave, but more distant from Kentucky, so only Comers Cave data supply the modern stratigraphic level in the Strait Canyon inferences. New Paris is far enough from both Comers Cave and Kentucky so that neither of the modern localities was used in trying to interpret population-level evolutionary patterns there. (However, both modern populations are depicted on all the plots, for the sake of other comparisons.)

Of the 12 plots (four traits for each of three localities), only one – posterior loop shape at Baker Bluff – shows a unidirectional trend (Fig-

Appalachian Localities

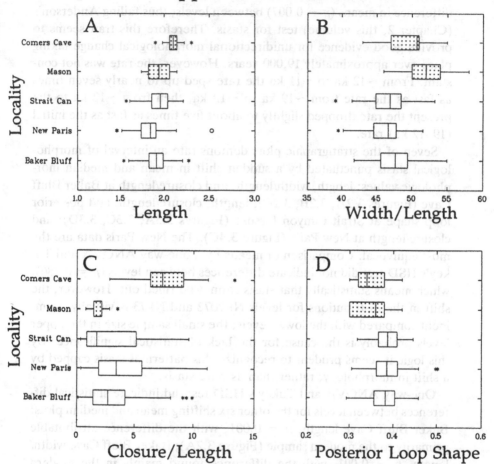

Figure 3.5. Box diagrams illustrating comparisons of lumped data for each fossil locality and modern Appalachian localities. The fossil data summarized in each box include all specimens from a given locality, irrespective of stratigraphic level. See Figure 3.2 for more explanation and Table 3.4 for sample sizes, means, and standard deviations.

ure 3.2D). This trend begins with a posterior loop that has a higher shape factor, that is, more closely approaches a circle, than is found as a mean or median value in any measured modern population (Barnosky, in press). It then proceeds unidirectionally toward a shape factor that characterizes both the modern Appalachian populations analyzed here (Comers Cave and Kentucky). The probability of this happening by chance alone is 0.03 (lumping both modern samples as one stratigraphic level). A one-way ANOVA and Tukey's HSD test indicated a significant difference in means ($p = 0.007$) between levels, thus failing Anderson's (Chapter 2, this volume) test for stasis. Therefore this trait seems to provide good evidence for unidirectional morphological change taking place over approximately 19,000 years. However, the rate was not constant: From ~12 ka to ~11 ka the rate sped up to nearly seven times as fast as the rate from ~19 ka to ~12 ka, then from ~12 ka to the present the rate dropped slightly to about five times as fast as the initial (19–12 ka) rate.

Seven of the stratigraphic plots demonstrate an interval of morphological stasis punctuated by a sudden shift in mean and median morphologic values: length, width/length, and closure/length at Baker Bluff Cave (Figures 3.2A, 3.2B, 3.2C); length, closure/length, and posterior loop shape at Strait Canyon Fissure (Figures 3.3A, 3.3C, 3.3D); and closure/length at New Paris (Figure 3.4C). The New Paris data are the most equivocal. Comparison of means by a one-way ANOVA and Tukey's HSD test did not indicate differences between levels ($p = 0.786$), which means statistically that stasis cannot be ruled out. However, the shift in the distributions for levels NP7072 and NP73 is visually prominent compared with the lower levels; the small sample size in the upper levels probably is the cause for the lack of statistical significance. By this logic it seems prudent to recognize this pattern as stasis capped by a shift in morphology, rather than as pure stasis.

One-way ANOVA and Tukey's HSD test did indicate statistical differences between levels for the other six shifting mean and median plots: Baker Bluff Cave length, $p = 0.001$, with the difference attributable primarily to the modern sample (Figure 3.2A); Baker Bluff Cave width/length, $p = 0.001$, with the difference found mainly in the modern specimens (Figure 3.2B); Baker Bluff Cave closure/length, $p = 0.001$, with the difference due mainly to the modern sample (Figure 3.2C); Strait Canyon length, $p = 0.045$, with the differences attributable to the modern and level SC+Q samples (Figure 3.3A); Strait Canyon closure/length, $p = 0.001$, with the modern sample being the different one (Figure 3.3C); Strait Canyon PLSHP, $p = 0.037$, with the modern sample containing most of the difference (Figure 3.3D). Visually prominent shifts that are not statistically significant, very likely because of the small sample size in the pertinent levels, also are evident in the Baker Bluff Cave plots for length and width/length from levels BB9 to BB6 (Figures 3.2A, 3.2B).

Most of the shifts in morphology take place between the highest fossil stratigraphic level and the present. These include length, width/length, and closure/length at Baker Bluff, and closure/length and posterior loop shape at Strait Canyon. At Baker Bluff, the highest reliable radiocarbon date, 10,560 ± 220 years BP, is from level BB5; the date from level BB6 is 11,640 ± 250 years BP. Assuming that the rates of deposition were similar from levels BB6 to BB3, level BB3 would date to around 8,400 years BP. Therefore, the tightest temporal bracket on the shift in morphology between level BB3 and the present is about 8,400 years. At Strait Canyon, the radiocarbon age on the highest stratigraphic level SC + Q is 18,420 ± 180 years BP; the changes in closure/length and posterior loop dimensions between level SC + Q and the present therefore took no more than about 18,420 years.

Observable (though not necessarily statistically significant by the techniques used in this chapter) changes in morphology also took place prior to the highest stratigraphic level in each site. Shifts in the values for length and width/length occurred at Baker Bluff Cave between levels BB9 and BB6. Level BB9 is radiocarbon-dated at 19,100 ± 850 years BP; comparison with the date for level BB6 indicates that these potential shifts in morphology span about 7,460 years. At Strait Canyon, length changed between level SC + A and SC + Q; this change cannot be bracketed tighter than about 11,000 years, because a radiocarbon date of 29,870 (+ 1,800/ − 1,400) years BP was obtained between the two levels (see Table 3.1). At New Paris, the change in length between level NP69 and NP7072 occurred within about 600 years, because a radiocarbon date from level NP69 was 11,910 ± 350 years BP, and one from near level NP7072 was 11,300 ± 1,000 years BP.

Of the seven sudden shifts in morphology, then, two occurred within about 18,000 years (Strait Canyon closure/length and posterior loop shape). One took place within about 11,000 years (Strait Canyon length). Five took place over a maximum of around 8000 years (Baker Bluff length and width/length twice and closure/length once). And one possibly occurred within 600 years (New Paris closure/length).

Three of the plots indicate morphological stasis throughout the time they represent: length, width/length, and posterior loop shape for New Paris (Figures 3.4A, 3.4B, 3.4D). None of these three plots for New Paris shows statistical significance between levels with a one-way ANOVA and Tukey's HSD test; they would pass Anderson's (Chapter 2, this volume) test for stasis. The length of time spanned by these relatively static traits is more than 1,300 years (assuming constant rates of deposition).

The Strait Canyon plot for width/length combines a period of stasis in levels SC-E, SC-C, and SC-B, with a sudden shift in mean and median at level SC-A, then a directional trend up to the modern width/length (Figure 3.3B). One-way ANOVA and Tukey's HSD test show a statistical difference between levels ($p = 0.003$), with the difference attributable to the contrast between levels SC-A and SC+A and modern Comers Cave. The shift at level SC-A is temporary, with overlying levels returning to medians and means that resemble those of levels below SC-A. Therefore this pattern reflects that the morphology fluctuates randomly around a mean through time but does not deviate from it to stay. Most workers would classify this pattern, taken as a whole, as stasis, even though it would not pass Anderson's test. Assuming constant rates of deposition at this site, admittedly a dangerous assumption in this case, the static part of the pattern (bottom three levels) would span about 6,500 years; the transition from SC-B to SC-A would span some 1,600 years; and the directional change from SC-A to the present would encompass some 46,000 years. [These calculations are based on the two known dates being separated by seven equally thick stratigraphic levels, equaling ~1,635 years per level, and level SC-A being 10 stratigraphic levels (i.e., 10 x 1,635 years) below level SC+J, from which the date of 29,870 years BP was obtained. SC+A, SC-A, SC-B, and SC-C are adjacent stratigraphic levels (listed in descending order), and SC-E is separated from SC-C by one stratigraphic level.]

Congruence among traits

At Baker Bluff the evolutionary patterns of two traits act in concert: length and width/length. Both show prominent shifts in median and mean from level BB9 to BB6 (taking ~8,000 years), indicating a lengthening of the tooth at that time, but not a corresponding widening (Figures 3.2A, 3.2B). A period of stasis prevails from level BB6 to BB3 (for ~3,000 years). Then between level BB9 and the present, length increases again, but this time it is accompanied by a proportionately even greater increase in width (Figures 3.2A, 3.2B); this change took place within ~8,000 years. Closure of triangles 1 and 2 (closure/length), meanwhile, remains static from level BB9 to BB3 (10,700 years), then jumps significantly between level BB3 and modern times (8,400 years). Posterior loop shape acts entirely independently, showing a gradual trend from top to bottom. Hence the overall sequence of change at this site is first an increase in length of the tooth (with no corresponding increase in width) that took place over ~8,000 years, then stasis for

~3,000 years, then concurrent increases in width, length, and opening of the triangles taking place in the ensuing ~8,000 years, all overlaid on a gradual increase in complexity of the posterior loop that took place continuously, but at varying rates, throughout the whole 19,000 years represented by the record.

At Strait Canyon, length and width/length also seem to march generally to the same beat, although they do get slightly out of step (Figures 3.3A, 3.3B). An increase in length accompanies a proportionately greater increase in width between levels SC + A and SC + Q; then stasis takes over from level SC + Q to the present. Prior to level SC + Q, stasis also prevails for both traits, although the minor decrease in length at level SC-A, then the increase to SC + Q, is not echoed by shifts in the length. The posterior loop shape (Figure 3.3D) and the ratio of closure/length (Figure 3.3C) show stasis through the entire fossil record, but means and medians for both traits shift between level SC + Q and the present. The overall pattern of change at this site, then, is stasis for the lower levels (perhaps as long as 8,000 years), then an increase in length and a proportionately greater increase in width between levels SC + A and SC + Q (taking place within 26,000 years, with no change after that for these traits), followed (some 26,000 years later?) by an increase in complexity of the posterior loop and widening confluence between triangles 1 and 2 between level SC + Q and modern times.

At New Paris, length, width/length, and posterior loop shape all act the same way: They show stasis throughout the record (Figures 3.4A, 3.4B, 3.4D). Closure/length appears to depart from the static pattern, showing jumps in mean and median between levels NP69 and NP7072 (Figure 3.4C). The overall pattern of change here is stasis in all characters for at least 1,300 years, then a possible widening of confluence between triangles 1 and 2 that may have been accomplished in as little as 200 years.

Correlation with climate change

The fossil records span times of noticeable climate change in the central Appalachian area, as documented by glaciostratigraphic and palynological evidence (Figure 3.6). Somewhere around 19 ka, climatic cooling effected the transition from interstadial (warm time within a glacial period) to full glacial conditions. A more pronounced climate change took place at the end of the Pleistocene, when the last glacial gave way to the present interglacial. Although in general that event took place

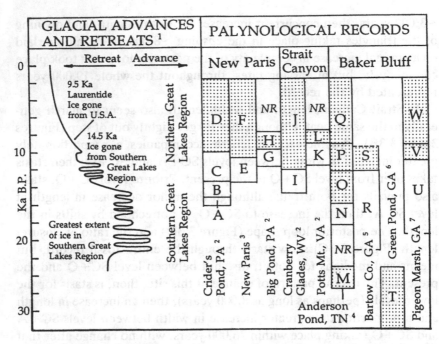

Figure 3.6. Correlations of data indicating climate changes near the fossil sites where teeth of *Microtus pennsylvanicus* were found. Information is from the following sources: [1]Mikelson et al. (1983); [2]Watts (1979); [3]Guilday et al. (1964); [4]Delcourt (1979); [5]Watts (1970); [6]Watts (1973); [7]Watts (1975). Stipple in bars that summarize pollen diagrams designates warm times, generally recognizable by a relatively high percentage of *Quercus* pollen; clear bars indicate cool times, generally recognizable by the presence of *Picea* in relatively high percentage and the absence or much-reduced frequency of *Quercus*. Capital letters mark vegetational zones characterized by high relative percentages of certain pollen types. For each zone, the following list summarizes the most important pollen types ordered from highest to lowest abundance: **A,** *Picea, Pinus, Betula*; **B,** *Picea, Pinus*; **C,** *Picea* decreases while *Pinus* increases; **D,** *Quercus* dominates; **E,** *Pinus, Picea*; **F,** *Alnus, Betula, Quercus*; **G,** *Pinus, Picea, Betula*; **H,** *Quercus, Pinus*; **I,** *Picea, Pinus*; **J,** *Quercus* dominates, with *Tsuga* early giving way to *Betula* later; **K,** *Quercus, Picea, Pinus*; **L,** *Quercus, Tsuga, Pinus, Betula*; **M,** *Pinus, Quercus, Picea*; **N,** *Pinus, Picea*; **O,** *Quercus, Picea, Pinus*; **P,** *Quercus, Picea, Ostrya-Carpinus, Fraxinus*; **Q,** *Quercus, Fraxinus*; **R,** *Pinus, Picea, Quercus*; **S,** *Quercus, Liquidambar, Pinus*; **T,** *Quercus, Carya, Pinus*; **U,** *Pinus, Quercus*; **V,** *Quercus, Fagus, Pinus*; **W,** *Quercus, Pinus, Carya*; **NR,** no record.

between 10,000 and 15,000 years ago, the exact timing varied for different latitudes and elevations as a function of distance from the ice sheet, atmospheric circulation patterns, and local climatic patterns. In view of the multiple climate changes that occurred after 30 ka and the control exerted by local influences, correspondences between climate change and morphological change can be identified securely only when the timing of morphologic change at a given site is compared with nearby paleoclimatic records.

At New Paris, fossil pollen profiles from the fissure fill itself (Guilday, Martin, and McCrady, 1964) and from two lake cores taken from Crider's Pond and Big Pond, Pennsylvania (Watts, 1979), provide a means of assessing when the climate changed, independent of the mammal fossils. The pollen profiles indicate a change from forests dominated by *Picea* and *Pinus* to those dominated by *Quercus* and other deciduous trees between 12 and 10 ka. The final waning of the Laurentide ice sheet, leading to its exit from North America by 9.5 ka, also corresponds (Mikelson et al., 1983). The only concurrent morphological change in *M. pennsylvanicus* teeth from New Paris is a widening of closure (Figure 3.4C). Therefore, three of the four analyzed traits showed no response to the terminal Pleistocene climate change (compare Figures 3.4C and 3.5).

Baker Bluff provides weak evidence for a correlation between climate change and morphological change. The paleoclimatic inferences derive from the nearest good palynological records, which still are farther from Baker Bluff than would be ideal: Anderson Pond, Tennessee (Delcourt, 1979), and four bogs in Georgia (Watts, 1970, 1973, 1975, 1979). Between 19 and 12 ka, changes in length and width/length ratios (Figures 3.2A, 3.2B) took place in the same interval that the full glacial apparently gave way to warmer times, between 16 and 12 ka. In the pollen records, this is marked by the change from assemblages that include relatively high proportions of *Pinus* and *Picea* to those dominated by *Quercus* and other deciduous trees. However, the interval between fossil tooth samples is so broad that it is impossible to document a close correspondence between morphological change and climate change. The two other traits (closure and posterior loop) did not change between 19 and 12 ka. Three of the four traits (length, width/length, closure) also change between 8.4 ka and the present, a time when the pollen records show possible evidence for climate change reflected by the pronounced decreases in *Fagus* (Pigeon Marsh) and *Ostrya-Carpinus* (Anderson Pond). The exact nature of this climate change and why it might affect *M. pennsylvanicus* are unknown; however, it may mark the shift from early Holocene moist cool-temperate conditions to the slightly drier conditions of the middle Holocene (Watts, 1975, p. 291; Delcourt, 1979, p. 271). Thus, at the Pleistocene–Holocene transition, two traits may have changed, and two were unaffected. And at a much less marked environmental change at 8.4 ka, three of four traits changed. Posterior loop shape apparently trended independent of all climate changes.

Detailed comparisons of the Strait Canyon record with climate changes are not possible because the dating of most levels is too imprecise, and because palynological records of appropriate age do not exist in the region. The pollen records from farther north and south indicate that climate changes did indeed take place between 18 ka and the present, when closure and posterior loop shape changed at Strait Canyon; therefore, changes in these traits *may* correspond with climate changes, but length and width/length did not change through the same interval. Length and width/length did change between 44 and 18 ka, and the changes *may* have corresponded with interstadial-stadial transitions in that time frame; however, the other two traits remained static then.

These relationships between climate change and morphological change are summarized in Table 3.5. An unequivocal morphological change coeval with climate change is present only in one case ("Yes" in Table 3.5). Correspondence with climate change cannot be rejected for nine other cases; neither can it be accepted in most of these cases, because either the sampling interval is too wide or the climate change is not well documented, or both ("Maybe" in Table 3.5). Taken at face value, these data do not contradict the idea that climate change may be important in triggering morphological change, because 9 of the 10 traits that changed at least conceivably did so at times of climate change. Another consistent example was provided by Martin and Prince (1990, p. 126), who interpreted dwarfing in *M. pennsylvanicus* from three stratigraphic levels in Bell Cave, Alabama, as a possible response to climatic cooling between 30 and 18 ka. However, in eight instances ("No" in Table 3.5), morphology remained static through climate change; one trait (posterior loop shape at Baker Bluff) demonstrated morphological change independent of climate change. Moreover, climate change does not affect every trait at a given time and site (16–12 ka at Baker Bluff, or 12–11 ka at New Paris). Nor does it affect the same trait every time that the climate changes (closure at Baker Bluff). Therefore, the data do not allow for generalizations about which traits will be affected by climate change, nor for the generalization that climate change drives morphological change. Considerably more work on detailed sequences of morphological change and climate change will be needed to establish whether or not general links exist.

Table 3.5. *Correspondence between morphological change and climate change in analyzed traits of* Microtus pennsylvanicus *dentition*

	Climate change: Did morphological change correspond with climate change?			In situ Evolution: Does the pattern suggest evolution or immigration?	
	>19 ka	<19 ka	<9 ka	Evol.	Imm.
Length					
SC VA	Maybe	No <18 ka	–	Maybe	No
BB TN	– [a]	Maybe 16-12 ka	Maybe 8.4 ka	Maybe	No
NP PA	–	No 11-12 ka	–	Maybe	Yes
Width/length ratio					
SC VA	Maybe	No <18 ka	–	Maybe	Yes
BB TN	–	Maybe 16-12 ka	Maybe 8.4 ka	Maybe	Yes
NP PA	–	No 11-12 ka	–	Maybe	Yes
Closure					
SC VA	No	Maybe <18 ka	–	Yes	No
BB TN	–	No 16-12 ka	Maybe 8.4 ka	No	No
NP PA	–	Yes 11-12 ka	–	No	No
Posterior loop					
SC VA	No	Maybe <18 ka	–	No	No
BB TN	–	No 16-12 ka[b]	No 8.4 ka[b]	Yes	No
NP PA	–	No 11-12 ka	–	No	No

[a]Undeterminable.
[b]Trend independent of climate change.

Discussion

Caveats

The main caveat in interpreting these data is the dating control, as is true in most studies of evolutionary rates and patterns. Whereas all of the fossil sites have at least two radiocarbon dates, they still fall short of the ideal of a date for each level. Therefore, to calculate time spanned from the bottom to the top of each deposit, one is forced into the assumption of constant sedimentation for strata without dates within each deposit, an assumption that can affect evolutionary interpretation (MacLeod, 1991).

In the case of Baker Bluff, this assumption probably is justified within the top or bottom parts of the sequence, but not when comparing the top two-thirds with the bottom one-third. The deposits at Baker Bluff are near the mouth of the cave and represent *Neotoma* midden deposits, in which "unlike rodent middens in the arid West . . . where plant remains are well preserved, most plants and pollen had decayed in [the] humid eastern environment, leaving a lag deposit of bones, teeth, snails and hackberry (*Celtis*) seeds. . . . Raptors, primarily owls, were responsible for the bulk of the small vertebrates in the deposit" (Guilday et al., 1978, p. 7). This depositional environment remained constant throughout the sequence, and the excavation shaft retained more or less constant dimensions from top to bottom (Guilday et al., 1978). However, sedimentation rates obviously do change in this sequence, shown by the three stratigraphic levels from BB9 to BB6, representing 7,460 years (rate = 2,486 years/level), whereas the adjacent levels BB6 and BB5 represent just 1,080 years (1,080 years/level). The slower rate is assumed for the upper two levels at Baker Bluff because during that time climatic conditions, which would contribute to sedimentation rate, more closely resembled those during BB6 than those during BB9.

New Paris is a fissure fill that retains relatively constant diameter throughout its depth and shows no signs of major sedimentary changes through the bone-bearing deposits. Small mammals apparently fell into this natural trap throughout the time the sequence represents. There is no objective evidence to contradict the assumption that the sediment accumulation rate was more or less constant through the bone-bearing interval. Guilday et al. (1964, p. 174) suggested that the bone-bearing part of the sequence spanned from about 11.3 ka to as much as 15 ka BP, on the basis of a radiocarbon date from the top of the deposit (4.6-m level) and pollen and faunal composition indicative of "boreal" conditions at the base of the excavation (8.5-m level). Since the report by Guilday et al. (1964), an additional radiocarbon date of 11.91 ka BP at 6.4 m provides evidence that the sedimentation rate was slower than Guilday and associates assumed for the lower part of the deposit. Projection of the rate calculated between the two dates places the bottom of the deposit at ~13 ka (Figure 3.4 and Table 3.2). However, radiocarbon dates at the bottom of the excavation and at the 5.2–6.4-m level were anomalously young (9,250 ± 120 and 9,540 ± 500 years BP, respectively), probably because of contamination by younger material from above. If similar contamination affected the 11.91-ka date, the bottom of the deposit probably would be older than 13 ka.

The assumption of a constant sedimentation rate at Strait Canyon Fissure may not be trustworthy, first because Strait Canyon is a narrow, irregularly shaped fissure where 1 m of sediment in a narrow section will represent less time than 1 m in a wide section, and second because a radiocarbon date at the bottom of the deposit indicated a younger stratum than the top two readings, suggesting either contamination by younger bones there or unknown problems in obtaining reliable dates (see Table 3.1). The more recent date at the bottom implies that if an error is being made in calculating the sedimentation rate, the error is in interpreting too slow a rate. Therefore the time spanned at Strait Canyon should be recognized as the maximum possible for the sequence.

Despite these problems, the dating control on these sequences still is more precise than that on any other stack of strata used to interpret evolutionary patterns.

The principle of studying evolution with fossils involves using morphological patterns through time to infer evolutionary patterns. Morphological change by itself, however, does not indicate evolution if the changes can also be explained by wholesale displacement of populations, as, for example, in front of retreating and advancing ice sheets in the case of late Pleistocene organisms. Detailed analyses of modern morphological patterns (Figure 3.1) and comparisons between them and the fossil patterns (Barnosky, 1990, in press) indicate that although displacement of populations cannot be ruled out for some traits, it can for others, as illustrated in the right-hand columns of Table 3.5. If displacement of populations caused the apparent morphological changes, then patterns of change consistent with immigration should show up for all characters ("Yes" in all cells). All of the traits are on the same tooth within each individual, after all. However, the only unequivocal agreement with the expectation of widespread population displacement lies in the length changes for New Paris and in the width/length ratios for all sites, and *in situ* evolution cannot be ruled out for these. On the other hand, morphological change resulting mainly from immigration can be ruled out in a number of cases (Barnosky, in press), and there is particularly good evidence for *in situ* evolution in two cases: toward less pronounced closure of triangles at Strait Canyon, and the marked trend toward more complex posterior loops at Baker Bluff. Therefore it seems warranted to interpret the morphological changes at these fossil sites as indicating *in situ* evolution.

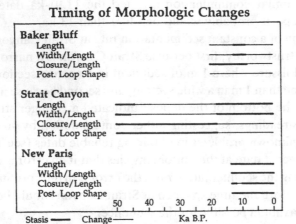

Figure 3.7. Summary of when morphological changes occurred in teeth of *Microtus pennsylvanicus*.

Evolutionary patterns and processes

The first thing that becomes apparent from the data presented here is that most of the morphological changes illustrated for each site indicate that a pattern of stairstep morphological change prevails within populations of *M. pennsylvanicus* on the time scale of 1,000 to 10,000 years. (The only exception is posterior loop shape at Baker Bluff.) However, only in the case of New Paris closure/length can the change be demonstrated to have taken place in less than 600 years (possibly as little as 200 years). For all of the other morphological jumps, the two data points are separated by at least 8,000 years. Considering that the data presented here represent some of the most precise dating of temporal differences between stratigraphic levels, even what appears as stairstep change in the fossil record can very easily be accommodated by microevolutionary processes.

The second important point is that in any population, several traits change through time, but not all at the same time and not all in the same way, as summarized in Figure 3.7. For example, at Baker Bluff, length and width/length changed twice, separated by a period of stasis; closure/length changed only once, after a longer period of stasis that lasted through the first change for length and width/length; and posterior loop shape gradually was becoming more complex throughout the times of both stasis and rapid change in the other traits.

Such a pattern could result only if the genes controlling each trait were not tightly linked; that is, one gene (or gene complex) might control length, another might control width, a third would control the amount of confluence between triangles, and a fourth would determine the shape of the posterior loop. Therefore, analysis of just one trait on the tooth might yield a very different evolutionary interpretation than would another trait. For example, looking only at length at Baker Bluff implies a process of punctuated change, whereas looking only at the posterior loop implies phyletic gradualism. The truth is that both were happening in a given animal at the same time.

The real process seems to involve a mosaic of change, whereby each of the four traits is varying essentially independently (in the genetic sense) through time, but with a selective "rubber band" that holds the combination of them within certain limits. In this model, selection is regarded as important and acts on the entire tooth morphology, not on just one of the traits, to cause evolutionary change. If, for example, the important advantage for microtine rodents is to increase the amount of abrasion their teeth can withstand (a logical assumption given the overall trend of increasing hypsodonty and lophodonty through time in all microtine rodents), they can do so by changing the four traits analyzed in this study in different ways. One way is simply to increase tooth length, which correspondingly increases the length of the enamel cutting band. But because tooth length is linked to the size of the animal (Martin, 1986, and Chapter 11, this volume; Damuth and MacFadden, 1990), that involves an energetic cost (= more food and more tooth abrasion), which may not be desirable. Another way to slow tooth wear is to increase complexity in the posterior loop; yet a third way is to leave the posterior loop alone, but increase the indentations between triangles 1 and 2 so that those two triangles become less confluent. Or they could increase complexity just slightly in the posterior loop *and* decrease the confluence between triangles. Essentially, a given length of cutting enamel can be achieved by adjusting the four traits analyzed in a variety of ways – doing all of the adjusting in one trait, or spreading it out over two or more. As long as the overall morphology does not break the selective rubber band, any combination of the traits is legal. If unlinked genes control different traits of a single tooth, the implication for analyzing evolutionary patterns is that looking at single traits will not provide all the information that is needed for reliably interpreting the evolutionary process in an organism. Instead whole character complexes must be analyzed.

The idea that the four traits can combine in different ways to produce the same end result for chewing explains why variation is pronounced among populations: Different parts of the tooth can change and yet still keep the total morphology within the selective constraints for a given area. However, the interesting question is why the differences arise in the first place. Does the morphology of the founding members of a given population remain in a given area, or do slightly different selective pressures create and maintain differences among sites, or does morphology at a given site change by randomly drifting around within the selective rubber band? At Baker Bluff, where the dating control is best, the data seem to argue for some combination of all three possibilities. There the change in posterior loop shape seems most readily explainable by a selective pressure acting over 19,000 years to produce a more complex loop and hence lengthen the enamel band that resists wear. This postulate is corroborated by the fact that, on average, all of the fossil material (including that from Strait Canyon and New Paris) has a more simple posterior loop than all of the modern material. Overlaid on the directional trend of increasing posterior loop complexity are spurt-like changes in length, width/length, and closure/length that may have been due to pronounced selection at certain times, to random drift, or to periodic influx of genes from immigrants (Figure 3.2). The Baker Bluff example may illustrate how character complexes can evolve by a mixture of selection acting over the long term on one part of the complex, with more minor, possibly neither advantageous nor disadvantageous, changes being superimposed every now and then by chance. One cannot totally dismiss the possibility that the trend at Baker Bluff was driven by chance rather than by selection, but the probability of the trend sustaining itself by chance for as many stratigraphic levels as it did is very low ($p = 0.003$).

Strait Canyon and New Paris show slightly different overall patterns of evolution than those at Baker Bluff, and they also differ from each other. Both lack directional trends, and New Paris indicates more stasis than prevailed at Baker Bluff or Strait Canyon (Figure 3.7). This indicates that the factors influencing morphological change were somewhat different at the various sites, even though all sites are in the Appalachians. At Strait Canyon, the difference could be temporal, because most of the sequence is somewhat older than the other two sites. However, New Paris and Baker Bluff overlap in age: Between ~13 ka and 10 ka these two populations were evolving in different ways, with New Paris essentially showing change in closure/length while Baker Bluff underwent stasis, and New Paris showing stasis in length while Baker

Bluff was demonstrating considerable change. It is unclear whether the slightly different timings of late Pleistocene climate change at these two localities influenced the morphological patterns, but whatever the ultimate cause, the contrasting patterns among sites indicate that morphological change probably is geographically localized when it takes place, rather than having selective pressures of wide geographic scale acting across large parts of a species range. Concordant with this idea are data from Martin and Prince (1990) showing decreasing tooth lengths of *M. pennsylvanicus* from ~27 to 12.5 ka BP, during the time the same dimension demonstrated an increase, then stasis, at Baker Bluff. These contrasting patterns among localities support models of speciation that favor local origin of new species, with subsequent expansion of the species range, or else local origin of a highly advantageous trait, with subsequent spread of that trait through adjacent populations if dispersal was rampant enough.

Conclusions

The analysis of these four different dental traits in fossil and modern specimens of *M. pennsylvanicus* demonstrates that stasis, episodic change, and gradual change are not mutually exclusive patterns of morphological response through time. All can be evident concurrently on different traits within the same character complex. One paradigm for why this takes place rests on the fact that selection acts not just on one trait, but on combinations of traits, and different combinations can result in the same solution to a morphological problem. Hence one population may find the solution to resisting tooth abrasion by increasing the complexity of the posterior loop, whereas another might solve the problem by leaving the posterior loop simple but increasing the entire tooth's width and decreasing confluence between triangles. In this paradigm for why different styles of morphological change can be apparent simultaneously in a given character complex, selection figures prominently in setting the limits through which morphology can fluctuate and sometimes in initiating selective changes, but the way in which morphology changes really depends on chance. In the case of the traits analyzed for *M. pennsylvanicus*, the selective part of the equation comes in if, say, vegetation changes such that locally only more abrasive food is available (perhaps a competing species moves into a geographic area and utilizes the more tender parts of grass shoots that otherwise would be available to the meadow vole, or succession or local climatic perturbations change the prevailing plants). The required morphological solution is to make the tooth more resistant to abrasion. The chance part of the equation comes in because the exact solution depends on which genes are available to initiate change in the character complex. If a few animals with genes for complex loops happen to exist in the area, those genes may rapidly spread through the population, and the problem is solved. On the other hand, if complex loops are very sparse in the initial members of the population, but some animals have the genetic potential to decrease confluence between triangles, that may be the preferred solution. While this response to the selective event is going on, other more or less random morphological changes may appear in other traits as immigrants enter the population and contribute a gene for, say, narrower teeth and initiate an apparently sudden change that is unrelated to the selective event. Even in the absence of a coeval selective event, these random morphological alterations can occur. The morphological change can take hold as long as it does not push the total character complex outside its required functional limits.

In this way, morphological change at the population level probably reflects a process of mosaic evolution, whereby different parts of the same character complex are simultaneously changing in different ways in response to the microevolutionary mechanisms of selection, immigration, and chance mutation. At this stage of the evolutionary process, not surprisingly, there is no evidence for a decoupled process of macroevolution, because macroevolutionary processes are supposed to become emergent only at the species level.

However, the analytic method used in this chapter very closely parallels that used for interpreting macroevolutionary patterns – drawing conclusions about evolutionary process from stratigraphic sequences of morphological change – but does so at temporal, geographic, and taxonomic scales that add insight to interpreting stratigraphic changes in morphology. First, inferring a speciation event just because two temporally separated populations are statistically different is dangerous. Numerous statistical differences exist between populations of *M. pennsylvanicus,* both across modern geographic space and through time at some of the fossil sites, and these traits (confluence of triangles, complexity of posterior loop) are those that might be (and have been) interpreted to indicate specific differences in the fossil record. Probably the most reliable way to recognize that speciation has occurred in the fossil record (by either anagenetic or cladogenetic change) is to find that differences between the fossil samples exceed the differences between living populations of the most closely related extant species (but see R. Martin, Chapter 11, this volume, for a contrasting view). This methodology has the advantages of (1) defining fossil and modern species in the same terms, that is, the amount of morphological distance in given traits that separates two samples, and (2) separating the concept of what species *are* from the concept of how they *arise* (Chandler and Gromko, 1989). Although it sounds simple, demonstrating similar morphological distances for fossil species and modern species usually involves a tremendous amount of work and therefore is seldom attempted in conjunction with paleontological studies that seek to infer evolutionary patterns. Nevertheless, it is a standard that must be achieved if we are to distinguish between microevolution and macroevolution.

Second, conclusions about evolutionary process based on examining only one character probably are not very reliable, because within a single character complex one can find evidence for stasis, episodic change, and gradual change (e.g., the Baker Bluff data), depending on the trait on which one focuses. The most reliable conclusions will come from analyzing multiple traits in the same character complex.

Third, morphological patterns taken from single geographic areas (one thick cliff sequence, for example) illustrate what is happening to a *population* through time, not to a whole *species*. What was happening to M. *pennsylvanicus* at Baker Bluff, for example, was different from what was happening at New Paris. In order to interpret patterns of change in a species, the sequence of local morphological changes from many different areas must be used to build a three-dimensional network of both temporal and geographic changes, because species do spread out over both time and space. Keeping in mind this distinction between evolutionary patterns in populations and evolutionary patterns in species is critical to sorting out whether or not microevolution is different from macroevolution, because the latter is supposed to be a species-level phenomenon, not a population-level phenomenon.

Finally, on a trait-by-trait basis, there is surprising similarity between the morphological patterns analyzed in this study (those at the population level on a time scale of 1,000 to 10,000 years) and longer-term (hundreds of thousands to millions of years) morphological patterns that have been used to argue for macroevolutionary patterns. In fact, if one simply took away the time scales on the M. *pennsylvanicus* diagrams and treated each diagram individually, they could be used to argue for any macroevolutionary process that was desired. This makes it evident that on both microevolutionary and macroevolutionary time scales, it is possible to observe examples of stasis, examples of punctuated change, and examples of gradual change. Our *modus operandi* as paleontologists and evolutionary biologists is to infer process from pattern: Can the same pattern at different time scales reflect different processes? Only if the answer is yes – and it may be, though considerably more work will be needed to find out – can we reasonably decouple macroevolution from microevolution.

Acknowledgments

Funding for this project was supplied by NSF grants EAR-8615373 and BSR-8916940. I thank Paul Barber, Elizabeth Barnosky, Felicia Keesing, William Lidicker, Jr., and Robert Martin for providing comments that improved the manuscript. This represents contribution no. 1548 from the Museum of Paleontology, University of California.

References

Barnosky, A. D. (1987). Punctuated equilibrium and phyletic gradualism: some facts from the Quaternary mammalian record. *Current Mammalogy*, 1:109–47.

(1990). Evolution of dental traits since latest Pleistocene in meadow voles (*Microtus pennsylvanicus*) from Virginia. *Paleobiology*, 16(3):370–83.

(in press). Defining climate's role in ecosystem revolution: clues from late Quaternary mammals. *Historical Biology*.

Chandler, C. R., and M. H. Gromko (1989). On the relationship between species concepts and speciation processes. *Systematic Zoology*, 38(2):116–25.

Damuth, J., and B. J. MacFadden (eds.) (1990). *Body Size in Mammalian Paleobiology: Estimation and Biological Principles*. Cambridge University Press.

Delcourt, H. R. (1979). Late Quaternary vegetation history of the eastern highland rim and adjacent Cumberland Plateau of Tennessee. *Ecological Monographs*, 49(3):255–80.

Delcourt, H. R., and P. A. Delcourt (1985). Quaternary palynology and vegetational history of the southeastern United States. In V. M. Bryant, Jr., and R. G. Holloway (eds.), *Pollen Records of Late-Quaternary North American Sediments* (pp. 1–37). American Association of Stratigraphic Palynologists Foundation.

Flynn, L. J. (1986). Species longevity, stasis, and stairsteps in rhizomyid rodents. In K. M. Flanagan and J. A. Lillegraven (eds.), *Vertebrates, Phylogeny, and Philosophy* (pp. 273–85). University of Wyoming Contributions to Geology, Special Paper 3.

Gingerich, P. E. (1985). Species in the fossil record: concepts, trends, and transitions. *Paleobiology*, 11(1):27–42.

Guilday, J. E., H. W. Hamilton, E. Anderson, and P. W. Parmalee (1978). The Baker Bluff cave deposit, Tennessee, and the late Pleistocene faunal gradient. *Bulletin of Carnegie Museum of Natural History,* 11:1–67.

Guilday, J. E., P. S. Martin, and A. D. McCrady (1964). New Paris no. 4: a Pleistocene cave deposit in Bedford Co., Pennsylvania. *Bulletin of the National Speleological Society,* 26(4):121–94.

Hall, E. R. (1981). *The Mammals of North America.* New York: Wiley.

Hoffman, A. (1989). *Arguments on Evolution: A Paleontologist's Perspective.* Oxford: Oxford University Press.

MacLeod, N. (1991). Punctuated anagenesis and the importance of stratigraphy to paleobiology. *Paleobiology,* 17(2):167–88.

Martin, R. A. (1986). Energy, ecology, and cotton rat evolution. *Paleobiology,* 12:370–82.

Martin, R. A., and R. H. Prince (1990). Variation and evolutionary trends in the dentition of late Pleistocene *M. pennsylvanicus* from Bell Cave, Alabama. *Historical Biology,* 4:117–29.

Mayr, E. (1970). *Populations, Species, and Evolution.* Cambridge, Mass.: Belknap Press.

Mikelson, D. M., L. Clayton, D. S. Fullerton, and H. W. Borns, Jr. (1983). The late Wisconsin glacial record of the Laurentide Ice Sheet in the United States. In H. E. Wright, Jr. (ed.), *Late Quaternary Environments of the United States,* Vol. 1 (Stephen C. Porter, ed.) (pp. 3–37). Minneapolis: University of Minnesota Press.

Rose, K. D., and T. M. Bown (1986). Gradual evolution and species discrimination in the fossil record. In K. M. Flanagan and J. A. Lillegraven (eds.), *Vertebrates, Phylogeny, and Philosophy* (pp. 119–30). University of Wyoming Contributions to Geology, Special Paper 3.

Stanley, S. M. (1985). Rates of evolution. *Paleobiology,* 11(1):13–27.

Vrba, E. S., and N. Eldredge (1984). Individuals, hierarchies and processes: towards a more complete evolutionary theory. *Paleobiology,* 10:146–71.

Vrba, E. S., and S. J. Gould (1986). The hierarchical expansion of sorting and selection cannot be equated. *Paleobiology,* 12:217–28.

Watts, W. A. (1970). The full-glacial vegetation of northwestern Georgia. *Ecology,* 51:17–33.

(1973). The vegetation record of a mid-Wisconsin interstadial in northwest Georgia. *Quaternary Research,* 3:257–68.

(1975). Vegetation record for the last 20,000 years from a small marsh on Lookout Mountain, northwestern Georgia. *Geological Society of America Bulletin,* 86:287–91.

(1979). Late Quaternary vegetation of central Appalachia and the New Jersey coastal plain. *Ecological Monographs,* 49(4):427–69.

4

Variogram analysis
of paleontological data

ANDREW P. CZEBIENIAK

Since Eldredge and Gould (1972) published their seminal paper on punctuated equilibrium, the field of evolutionary paleontology has been wracked by sometimes bitter disputes over the apparent reality or unreality of evolutionary *stasis* (Gould and Eldredge, 1977; Gingerich, 1985). Arguments have been advanced for stasis in lineages of protists (Wei and Kennett, 1988), invertebrates (Stanley and Yang, 1987), and vertebrates (Lich, 1990); see Barnosky (1987) and Levinton (1988) for additional references. Many, if not most, studies of morphological stasis in the fossil record have relied on qualitative and/or quantitative inferences about the behavior of *location* parameters (e.g., means and medians) as functions of geological time (usually estimated from stratigraphic and/or radiometric information). The simplicity and robustness of such tests cannot be gainsaid, but they do leave much unsaid. Figure 4.1 illustrates an artificial data set for which we would fail to reject the null hypothesis of evolutionary stasis when using such statistical tests as the nonparametric runs test (Sokal and Rohlf, 1981) or Lande's (1986) recommended test regarding the overlap of standard deviations about the means. Despite this conclusion we can see an "obvious" evolutionary trend in Z, the variable of interest. An alternative to such procedures is to examine the behavior of *dispersion* parameters. The purpose of this chapter is to introduce the geostatistical technique of *variogram analysis* to evolutionary paleontology as a way to supplement location inferences with dispersion inferences, specifically in the case of the Pliocene arvicoline rodent *Cosomys primus* (Lich, 1990; Anderson, Chapter 2, this volume).

The variogram is related to the (auto)covariogram and (auto)cor-

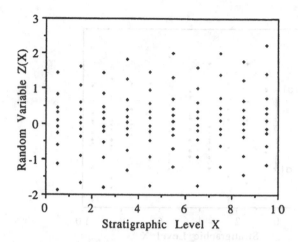

Figure 4.1. An artificial data set from a normal distribution that would satisfy Lande's (1986) test for stasis and a nonparametric runs test for stasis, yet is not static.

relogram, but unlike them has no apparent analogue in time-series analysis (Cox and Lewis, 1966; Box and Jenkins, 1976). First developed in the field of ore-reserve estimation (Matheron, 1963), the practice of variogram analysis, or "variography" as it is sometimes called, has spread to the petroleum and groundwater industries. The variable of interest may be the porosity or permeability of rocks, the concentration of ore or contaminants, local rainfall, piezometric height, soil moisture, or any of innumerable other quantities; see Journel and Huijbregts (1978) for references. Usually the interest is in two-dimensional or three-dimensional spatial variability, which may explain why the technique has not already been adopted by paleontologists for our one-dimensional problem of temporal variability. However, by regarding stratigraphic information as a proxy for time, we can examine the variability of fossils as a function of the stratigraphic distance between them.

Statistical foundations

The variogram of a spatiotemporal data set is a measure of the variance of data points as a function of the distance separating them. For k samples of data from a normal (Gaussian) random function with mean μ_i and variance σ^2_i we would write

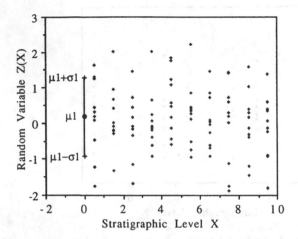

Figure 4.2. An artificial data set from a stationary normal random function.

$$Z(X_i) \sim N(\mu_i, \sigma^2_i), \qquad i = 1, \ldots, k,$$
$$\Sigma[Z(X_i)] = \mu_i, \qquad i = 1, \ldots, k,$$
$$\mathrm{var}[Z(X_i)] = \sigma^2_i, \qquad i = 1, \ldots, k,$$

where X_i represents the positions of all data points in the ith sample (Figure 4.2). For two data points separated by time or spatial distance h, we define the variogram function $2\gamma(X_i, h)$ as

$$2\gamma(X_i, h) = \mathrm{var}[Z(X_i) - Z(X_i + h)] = \mathrm{var}[W(X_i, h)],$$

where $W(X_i, h) = Z(X_i) - Z(X_{i+h})$ is normally distributed, with mean $\mu_i - \mu_j$ such that $\mu_j = \Sigma[Z(X_j)] = \Sigma[Z(X_i + h)]$. If the mean and variance of each sample are independent of X_i, then $Z(X_i)$ is called an *intrinsic* random function (Matheron, 1973), and the variogram is also independent of X_i. [*Stationary* random functions, which have stricter conditions (Vanmarcke, 1983), are the subsets of intrinsic random functions with which we are most concerned.] This permits reexpression as

$$2\gamma(h) = 2\gamma(X_i, h) = \Sigma[\{W(X_i, h)\}^2] = \Sigma[\{Z(X_i) - Z(X_i + h)\}^2],$$

which is more easily computed. [The function $2\gamma(h)$ is properly referred to as the semi-variogram, but many authors (e.g., David, 1977) refer to $2\gamma(h)$ as the variogram. This is a matter of personal preference and is unimportant in the course of research so long as the factor 2 is not forgotten.] Because each variogram value $2\gamma(h)$ [of normal data $Z(X_i)$]

is a sum of squares of normal random variables [$W(X_i, h)$], it behaves as a multiple of a χ^2 random variable (Cressie and Hawkins, 1980). For a stationary random function, the variogram will behave as a *central* χ^2 random variable. If the mean is somehow dependent on X_i, the random function is said to be *nonstationary*, and the variogram will behave as a *noncentral* σ^2 random variable. However, regardless of stationary conditions, if the data in each of the k samples are normally distributed, we can calculate confidence envelopes for the values of the variogram function using the central and noncentral χ^2 distributions (Czebieniak, 1991) according to the equation

$$\Pr[2n_h\gamma^*(h)/\chi^2_{n_h,1-\alpha} < 2\gamma(h) < 2n_h\gamma^*(h)/\chi^2_{n_h,\alpha}] = (1 - 2\alpha),$$

where $2\gamma^*(h)$ is the experimental variogram (following David's [1977] notation), calculated from the input data as

$$2\gamma^*(h) = \Sigma\{Z(X_i) - Z(X_i + h)\}^2/n_h,$$

n_h being the number of pairs of data points separated by the distance h. This immediately suggests a test for stationarity. If we calculate the experimental variogram from our data, then we can propose the test

H_0: stationarity (= evolutionary stasis)

H_A: nonstationarity

Under the null hypothesis, each value of the variogram function should behave as a central χ^2 random variable, and we should be able to draw a single horizontal line from the y-axis (at $h = 0$) through the $(1 - 2\alpha)\%$ confidence envelope without piercing the envelope. If we cannot draw such a line, then we shall reject the null hypothesis at the $(1 - 2\alpha)\%$ confidence level. Preliminary work (Czebieniak, 1991, unpublished data) indicates that the low correlation of consecutive values of the variogram function [i.e., $2\gamma^*(h)$ and $2\gamma^*(h + 1)$] under H_0 results in the experimental variogram falling entirely within the 95% confidence envelope only 70% of the time, and within the 99% envelope only 93% of the time. This situation is analogous to having two consecutive $(1 - 2\alpha)\%$ tests whose combined confidence is $[(1 - 2\alpha)2]\%$. Therefore we shall calculate the "95%" confidence envelope using the 99.9% points of the central χ^2 distribution.

Figure 4.3 shows two cases illustrating the different possible outcomes. In Figure 4.3A, derived from an artificial stationary data set, the crosses represent the calculated values of the experimental semi-variogram, hereafter referred to as the "track" of the semi-variogram function, and the dotted lines represent the (expected) 95% confidence envelope for

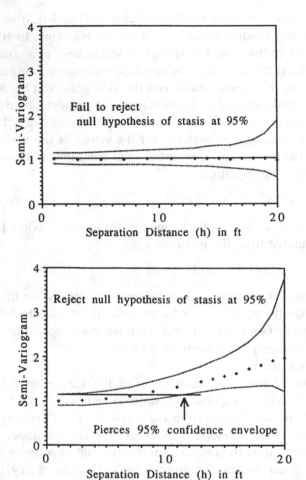

Figure 4.3. Semi-variogram functions (crosses) and 95% confidence envelopes (dotted lines) for two different evolutionary models. Horizontal solid lines illustrate test for stasis. (A) Evolutionary stasis (without punctuation events). (B) "Phyletic gradualism" (total increase of 1 SD).

the track. The horizontal solid line is entirely within the envelope, and therefore we would fail to reject the null hypothesis of evolutionary stasis at the 95% confidence level. However, in Figure 4.3B, derived from a nonstationary data set in which the mean increases gradually by one standard deviation over the 20-foot sampled interval, we can reject the null hypothesis of stasis at the 95% confidence level because the horizontal solid line pierces the envelope. The "flaring of the bell" with increasing h is a result of the decreasing sample size (i.e., there are fewer points very far apart), and variogram values supported by fewer than 30 pairs of data points are regarded as unreliable (Knudsen et al., 1984).

Actual calculation of the experimental variogram is done on a computer using such FORTRAN programs as MAREC2 (David, 1977), GAMA3 (Journel and Huijbregts, 1978), and GAMUK (Knudsen and Kim, 1978; Knudsen et al., 1984). For GAMUK the data are input as columns, such as

X-Dir.	Y-Dir.	Elev. (m)	Z-Variable
0.0	0.0	991.0	3.10
⋮	⋮	⋮	⋮
0.0	0.0	1004.0	2.85

(The X and Y columns usually include longitudinal and latitudinal information in the two- and three-dimensional cases, but serve only as placeholders in the one-dimensional paleontological case.)

The user also inputs a "class size" or "lag" H that groups together those values of the variogram function for h between integer multiples of H. [This presupposes that $2\gamma^*(100.0)$ is not significantly different from $2\gamma^*(101.0)$, etc.] A larger class size will smooth the track of the experimental variogram, possibly enhancing or obscuring any pattern, so great care must be exercised when choosing the class size. The value of the variogram function at $h = 0$ is referred to as the "nugget" and (for paleontological data) represents a measure of intrapopulational variability.

Variogram analysis of a fossil lineage

Lich (1990; Anderson, Chapter 2, this volume) examined the morphometrics of the M_1 in the Pliocene (Blancan) arvicoline rodent *Cosomys primus* from the Glenns Ferry Formation of Idaho. Her conclusion, based on a one-way analysis of variance, an ANOVA power test, a nonparametric runs test, and Bookstein's (1987) test against a null hypothesis of random walk, is that molar size in *C. primus* is static over the measured interval. Performing a variogram analysis on her unpublished raw data yields interesting results. Figure 4.4 illustrates the semi-variogram functions for $\log_{10}(M_1$ length) and $\log_{10}(M_1$ width) at class sizes of 0.3, 1.5, and 3.0 m. What little smoothing occurs is a result

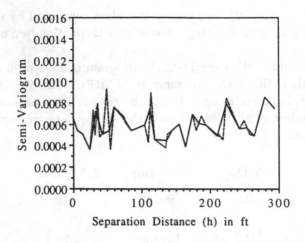

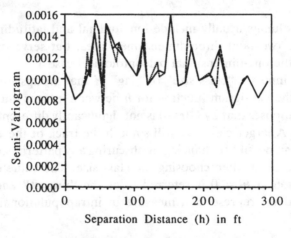

Figure 4.4. Semi-variogram functions for *Cosomys primus* tooth metrics at class sizes of 1 foot (dotted line), 5 feet (dashed line), and 10 feet (solid line). (A) Semi-variogram for $\log_{10}(M_1$ length) in *Cosomys primus*. (B) Semi-variogram for $\log_{10}(M_1$ width) in *Cosomys primus*.

of combining experimental variogram values with poor support. The "spikiness" that remains is presumed to be real and not artifactual. Notable is the lack of apparent trend in the track of the variogram, which might be interpreted as supporting the null hypothesis of stasis. However, the confidence envelopes in Figure 4.5 tell a profoundly different story. Around the track of the variogram the dashed line represents the confidence envelope calculated from the 95% points of the central χ^2 distribution; the dotted line is from the 99.9% points. The

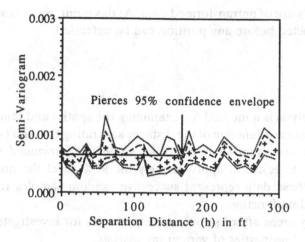

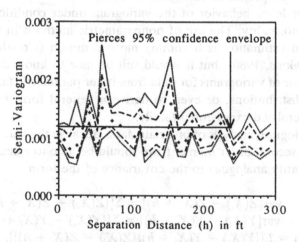

Figure 4.5. Semi-variogram functions (crosses), 95% confidence envelopes (dashed lines), and 99.9% confidence envelopes (dotted lines) for *Cosomys primus* tooth metrics at a class size of 10 feet. Horizontal solid lines illustrate test for stasis. (A) Semi-variogram for $\log_{10}(M_1$ length) in *Cosomys primus*. (B) Semi-variogram for $\log_{10}(M_1$ width) in *Cosomys primus*.

outer envelope is pierced for both measured variables, suggesting that we should reject the null hypothesis of evolutionary stasis, although our choice of an alternative hypothesis is unclear. Whether or not the pattern is compatible with the variogram of data generated by a random walk has not yet been determined. The data may simply be inadequate, although the sample is larger than the one that produced Figure 4.3. Perhaps the log-transformation has permitted us to see a pattern obscured by Lich's use of untransformed data. At this point, more research must be conducted before any position can be defended.

Conclusions

Variogram analysis is a method for examining the spatial and temporal variability of data as a function of the distance separating the data points. Application of variogram analysis to Lich's (1990) *C. primus* data set allows us only to reject at approximately the 95% level the null hypothesis that these data represent successive random samples from a stationary random function.

Two primary areas afford significant opportunity for investigation.

1. General properties of variogram analysis.
   a. Confidence behavior of the variogram under conditions of nonnormality. The use of nonparametric methods in variogram estimation is becoming more common (Cressie and Hawkins, 1980), but it would still be nice to know the behavior of variograms for data from other parametric families of distributions, or even to derive a general form for the generalized exponential family.
   b. Variogram analysis of bivariate data. Because the variogram is a second-order moment, we should be able to calculate a quantity analogous to the covariance of the form

$$2\gamma_Z(h) = \text{var}[Z(X_i) - Z(X_i + h)] = \Sigma[\{Z(X_i) - Z(X_i + h)\}^2],$$
$$2\gamma_Y(h) = \text{var}[Y(X_i) - Y(X_i + h)] = \Sigma[\{Y(X_i) - Y(X_i + h)\}_2],$$
$$2\gamma_{YZ}(h) = E[\{Y(X_i) - Y(X_i + h)\}\{Z(X_i) - Z(X_i + h)\}].$$

   Because the term "covariogram" is already taken, Journel and Huijbregts (1978) suggest the term "cross-(semi-) variogram," but we might use a less cumbersome term such as "bivariogram."
   c. Use of the central and noncentral F distribution for confidence envelopes around the relative variogram, $2\gamma(h)/\sigma^2$, where σ^2 is the total sample variance. (This would permit greater use of standardized tables, which are more common for the F distributions.)
2. Variogram analysis of paleontological data:
   a. Variogram analysis of different types of nonstationary data,

        including random walk, directional selection, doubly sto-
        chastic processes, shot-noise processes, chaotic behavior,
        and so forth.

  b. "Bivariography" of data such as principal components, Four-
        ier coefficients, coevolutionary data, data pertaining to mos-
        aic evolution, and so forth.

  c. Characterization of geographic variation in fossil and Recent
        species.

  d. Investigations of heterochrony using ontogenetic indepen-
        dent (or pseudoindependent) variables in place of spati-
        otemporal ones.

  e. Effects of sampling schemes (including the vagaries of ta-
        phonomy) on the appearance of experimental variograms
        from the same data set.

Variogram analysis appears to hold much promise for the field of
evolutionary paleontology, provided we can find its limitations and adhere to its rules.

Acknowledgments

I would like to thank Dr. Philip Gingerich for requesting that I speak
on variogram analysis at the symposium on morphological change and
for reviewing this manuscript; Dr. James Sprinkle, in whose seminar I
first broached this topic, for encouraging me to submit this manuscript;
and the Reservoir Characterization Research Laboratory at the Bureau
of Economic Geology in Austin, Texas, for allowing me to use their
variogram computer programs. Thanks also to my parents for paying
my travel costs on such short notice, and to Dr. Deborah K. Lich for
promptly and enthusiastically providing her raw unpublished data on
Cosomys primus.

References

Barnosky, A. D. (1987). Punctuated equilibrium and phyletic gradualism: some
    facts from the Quaternary mammalian record. *Current Mammalogy*, 1:109–
    47.

Bookstein, F. L. (1987). Random walk and the existence of evolutionary rates.
    Paleobiology, 13:446–64.

Box, G. E. P., and G. M. Jenkins (1976). *Time Series Analysis: Forecasting
    and Control*. Oakland: Holden-Day.

Cox, D. R., and P. A. W. Lewis (1966). *The Statistical Analysis of Series of Events*. London: Chapman and Hall.

Cressie, N., and D. M. Hawkins (1980). Robust estimation of the variogram: I. *Mathematical Geology*, 12:115–25.

Czebieniak, A. P. (1991). Exploratory variography of a fossil lineage. Unpublished M.S. report in statistics, University of Texas at Austin.

David, M. (1977). *Geostatistical Ore Reserve Estimation: Developments in Geomathematics 2*. Amsterdam: Elsevier Scientific.

Eldredge, N., and S. J. Gould (1972). Punctuated equilibria: an alternative to phyletic gradualism. In T. J. M. Schopf (ed.), *Models in Paleobiology* (pp. 82–115). San Francisco: Freeman.

Gingerich, P. D. (1985). Species in the fossil record: concepts, trends, and transitions. *Paleobiology*, 11:27–41.

Gould, S. J., and N. Eldredge (1977). Punctuated equilibria: the tempo and mode of evolution reconsidered. *Paleobiology*, 3:115–51.

Journel, A. G., and C. J. Huijbregts (1978). *Mining Geostatistics*. San Diego: Academic Press.

Knudsen, H. P., and Y. C. Kim (1978). A short course on geostatistical ore reserve estimation. University of Arizona Department of Mining and Geological Engineering.

Knudsen, H. P., Y. C. Kim, E. Y. Baafi, and B. L. Barua (1984). Computer program GAMUK for variogram computation. University of Arizona Department of Mining and Geological Engineering.

Lande, R. (1986). The dynamics of peak shifts and the pattern of morphological evolution. *Paleobiology*, 12:343–54.

Levinton, J. S. (1988). *Genetics, Paleontology, and Macroevolution*. Cambridge University Press.

Lich, D. K. (1990). *Cosomys primus:* a case for stasis. *Paleobiology*, 16:384–95.

Matheron, G. (1963). Principles of geostatistics. *Economic Geology*, 58:1246–66.

(1973). The intrinsic random functions and their applications. *Advances in Applied Probability*, 5:439–68.

Sokal, R. R., and F. J. Rohlf (1981). *Biometry: The Principles and Practice of Statistics in Biological Research*, 2nd ed. San Francisco: Freeman.

Stanley, S. M., and X. Yang (1987). Approximate evolutionary stasis for bivalve morphology over millions of years: a multivariate multilineage study. *Paleobiology*, 13:113–39.

Vanmarcke, E. (1983). *Random Fields: Analysis and Synthesis*. Cambridge, Mass.: MIT Press.

Wei, K. Y., and J. P. Kennett (1988). Phyletic gradualism and punctuated equilibrium in the late Neogene planktonic foraminiferal clade *Globoconella*. *Paleobiology*, 14:345–63.

5

Morphological change in Quaternary mammals: a role for species interactions?

TAMAR DAYAN, DANIEL SIMBERLOFF, AND
EITAN TCHERNOV

Different scientists (e.g., Guthrie, 1984; Martin, 1986) have adduced various evolutionary interpretations of morphological change in different species of Quaternary mammals. "Mere" size fluctuations [to the extent that changes in size can be dissociated from changes in shape (Mosimann and James, 1979)] have often been interpreted as responses to paleoclimatic change in accord with Bergmann's rule, which states that the body sizes of homeotherms from cooler regions tend to exceed the body sizes of those from warmer regions (Bergmann, 1847). This rule was later reformulated by James (1970) to include effects of changes in humidity. Although the physiological interpretation advanced by Bergmann (1847) has been debated (e.g., Scholander, 1955, 1956; Irving, 1957; Hayward, 1965), the notion that morphological clines parallel climatic ones is widely accepted. Climatic fluctuations in the Quaternary have been dramatic, and much research is dedicated to studying them. Consequently, fluctuations in sizes of fossil mammals have often been related to climatic change (Tchernov, 1968; Davis, 1977, 1981; Klein, 1986), although deviations from this relationship, in both time and space, are numerous (Dayan et al., 1991).

The possible evolutionary role of interspecific competition, on the other hand, has been largely neglected in the Quaternary fossil literature. This neglect may reflect the debate over the evolutionary role of interspecific interactions in the neontological literature, where the role of competition in structuring ecological communities and in affecting the morphology of competing animals has been a major source of dispute in evolutionary ecology (Lewin, 1983; Pimm and Gittleman, 1990).

For many years numerous studies interpreted morphological size re-

71

lationships among sympatric animal species of similar ecological requirements as resulting from competition; see references cited by Roth (1981), Simberloff and Boecklen (1981), and Greene (1987). Brown and Wilson (1956) adduced several examples in which species whose ranges partially overlap have enhanced morphological differences in the zone of overlap. They termed this phenomenon "character displacement." Several years later Hutchinson (1959) suggested that competition for food imposes a limit on the similarity between animal species; he argued that in order for two or more species to coexist, a size ratio of at least 1.3 in the linear dimension of the trophic apparatus is necessary. A similar pattern often depicted [starting with Holmes and Pitelka (1968)] is that of apparent size-ratio equality between sequential pairs among three or more size-ranked species using the same class of resources in a similar way – that is, species that are guild (Root, 1967) associates. This phenomenon was later termed "community-wide character displacement" (Strong, Szyska, and Simberloff, 1979). Such morphological patterns were generally interpreted as a competitively induced coevolutionary response of the different species to one another.

In recent years much of the literature on size ratios and size relationships has been reconsidered and contested. Grant (1972) showed that many claimed cases of ecological character displacement do not bear scrutiny and that the phenomenon is, at best, quite rare. In particular, studies claiming size-ratio equality were subjected to rigorous statistical analysis (Simberloff and Boecklen, 1981), and it appeared that, in the vast majority of published claims, the data do not falsify a hypothesis of independence between mean sizes. Some debate has existed over the relative power of various statistical tests for equality of size ratios (e.g., Bush and Holmes, 1983; Hopf and Brown, 1986; Tonkyn and Cole, 1986). The crux of the matter is that all these tests appear not to be very powerful; that is, one might be led to accept a false null hypothesis of ratio inequality when, in fact, ratios are more nearly equal than a random set would be. Thus one should be wary of rejecting a claim of significance for a borderline result. No one has comprehensively studied the relative power of these tests. Throughout previous papers, and in the present analysis, we used a test by Barton and David (1956).

Subsequently much skepticism has arisen about the role of interspecific interactions both in structuring ecological communities and in affecting the morphology of the constituent species. In the fossil Quaternary literature, likewise, the possible influence of competition in generating different-sized species has generally been deemphasized (e.g., Davis, 1977; Klein, 1986).

A review of some evidence from Recent mammals

The theory of limiting similarity (Hutchinson, 1959) dealt with minimum size ratios between the trophic apparati of potentially competing species. Trophic apparati of different sizes were seen as enabling different species of animals to use different sizes of food particles. For mammals, Hutchinson suggested that skull length provides an appropriate measure, and later studies followed in the use of this trait or else took other measures of body size, such as head plus body length. In recent years the realization that trophic apparati more directly related to feeding ecology should be studied has led to several studies of characters such as mandible length, gape, and bite force (e.g., Kiltie, 1984, 1988; Malmquist, 1985), but the evidence for character displacement has remained equivocal. Dayan et al. (1989a,b, 1990, 1992a) suggested that for carnivores, the trophic apparati most directly related to the feeding ecology of the different species may in fact be their specialized dentitions. In carnivores the emphasis is on the canine and carnassial teeth. Canines are used to seize and kill live prey; carnassials, an evolutionary adaptation of all carnivorous mammalian orders, shear prey in a scissors-like fashion.

Dayan et al. (1989a,b, 1990, 1992a) studied the specialized dentitions of several guilds of extant carnivores. A guild is defined as a group of species using the same class of resources in similar ways (Root, 1967). The locomotor abilities of a carnivore indicate much about its way of life – its preferred habitat, foraging behavior, and escape strategy (Van Valkenburgh, 1984). Thus, Dayan et al. (1989a,b, 1990, 1992a, in press) used morphological characteristics of limb morphology to divide carnivores into guilds:

1. Mustelids-viverrids: elongated, short-limbed, fairly slow pentadactyl plantigrades to semiplantigrades (e.g., Novikov, 1962; Harrison, 1968; Osborn and Helmy, 1980). The ulna is not reduced, and the radius retains the primitive ability to rotate around it.
2. Felids: swift, stalking species with digitigrade feet. The radius retains the ability to rotate around the ulna, thus enabling cats to grip with their forepaws, a movement essential for climbing, and also to manipulate prey with their forepaws.

3. Canids: highly adapted cursorial carnivores of open areas. Feet are digitigrade, and the ulna and radius are partly fused, thus reducing the ability of canids to rotate their forepaws.

Dayan et al. (1989a,b, 1990, 1992a) hypothesized that these differences in foraging and killing behavior reduce ecological overlap between the different groups of species to an extent that justifies their grouping into separate guilds; species of the same guild are more likely to encounter and pursue the same class of prey species than are species from different guilds.

Mustelids, viverrids, and felids kill their prey by a specialized bite: they insert their upper canines between two vertebrae and dislocate them, thus lacerating the spinal cord (e.g., Leyhausen, 1965, 1979; Macdonald, 1984). This stylized behavior suggests that for each species, canines may well reflect the typical size of their prey; they would be adapted to the particular vertebral size and structure most frequently encountered (Leyhausen, 1965, 1979). Partitioning of resources through the use of different-sized prey would be reflected in canine size differences between coexisting species (Dayan et al., 1989a,b, 1990, 1992a).

These guilds generally consist of highly dimorphic species. Canines are even more dimorphic than carnassial lengths and skull lengths (Dayan et al., 1989a, 1990). In previous studies of highly dimorphic species, means or midpoints for both sexes lumped together were taken (e.g., Kiltie, 1984; Schoener, 1984), or else males were compared to males, and females to females, separately (e.g., Rosenzweig, 1968; McNab, 1971). Dayan et al. (1989a, 1990) hypothesized that if competition is a prime selective pressure, each sex of each species is competing against its conspecific sex as well as the two sexes of each other species. They therefore viewed each sex as a "morphospecies" and analyzed all morphospecies at once, thus testing for the possible effects of both interspecific and intraspecific competition.

In the mustelid-viverrid and small felid guilds of Israel, in the small-felid guild of the Sind desert, and in North American weasels (which comprise the smaller species of the North American mustelid guild), the means for the canine diameters of the different morphospecies were found to be regularly spaced so as to form a nonrandom series (Dayan et al. 1989a, 1990). This pattern, communitywide character displacement, suggests the existence of coevolutionary morphological responses of the different species (and sexes) within each guild to one another.

For North American weasels, Dayan et al. (1989a) demonstrated this pattern in several geographic regions. These species, in particular the ermine (*Mustela erminea*), exhibit considerable geographic variation in size, yet in seven of eight localities a regular pattern obtained. The Old World small felids vary in size between Israel and the Sind desert, yet both localities have a regular pattern of equal size ratios. This fact indicates that the regular pattern exhibited does not result simply from species sorting by size, with some species from the regional pool locally excluded by virtue of similarity to others, but from actual character displacement.

An extinct species of the Israeli mustelid guild conforms to the general pattern. In the Israeli fossil record, a small carnivore, the common weasel (*Mustela nivalis*) has been found (Dayan and Tchernov, 1988) that by limb morphology should be grouped with the mustelids-viverrids. The size of the fossil specimens appears quite similar to that of the modern Egyptian population. In canine diameter the modern Egyptian population fits well with the general trend (Dayan et al., 1989a), a pattern of regularly spaced means that is significantly different from random.

Canids display less specialized killing behavior than do felids and mustelids. They are generally more omnivorous, and they possess no morphological specializations that would enable them to bring down relatively large prey as quickly and efficiently do as cats and mustelids (Kleiman and Eisenberg, 1973). Their canine teeth are large, but not highly specialized; they are not particularly sharp and not much flattened. They are good all-purpose weapons, but are not specifically adapted to delivering a highly oriented death bite (Ewer, 1973), nor are they as strong as the canines of felids (Van Valkenburgh and Ruff, 1987). Canids kill their prey through a series of slashing bites (Ewer, 1973; Leyhausen, 1979). They kill small prey by violent shaking (Seitz, 1950). The rostrum of the canid is long, so the shearing effect of the jaws is somewhat reduced, except near the angle of the jaw itself (Kleiman and Eisenberg, 1973). This morphology dictates less bite strength of the canines relative to the shearing power of the carnassials.

Dayan et al. (1989b, 1992a) therefore hypothesized that differences in canine size would be less significant as a means for niche partitioning in canids than in felids, mustelids, and viverrids. They thought the carnassials might be more important in this group for holding, multiple slashing bites, and cutting up prey and thus that competition might generate differences in carnassials (Dayan et al., 1989b, 1992a).

That expectation was only partly met by data on Middle Eastern and North African canids (Dayan et al., 1989b, 1992a). Dayan and associates studied mixed samples of males and females because sexual size di-morphism in canids is less than in mustelids and felids. They found character displacement and communitywide character displacement to be manifested by carnassial lengths, but to almost the same extent by both canine diameter and condylobasal skull length (Dayan et al., 1989b, 1992a). This result contrasted with those from felids and mustelids-viverrids in Israel, where only the means for canine diameter of the different morphospecies produced a regular pattern of equal size ratios (Dayan et al., 1989a, 1990). Among the small cats of the Sind desert and among weasels in several localities throughout North America, they did find size ratios tending toward equality also in skull lengths. Because size-ratio equality was a consistently repeated pattern for canines alone in these guilds, they hypothesized that this pattern in skull length, where it occurs, is simply a passive correlated response. In canids it is still difficult to ascertain whether resource partitioning directly affects the specialized dentitions, or only carnassial lengths, and only indirectly affects skull lengths, or whether it affects all three traits equally.

It has been suggested that a possibly similar pattern in skull length may have been distorted by greater variation in cranial characters than in dental characters, because cranial characters are generally considered to be more susceptible to both environmentally induced variability and age-related variability. However, this hypothesis seems unlikely in view of the fact that for all carnivores studied, dental characters appear to be less variable than cranial characters, and in particular skull length displays very low variability (Dayan, Wool, and Simberloff, 1992b).

Communitywide character displacement manifested by teeth is by no means limited to carnivores. A recent study of heteromyid rodents, granivores of North American deserts, revealed a similar morphological pattern (Dayan and Simberloff, 1992). Heteromyids consist of bipedal and quadrupedal species; these two groups differ from one another in their foraging methods, thus reducing ecological overlap between them. Therefore, Dayan and Simberloff (1992) divided them into two guilds, based on differences in limb morphology, much the same as in carni-vores. These species are not highly dimorphic, so Dayan and Simberloff (1992) studied the means for mixed sex populations. Within each guild they found a regular pattern of equal size ratios in widths of the upper incisors (= cutting edge). These incisors are used for husking seeds that are carried in external cheek pouches. Some of the husking occurs above ground, where predation risk is high. Therefore husking speed may be critical. For each species, they hypothesized an optimal seed size, too large to be pouched without husking, but small enough to be husked efficiently and quickly enough to outweigh the risk of predation. They suggested that, in selective terms, it may be well worth having an incisor width specialized for this seed size (Dayan and Simberloff, 1992). The pouch volumes of these rodents show a similar pattern of equal size ratios and a high correlation with incisor width. These observations support the hypothesis. There is no similar pattern for skull length, body weight, or teeth other than upper incisors. Whatever the selective mechanism, this pattern attests to a coevolutionary morphological response among the different heteromyid rodent species in each guild (Dayan and Simberloff, 1992).

An interesting phenomenon is that these species of rodents and carnivores do partition habitats to some extent, although some overlap occurs. This partitioning has been observed for carnivores and has been carefully studied for heteromyid rodents, as reviewed by Brown (1987) and Kotler and Brown (1988), but a morphological pattern arises in spite of the partial habitat partitioning. It does not occur where species have completely disjunct habitats and cannot be construed as guild associates; for example, the otter (*Lutra lutra*), which finds its prey in aquatic habitats, deviates from this regular pattern (Dayan, 1989). The otter's specialized habitat and dietary requirements, reflected in its skeletal morphology, clearly separate it from other Israeli carnivores. When taken together with the mustelid-viverrid guild of Israel, this species distorts the regular pattern of equal size ratios exhibited by all other species (Dayan, 1989).

Some morphological patterns in Quaternary canids

Studies of character displacement in Quaternary fossil mammals are still preliminary. So far, only canids have been studied, and the results outlined here conform well to expectations generated by studying modern carnivore guilds.

In Israel, Dayan et al. (1992a) described a pattern of equal size ratios between the Recent populations of canids (*Canis lupus,* gray wolf; *C. aureus,* golden jackal; *Vulpes vulpes,* red fox; *V. ruppelli,* Ruppell's sand fox; and *V. cana,* Blanford's fox) in all three traits studied: skull length, lower carnassial length, and canine diameter. In the fossil record only the three larger species (wolf, golden jackal, red fox) have been found. In Middle Paleolithic deposits, remains of all three species have been uncovered, and all three were larger then. For two of these species, the wolf and the red fox, the size increase has been interpreted as a response to paleoclimatic change: That period was cooler and wetter than today (Kurtén, 1965; Davis, 1977, 1981). These two species currently exhibit a Bergmannian size gradient. The modern populations of golden jackal exhibit no such spatial pattern, so the large Middle Paleolithic form was taken to be a distinct species (Bate, 1937; Kurtén, 1965). Dayan et al. (1992a), however, found equal size ratios between carnassial lengths of the three larger canids also in the Middle Paleolithic, and these ratios were almost identical with those occurring between the modern populations. It appears that these three species retained the same regular pattern of equal size ratios despite their temporal fluctuations in size. Dayan et al. (1992a) suggested that the large size of the Paleolithic golden jackal populations may have resulted from either a Bergmannian response to paleoclimatic change or character displacement in response to the shift in size of both the wolf and fox. Therefore a different specific status for the Paleolithic jackal of Israel is not warranted.

A comparative study of North American Quaternary canids is now under way. Three large canids are found at Rancho la Brea: the dire wolf (*Canis dirus*), the largest of the three; the gray wolf (*C. lupus*); and the coyote (*C. latrans*), the smallest of the three. The dire wolf and the coyote are remarkably abundant, whereas the gray wolf is relatively rare. This pattern may indicate some reduction in habitat overlap between the gray wolf and its larger and smaller congeners. However, for modern populations of both carnivores and rodents we have seen regular morphological patterns among species that display partial habitat segregation. Because these three North American canids do, in fact, overlap geographically, and at least partially in habitat, it is a tenable hypothesis that ecological character displacement would occur among them, if competition is an important evolutionary force. The remarkable preservation of the fossil material of Rancho la Brea permits a detailed morphological analysis of the relationships among the three species.

We reanalyzed some data published by Nowak (1979). For the greatest total length of the skull and for upper carnassial length, enough measurements are available, and they demonstrate that the three species do indeed display equal size ratios in upper carnassial length; recall that the low power of ratio equality tests suggests significance at a higher than nominal level. There was no such indication of equality for skull length ratios (Table 5.1). One should still treat these results cautiously, because these data were published before ^{14}C datings were available for

Table 5.1. *Sample statistics and Barton-David (B-D) statistics for the greatest total length of the skull and for upper carnassial length for the dire wolf, gray wolf, and coyote from Rancho la Brea*

Parameter	N	Mean	SD	Ratio between means	Two-tailed probability of observed B-D statistic
Greatest length of skull					
C. dirus	62	294.8	11.31		
				1.18	
C. lupus	6	250.7	14.77		p = 0.103
				1.22	
C. latrans	44	205.5	9.03		
Upper carnassial length					
C. dirus	62	31.75	1.38		
				1.21	
C. lupus	11	26.19	2.03		p = 0.061
				1.24	
C. latrans	43	21.05	1.29		

Source: Adapted from Nowak (1979).

the La Brea fossils. The vast majority of Nowak's (1979) canids clearly are a fairly homogeneous group from the later stages of the site, but the pattern may still change, though slightly, when the new datings are taken into consideration. A multivariate analysis of this currently dated material is under way.

Morphological and paleontological implications and perspectives

The relationship between competition and morphology has profound evolutionary significance for understanding the evolution of competing species. How does it affect our understanding of morphological change in fossil mammals?

It appears that teeth offer uniquely appropriate characters for study in mammalian character displacement research. This fact is very significant to paleontological systematic and evolutionary research that centers on teeth. Teeth are generally the best-preserved elements among mammalian fossil remains and carry highly diagnostic characters. Therefore, understanding the evolutionary forces affecting mammalian dentitions is of special importance to paleontological research.

The sizes and shapes of the specialized dentitions reflect the feeding ecology of mammals, so resource partitioning through use of different sizes of food particles may well be reflected in different sizes of teeth. Body size, on the other hand, may be more affected by other selective forces, some at least pertaining to the physiology of mammals, as envisioned in Bergmann's rule. Skull length may well be affected also by other functions, such as carrying the brain and the chief senses. Perhaps the controversy in the wide literature on character displacement stems from the use of characters influenced by factors other than feeding (Dayan et al., 1989a).

Size changes in fossil mammals should also be studied in the context of changes in community composition. Such studies may explain in part the all-too-common deviations from expectations based on Bergmann's rule, as well as the not-infrequent temporal size trends that do not conform to independently recorded paleoclimatic change (Dayan et al., 1991). That different character sets are under different selective pressures implies changes in proportions (shape) as well as changes in size under different synecological pressures. The complex interrelationships of various selective pressures on dental and cranial/skeletal characters are particularly critical for paleontological research, which relies rather heavily on the study of teeth.

The fact that character displacement and communitywide character displacement occur between species that show fine differences in habitat preference enables us to search for these patterns also among fossil mammals, whose ecological and spatial relationships are not as well known as those of extant mammals. Extreme ecological adaptations are generally reflected in gross morphological differences, and these usually can be recognized in fossils.

Acknowledgments

We thank A. Barnosky, R. Martin, and K. Seymour for helpful suggestions and comments on this manuscript.

References

Barton, D. E., and F. N. David (1956). Some notes on ordered intervals. *Journal of the Royal Statistical Society*, B18:79–94.

Bate, D. M. A. (1937). Palaeontology: the fossil fauna of the Wadi el-Mughara caves. In D. A. E. Garrod and D. M. A. Bate (eds.), *The Stone Age of Mount Carmel, Excavations at the Wadi el-Mughara* (pp. 137–233). Oxford: Clarendon Press.

Bergmann, C. (1847). Ueber die Verhaltnisse der Warmekonomie der Thiere zu ihrer Grosse. *Gottingen Studien*, 3:595–708.

Brown, J. H. (1987). Variation in desert rodent guilds: patterns, processes, and scales. In J. H. R. Gee and P. S. Giller (eds.), *Organization of Communities: Past and Present* (pp. 185–203). Oxford: Blackwell Scientific.

Brown, W. L., and E. O. Wilson (1956). Character displacement. *Systematic Zoology*, 5:49–64.

Bush, A. O., and J. C. Holmes (1983). Niche separation and the broken-stick model: use with multiple assemblages. *American Naturalist*, 122:849–55.

Davis, S. (1977). Size variation of the fox, *Vulpes vulpes*, in the Palaearctic region today, and in Israel during the late Quaternary. *Journal of Zoology (London)*, 182:343–51.

(1981). The effects of temperature change and domestication on the body size of the Pleistocene to Holocene mammals of Israel. *Paleobiology*, 7:101–14.

Dayan, T. (1989). The succession and community structure of the carnivores of the Middle East in space and time. Unpublished Ph.D. dissertation, Tel Aviv University, Israel (in Hebrew, English summary).

Dayan, T., and D. Simberloff (1992). Morphological relationships among coexisting heteromyids: an incisive dental character. Unpublished manuscript.

Dayan, T., D. Simberloff, E. Tchernov, and Y. Yom-Tov (1989a). Inter- and intra-specific character displacement in mustelids. *Ecology*, 70:1526–39.

(1990). Feline canines: community-wide character displacement in the small cats of Israel. *American Naturalist*, 136:39–60.

(1991a). Calibrating the paleothermometer: climate, communities, and the evolution of size. *Paleobiology*, 17:189–99.

(1992a). Canine carnassials: character displacement among the wolves, jackals, and foxes of Israel. *Biological Journal of the Linnaean Society*, 45:315–31.

Dayan, T., and E. Tchernov (1988). On the first occurrence of the common weasel (*Mustela nivalis*) in the fossil record of Israel. *Mammalia*, 52:165–8.

Dayan, T., E. Tchernov, D. Simberloff, and Y. Yom-Tov (1991b). Tooth size: function and coevolution in carnivore guilds. In P. Smith and E. Tchernov (eds.), *Structure, Function and Evolution of the Teeth*. Jerusalem: Freund Publishing.

Dayan, T., E. Tchernov, Y. Yom-Tov, and D. Simberloff (1989b). Ecological character displacement in Saharo-Arabian *Vulpes:* outfoxing Bergmann's rule. *Oikos*, 55:263–72.

Dayan T., D. Wool, and D. Simberloff (1992b). Teeth: skeletons in the closet of paleontology? Unpublished manuscript.

Ewer, R. F. (1973). *The Carnivores*. Ithaca, N.Y.: Cornell University Press.

Grant, P. R. (1972). Convergent and divergent character displacement. *Biological Journal of the Linnaean Society*, 4:39–68.

Greene, E. (1987). Sizing up size ratios. *Trends in Ecology and Evolution*, 2:79–81.

Guthrie, R. D. (1984). Alaskan megabucks, megabulls, and megarams: the issue of Pleistocene gigantism. In H. H. Genoways and M. R. Dawson (eds.), *Contributions in Quaternary Vertebrate Paleontology: A Volume in Memorial to John E. Guilday* (pp. 482–510). Carnegie Museum of Natural History, Special Publication 8.

Harrison, D. L. (1968). *The Mammals of Arabia*. Vol. 2. *Carnivora, Artiodactyla, Hyracoidea*. London: E. Benn.

Hayward, J. S. (1965). Metabolic rate and its temperature-adaptive significance in six geographic races of *Peromyscus*. *Canadian Journal of Zoology*, 43:309–23.

Holmes, R. T., and F. A. Pitelka (1968). Food overlap among coexisting sandpipers on northern Alaska tundra. *Systematic Zoology*, 17:305–18.

Hopf, F. A., and J. H. Brown (1986). The bull's-eye method for testing randomness in ecological communities. *Ecology*, 67:1139–55.

Hutchinson, G. E. (1959). Homage to Santa Rosalia, or Why are there so many kinds of animals? *American Naturalist*, 93:145–59.

Irving, L. (1957). The usefulness of Scholander's views on adaptive insulation of animals. *Evolution*, 11:257–9.

James, F. C. (1970). Geographic size variation in birds and its relationship to climate. *Ecology*, 51:365–90.

Kiltie, R. A. (1984). Size ratios among sympatric Neotropical cats. *Oecologia (Berlin)*, 61:411–16.

   (1988). Interspecific size regularities in tropical felid assemblages. *Oecologia (Berlin)*, 76:97–105.

Kleiman, D. G., and J. F. Eisenberg (1973). Comparison of canid and felid social systems from an evolutionary perspective. *Animal Behaviour*, 21:637–59.

Klein, R. G. (1986). Carnivore size and Quaternary climatic change in southern Africa. *Quaternary Research*, 26:13–170.

Kotler, B. P., and J. S. Brown (1988). Environmental heterogeneity and the coexistence of desert rodents. *Annual Review of Ecology and Systematics*, 19:281–307.

Kurtén, B. (1965). The Carnivora of the Palestine caves. *Acta Zoologica Fennica*, 107:1–74.

Lewin, R. (1983). Santa Rosalia was a goat. *Science*, 221:636–9.

Leyhausen, P. (1965). Ueber die Funktion der relativen Stimmungshierarchie dargestellt am Beispiel der phylogenetischen und ontogenetischen Entwicklung des Beautefangs von Raubtieren. *Zeitschrift für Tierpsychologie*, 22:412–94.

   (1979). *Cat Behavior*. New York: Garland STPM.

Macdonald, D. (ed.) (1984). *The Encyclopedia of Mammals*. London: Allen & Unwin.

McNab, B. (1971). On the ecological significance of Bergmann's rule. *Ecology*, 52:845–54.

Malmquist, M. G. (1985). Character displacement and biogeography of the pygmy shrew in northern Europe. *Ecology*, 66:372–7.

Martin, R. A. (1986). Energy, ecology, and cotton rat evolution. *Paleobiology*, 12:370–82.

Mosimann, J. E., and F. C. James (1979). New statistical methods for allometry with application to Florida red-winged blackbirds. *Evolution*, 33:444–59.

Novikov, G. A. (1962). *Carnivorous Mammals of the Fauna of the U.S.S.R.* Jerusalem: Israel Program for Scientific Translations.

Nowak, R. M. (1979). *North American Quaternary Canis*. Monographs of the Museum of Natural History, University of Kansas, no. 6.

Osborn, D. J., and I. Helmy (1980). The contemporary land mammals of Egypt (including Sinai). *Fieldiana Zoology, N.S.*, 5:1–579.

Pimm, S. L., and J. L. Gittleman (1990). Carnivores and ecologists on the road to Damascus. *Trends in Ecology and Evolution*, 5:70–3.

Root, R. B. (1967). The niche exploitation pattern of the blue-gray gnatcatcher. *Ecological Monograph*, 37:317–50.

Rosenzweig, M. L. (1968). The strategy of body size in mammalian carnivores. *American Midland Naturalist*, 80:299–315.

Roth, V. L. (1981). Constancy of size ratios of sympatric species. *American Naturalist*, 118:394–404.

Schoener, T. W. (1984). Size differences among sympatric, bird-eating hawks: a worldwide survey. In D. R. Strong, Jr., D. Simberloff, L. G. Abele, and A. B. Thistle (eds.), *Ecological Communities: Conceptual Issues and the Evidence* (pp. 254–81). Princeton, N.J.: Princeton University Press.

Scholander, P. F. (1955). Evolution of climatic adaptation in homeotherms. *Evolution*, 9:15–26.

(1956). Climatic rules. *Evolution*, 10:39–40.

Seitz, A. (1950). Untersuchungen über angeborene Verhaltensweisen bei Caniden. I und II. *Zeitschrift für Tierpsychologie*, 7:1–46.

Simberloff, D., and W. J. Boecklen (1981). Santa Rosalia reconsidered: size ratios and competition. *Evolution*, 35:1206–28.

Strong, D. R., Jr., L. A. Szyska, and D. S. Simberloff (1979). Tests of community-wide character displacement against null hypotheses. *Evolution*, 33:897–913.

Tchernov, E. (1968). *Succession of Rodent Faunas during the Upper Pleistocene of Israel*. Berlin: Parey.

Tonkyn, D. W., and B. J. Cole (1986). The statistical analysis of size ratios. *American Naturalist*, 128:66–81.

Van Valkenburgh, B. (1984). A morphological analysis of ecological separation within past and present predator guilds. Ph.D. dissertation, Johns Hopkins University, Baltimore.

Van Valkenburgh, B., and C. B. Ruff (1987). Canine tooth strength and killing behaviour in large carnivores. *Journal of Zoology, (London)*, 212:379–97.

6

Rates of evolution in Plio-Pleistocene mammals: six case studies

PHILIP D. GINGERICH

Evolution, in general, means change over time: The end is different from the beginning. Change is not inevitable, but it seems to be common. And some change can almost always be observed, given enough time. When do we see change? How much time does it take? To study evolution we have to measure change and time, and to understand evolution we have to know how the two are related.

Evolution in biology is often described as a process, but it is really a collection of processes – including the processes of mutation, dispersion, drift, and selection – that together produce whatever change we see. These processes may each be complicated individually, and their interactions produce additional complexity. Evolution is widely acknowledged as a fundamental concept in biology, and yet we have a surprisingly anecdotal and casual knowledge of how evolutionary change takes place. One way to improve this understanding is to quantify change over time in terms of rate. Evolutionary rates are quantitative expressions relating change to time.

Evolution takes place on many scales of time. Field and laboratory experiments can be designed to study change on short time scales, and fossils provide the most direct and best information about evolution on long time scales. The principal problem with the fossil record is that the time scales involved, typically millions of years, are so long that they are difficult to relate to the time scales of our lifetimes (and those of other organisms). Biologists as a group have a surprisingly poor understanding of evolution on a geological scale of time, and paleontologists as a group have a surprisingly poor understanding of evolution on a biological scale of time. One reason for this is that we have almost no

84

record of changes on intermediate scales of time, scales of hundreds or thousands of years, that would permit evolution on a laboratory scale of time to be related to evolution on a geological scale.

Fortunately biologists can sometimes study evolution on a scale of hundreds or thousands of years by taking advantage of accidents and events that have affected organisms, (e.g., chance colonization events) and have been documented in human history. And in favorable circumstances, paleontologists, too, can resolve time on scales of hundreds or thousands of years. It is generally true that time is easier to resolve the less far back we are in geological time, and the Pleistocene, spanning much of the past 2 million years, is important in this regard. The six Plio-Pleistocene studies discussed here illustrate a paleontological approach to understanding evolution on a wide range of time scales.

Quantification of evolutionary rates

Quantification of evolutionary rates is still in its infancy, and the history of ideas on this can be summarized rather easily. Time is uniquely important in paleontology, and it comes as no surprise that paleontologists figure prominently in the quantification of evolutionary rates. George Gaylord Simpson provided the first substantial treatment of rates in his book *Tempo and Mode in Evolution* (1944). At that time the geological time scale was poorly calibrated numerically, and Simpson was unwilling to trust estimates of the durations of epochs. He published a table of various measures of tooth size in horses and showed graphically that if the length of the ectoloph evolved at a constant rate, then the height of the paracone did not (Simpson, 1944, Figure 2). Simpson also provided a graph of these two characteristics plotted against geological time, showing that in all likelihood neither had evolved at a constant rate (Simpson, 1944, Figure 4). He noted that the slopes of lines in his graphs were proportional to rates of evolution, but did not attempt to quantify these. Simpson's most noteworthy contribution to quantification was recognition from the beginning that proportional change, not absolute change, is the quantity of interest, and consequently that a proportional (logarithmic) scale is the appropriate measurement scale for morphology.

J. B. S. Haldane (1949) quantified evolutionary rates in terms of proportional change and proposed the *darwin*, defined as change by a factor of e per million years (where e is the base of natural logarithms), as a convenient rate unit. Haldane calculated rates of evolution in horses for Simpson's two linear measures and for their ratio (Haldane's Table 1), but seems not to have recognized the incomparability of rates quantified for measures of different dimensions. Haldane mentioned the desirability of measuring morphological change in standard deviation units, and he also mentioned the desirability of measuring time in generations, suggestions developed further in this chapter.

Björn Kurtén (1959) measured rates of evolution in various ways in mammals that lived during the pre-Pleistocene (Tertiary), the Pleistocene, and the post-Pleistocene (Holocene), finding that each time period had its own characteristic range. Pre-Pleistocene mammals had rates ranging from 0.003 to 0.2 darwins (mean 0.02 d), Pleistocene mammals had rates ranging from 0.12 to 2.3 d (mean 0.5 d), and post-Pleistocene mammals had rates ranging from 3.7 to 43 d (mean 12.6 d). This led Kurtén to suggest that rates for Quaternary (Pleistocene and post-Pleistocene) mammals may have been higher than those for Tertiary (pre-Pleistocene) mammals, because environmental change in the Quaternary was rapid and revolutionary, whereas environmental change in the Tertiary was slow and gradual. In addition, Kurtén suggested the possibility that slow Tertiary rates might be partially or even wholly spurious because they are based on samples millions of years apart, with intervening histories that may have contained any amount of fluctuation at higher rates. Rates measured over long intervals of time are generally lower than rates measured over shorter intervals, and this is due in part to time-averaging of "fluctuating" higher rates (Gingerich, 1983). This does not mean that any are spurious, but simply that the effect of interval length must be considered when rates are compared.

Log rate–log interval graphs

Evolutionary rates do not exist independently of the interval over which they are measured. This observation is easily accommodated by plotting rates in an interval context. The graphs employed here have interval length plotted on a proportional scale on the abscissa, and the absolute value of evolutionary rate plotted on a proportional scale on the ordinate ($\log_{10}$ is used for convenience, because 10 is familiar as the base of our numbering system, but any base could be used). In the following discussion these "log rate–log interval" graphs are referred to as LR-LI graphs. Absolute values of rates are plotted on the assumption that rates of proportional increase are no different than rates of proportional decrease, and assignment of direction is arbitrary in any case.

Processes

Any real process producing systematic change (e.g., natural selection for small body size) does so at some characteristic intrinsic rate that is independent of interval length, meaning that the simple result of a single real process carried out long enough will plot as a horizontal distribution of average slope zero (0.0) on a rate-versus-interval graph (and on an LR-LI graph). The result is a distribution, because intrinsic rates and net rates (discussed later) have variance, and the horizontal line labeled "Process" (for process change) in the inset key in the following LR-LI graphs represents the zero slope of this possibility.

Stasis

Alternatively, a simple process operating at zero rate, or two or more processes that cancel each other out (as well as absence of a process), will produce no systematic change over time (stasis), meaning that differences observed by chance will tend to be constant while interval length increases. Rates calculated from these necessarily will decrease in inverse proportion to interval length, and in this alternative case, if carried out long enough, rates will plot as a distribution with an average slope of negative unity (-1.0) on an LR-LI graph. Again, the result is a distribution, because intrinsic and net rates have variance (lacking variance, stasis at zero rate could not be plotted on an LR-LI graph). The line of slope -1.0 labeled "Stasis" in the inset key of the following LR-LI graphs represents the slope of this possibility.

Pure process and pure stasis are two extreme possibilities in comparing rates measured over a range of time intervals, but neither necessarily happens all the time. Process and stasis can be combined in any proportion, and the slope of the resulting rate distribution on an LR-LI graph will reflect the proportion. The slope will be near zero if process predominates, and near -1.0 if stasis predominates.

Random change

One additional possibility, random change, deserves mention. Appearance of randomness usually is taken to mean that real component factors or component processes interact in complex ways that defy separation. For some purposes, these can be adequately modeled together rather than studied individually. Random walks can mimic process, and they can mimic stasis, but carried out long enough most will plot as a distribution with a slope of -0.5 on an LR-LI graph. The line of slope -0.5 labeled "Random" in the inset key of the following LR-LI graphs represents the slope of this possibility. Random change is an intermediate null model against which both process change and stasis must be compared to evaluate their possible significance (Gingerich, 1992).

Intrinsic rates and net rates

All stepped processes (including random drift) have two rates: an *intrinsic rate* that is the average rate observed at each step of the process (the generation-by-generation rate in evolution), and a *net rate* that is the average rate calculated over one or more steps (generations). The net rate can never exceed the intrinsic rate, although net rates approximate intrinsic rates in any real process. Net rates typically decrease in proportion to the square root of step number for random processes (hence the -0.5 slope on an LR-LI graph), and net rates decrease in proportion to step number in stasis (hence the -1.0 slope on an LR-LI graph).

Empirical limits

Empirically it turns out that rates of evolution have an upper bound and a lower bound, and both decrease systematically with interval length. Log rate of the upper bound and log rate of the lower bound decrease in inverse proportion to log interval, and each plots as a straight line of -1.0 slope. On an LR-LI graph, the upper bound is shown as a solid line because it reflects both a structural limit and an artifice of limited perception, and the lower bound is shown as a dotted line because it is purely an artifact of limited measurement.

The empirical upper and lower bounds of evolutionary rates define the limits of a broad distribution of possible rates having the -1.0 slope characteristic of stasis, which indicates that stasis, rather than process change or drift, predominates in long-term evolution.

Temporal scaling of evolutionary rates

The inverse relationship of evolutionary rates to the interval of time over which they are calculated means that rates can be compared directly only if they are calculated over intervals of the same length. When a distribution of rates calculated over a range of intervals is available, it is sometimes possible to use the relationship of rate to interval within this distribution to predict rates for some interval outside the original range of intervals. This *temporal scaling* is useful when comparing different distributions, and it is particularly useful for predicting intrinsic rates on a scale of one generation or a scale of 1 year.

Confidence on other time scales

Inference concerning the relationship of evolutionary rates and time intervals is most easily carried out in an LR-LI context. Log rate is a derived variable dependent on log interval; however, ordinary least-squares regression cannot be used for extrapolation and prediction in this context because distributions of derived log rates are usually negatively skewed, which means that medians are better than mean values as measures of location and that squaring deviant outliers would bias regression. The method of slope-and-intercept computation employed here is *robust maximum likelihood* (RML) estimation minimizing absolute deviations. The distribution of residual variation about a line with the computed slope and intercept is shown as an inset histogram in the lower left corner of all LR-LI graphs. Confidence intervals are calculated by bootstrapping, following Efron and Tibshirani (1986). A linear model is used for extrapolation and prediction because of its simplicity, recognizing that any relationship of log rate to log interval may not be linear over its entire range.

Rate units

Haldane (1949) proposed a standard unit for rates of morphological evolution called the *darwin*, quantifying rates of morphological evolution in terms of factors of *e* and time in millions of years:

$$\text{rate } (d) = \frac{\ln x_2 - \ln x_1}{t_2 - t_1}, \tag{6.1}$$

where $\ln x_1$ and $\ln x_2$ are individual natural log (ln) measurements or sample means of ln measurements at times t_1 and t_2, respectively. This is useful when comparing changes in measures of the same dimension (linear, areal, or volumetric) per unit time, but it is not generally useful when comparing changes in length with changes in area, or changes in length with changes in volume, or changes in length with changes in any compound ratio (shape).

Haldane mentioned the desirability of measuring morphological change in standard deviation units, and our colleague Björn Malmgren has suggested a rate unit called the *simpson*, quantifying rates of morphological evolution in terms of standard deviations per year or per million years. This has the advantage that standard deviations are measured in the original measurement units, making rates in standard deviation units independent of original dimension and thus more widely comparable.

Haldane also mentioned the desirability of measuring time in generations, and another rate unit of general interest is what I call the *haldane* (h), quantifying rates of morphological evolution in terms of standard deviations per generation. Morphological change still has to be corrected for proportion, and we are thus interested in change in natural logarithms (ln) of measured quantities, expressed in standard deviations of the natural log measurements. That is,

$$\text{rate } (h) = \frac{(\ln x_2 - \ln x_1)/s_{\ln x}}{t_2 - t_1} = \frac{z_2 - z_1}{t_2 - t_1}, \tag{6.2}$$

where x and t are defined as in equation (6.1), and $s_{\ln x}$ is the pooled standard deviation of $\ln x_1$ and $\ln x_2$. Natural logarithms are convenient here because the coefficients of variation commonly reported in the literature are a good approximation of the standard deviation of ln measurements (Lewontin, 1966). The term $\ln(\text{mean } x)$ is used to approximate the mean of $\ln x$ when necessary. The notation h_0 refers to intrinsic evolutionary rate in haldanes estimated at a scale of one generation, that is, standard deviations per generation at a scale of one generation (which is simply the rate intercept where log interval $= 0$ on any LR-LI graph).

Generation time (G) is related to body mass (M) in mammals by the power function

$$G = 0.16M^{0.26}, \tag{6.3}$$

where G is measured in years, and M is measured in grams. This equation is derived by regression of log G on log M ($N = 66$, $r = 0.82$), and it can be used to estimate generation time from body mass when body mass is known. Data used in the regression come from Eisenberg (1981), with generation time taken as the sum of age at first mating plus gestation time. Equation (6.3) yields an approximation that should be rounded downward for "*r*-selected" mammals and upward for "*K*-selected" mammals, especially those living in seasonal climates.

Rates in Plio-Pleistocene mammals: six case studies

Strait Canyon Microtus

The microtine rodent *Microtus pennsylvanicus* (meadow vole) was a common constituent of Pleistocene mammalian faunas of North America (and it is still widely distributed today). A. Barnosky studied samples of *M. pennsylvanicus* from 7 of 46 sampled intervals in Strait Canyon Fissure in Highland County, Virginia (Barnosky, 1990, and personal communication). Two of these sampling intervals 1.35 m apart have radiocarbon ages of 18.420 and 29.870 ky BP. Interpolation and extrapolation suggest that the seven *Microtus*-bearing intervals range in age from about 47.680 ky BP to about 20.960 ky BP. Samples are as few as 1.270 ky or as many as 26.720 ky apart in time. The *M. pennsylvanicus* weigh about 30 g and have a life span of about 1 year. Substitution in equation (6.3) indicates that *Microtus* probably reproduced at a rate of about three generations per year in the Pleistocene.

Rates of change of the length of the last upper molar (M^3) in Strait Canyon *Microtus* are shown in the LR-LI graph in Figure 6.1. Twenty-one rates can be calculated for six different intervals of time ranging from 3,810 generations to 80,160 generations. Twenty of the rates are nonzero. Thirteen rates are positive, indicating evolution toward larger size, and seven are negative, indicating evolution toward smaller size. The median rate is 0.000016 standard deviation per generation ($10^{-4.798}$), on a median time scale of about 33,000 generations ($10^{4.519}$), numbers that mean little by themselves. Fortunately, the *M. pennsylvanicus* data contain additional information.

All of the *Microtus* rates taken together have an RML slope of $-.512$, which suggests random change over time. The mean absolute deviation is 0.245. There are few enough data points that a bootstrapped 95% confidence interval on the slope ranges from -0.116 to -1.250, ruling out process change (slope 0.0), but including both randomness (slope -0.5) and stasis (slope -1.0). The RML intercept estimate is $h_0 = 0.003$ ($10^{-2.487}$), which suggests an intrinsic evolutionary rate of 0.003 standard deviation per generation on a time scale of one generation. Artificial selection experiments achieve intrinsic rates on the order of

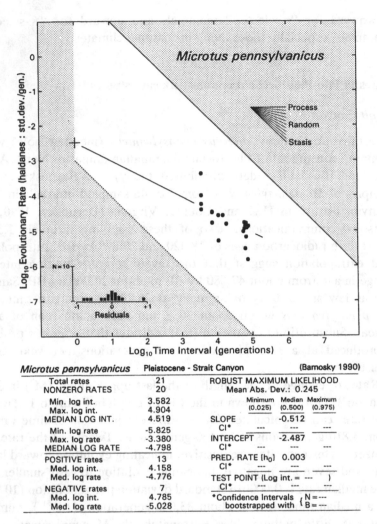

Microtus pennsylvanicus	Pleistocene - Strait Canyon		(Barnosky 1990)
Total rates	21	ROBUST MAXIMUM LIKELIHOOD	
NONZERO RATES	20	Mean Abs. Dev.: 0.245	

		Minimum (0.025)	Median (0.500)	Maximum (0.975)
Min. log int.	3.582			
Max. log int.	4.904			
MEDIAN LOG INT.	4.519	SLOPE	-0.512	
		CI*	---	---	---
Min. log rate	-5.825			
Max. log rate	-3.380	INTERCEPT	-2.487	
MEDIAN LOG RATE	-4.798	CI*	---	---	---
POSITIVE rates	13	PRED. RATE [h₀]	0.003	
Med. log int.	4.158	CI*	---	---	---
Med. log rate	-4.776	TEST POINT (Log int. = ---)		
NEGATIVE rates	7	CI*	---	---	---
Med. log int.	4.785	*Confidence Intervals	N = ---	
Med. log rate	-5.028	bootstrapped with	B = ---	

Figure 6.1. LR-LI graph for measurements of M^3 length in seven samples of late Pleistocene *Microtus pennsylvanicus* from Strait Canyon Fissure in Virginia. Solid circles are individual rates. Samples range in age from about 47.680 to 20.960 ky BP. *Microtus* is assumed to have had three generations per year, and rates are calculated over intervals ranging from 3,810 to 80,160 generations. Slope of rate distribution is −0.512. Intercept is −2.487, yielding an intrinsic rate $h_0 = 0.003$ standard deviation per generation on a time scale of one generation. This sample of rates is small enough that confidence intervals on the slope, extrapolated intercept, and h_0 (not shown) are very broad. Vertical scale bar represents $N = 10$ on inset histogram of residuals (mean absolute deviation is 0.245).

0.200 standard deviation per generation, and this *Microtus* result is about two orders of magnitude less than rates imposed by strong selection. However, here again there are few enough data points in the original rate sample that a bootstrapped 95% confidence interval for the predicted value of h_0 ranges from 0.00005 to 8.300 standard deviations per generation.

This example illustrates how net rates of evolution calculated on different time scales can be used to predict an intrinsic rate of change on a time scale of one generation, but the result cannot be distinguished from random change over time, and the confidence interval on the prediction is very broad.

Sandalja II Equus

A. Forsten (1990) studied late Pleistocene samples of the horse *Equus germanicus* from 7 of 12 sample intervals in the Sandalja II cave near Pula, in Istria, northwestern Yugoslavia. Two levels in the cave have radiocarbon ages, one near the bottom of the *Equus*-bearing sequence (25.340 ky BP), and one near the top (10.830 ky BP). Extrapolation and interpolation suggest that the seven *E. germanicus*-rich levels have ages of approximately 27.140, 24.740, 19.385, 14.675, 12.320, 11.575, and 10.830 ky BP. Successive samples differ by as few as 745 years and by as many as 5.355 ky, and *Equus* is assumed to have had a generation time of about 3 years in the Pleistocene.

Rates of change of measures of tooth size taken at the base of the crown and at the occlusal surface in Sandalja II *Equus* are shown in the LR-LI graph in Figure 6.2. A total of 714 rates can be calculated for intervals of time ranging from about 248 to 5,432 generations, and 666 of these rates are nonzero. Of the 666, 274 rates are positive, and 392 are negative, indicating a predominance of evolution toward smaller size. The median rate is 0.00021 standard deviation per generation ($10^{-3.686}$), on a median time scale of 2,466 generations ($10^{3.392}$). This median rate is about 13 times that for Strait Canyon *Microtus,* but the *Equus* rates are measured on a shorter median time scale.

All of the *Equus* rates taken together have an RML slope of -0.797, which lies on the stasis side of random. Here there are enough data points to constrain confidence intervals, and bootstrapped confidence intervals for the slope, intercept, and predicted intrinsic rate h_0 are shown in Figure 6.3. The 95% confidence interval on the slope ranges from -0.731 to -0.894. This narrow confidence interval rules out both

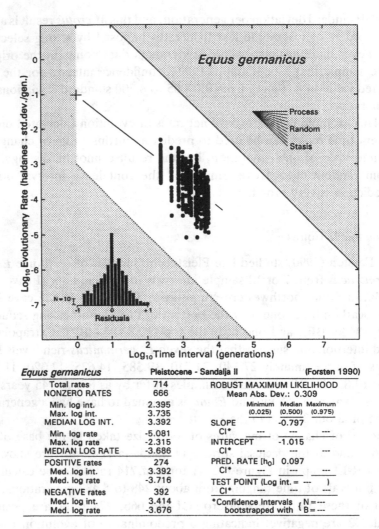

Figure 6.2. LR-LI graph for measurements of 34 dental characteristics in seven samples of late Pleistocene *Equus germanicus* from Sandalja II cave in Istria (Yugoslavia). Solid circles are individual rates. Samples range in age from about 27.140 to 10.830 ky BP, and rates are calculated over intervals ranging from 248 to 5,432 generations. Slope of rate distribution is −0.797. Intercept is −1.015, yielding rate $h_0 = 0.097$ standard deviation per generation on a time scale of one generation. Bootstrapped confidence intervals on the slope and intercept are shown in Figure 6.3. Vertical scale bar represents $N = 10$ on inset histogram of residuals (mean absolute deviation is 0.309).

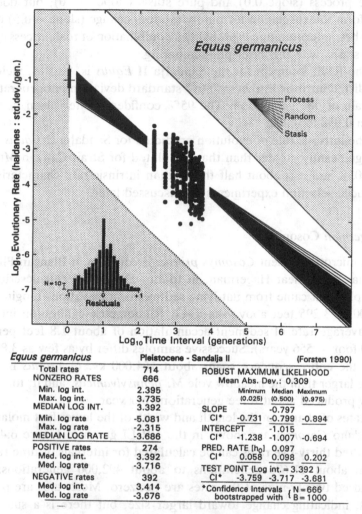

Figure 6.3. LR-LI graph for measurements of 34 dental characteristics in seven samples of late Pleistocene *Equus germanicus* from Sandalja II cave in Istria (Yugoslavia). Same information as in Figure 6.2, with addition of bootstrapped 95% confidence intervals for slope, intercept, and predicted intrinsic rate h_0 based on 1,000 resamplings of original 666 nonzero rates. Confidence intervals on slopes range from -0.731 to -0.894. This range of slopes excludes pure process change (0.0) and pure stasis (-1.0), and it does not conform to expectation for purely random change either (-0.5). Confidence interval on intrinsic rate h_0 ranges from 0.058 to 0.202 standard deviation per generation on a time scale of one generation.

pure process (slope 0.0) and pure stasis (slope -1.0), but does not conform to expectation for purely random change (slope -0.5) either. The best interpretation is probably a combination of real process change and stasis, with the latter predominating.

The RML intercept for the Sandalja II *Equus* is -1.015, yielding a predicted intrinsic rate $h_0 = 0.097$ standard deviation per generation at a scale of one generation. The 95% confidence interval on h_0 ranges from 0.058 to 0.202.

The intrinsic rate of evolution predicted for Sandalja II *Equus* (0.01) is significantly greater than that calculated for Strait Canyon *Microtus* (0.003), and it is about half the median intrinsic rate characteristic of artificial selection experiments, as discussed later.

Hagerman Cosomys

The microtine rodent *Cosomys primus* is common in Blancan Pliocene faunas found near Hagerman, in Idaho. D. Lich (1990) described 10 samples that came from flat-lying sediments at elevations ranging from 3,000 to 3,295 feet above sea level. Radiometric calibration indicates an average rate of sediment accumulation of about 1.8 feet per 1 ky (1.0 foot = 556 years). Successive samples differ by as few as 3.890 ky, and the entire sequence spans about 164.000 ky. *C. primus* is only a little larger than the meadow vole *M. pennsylvanicus,* and it, too, probably produced about three generations per year.

Rates of change of the length and width of the lower first molar (M_1) in Idaho *Cosomys* are shown in the LR-LI graph in Figure 6.4. One hundred thirty-five rates can be calculated for intervals of time ranging from about 11,670 generations to about 492,000 generations. One hundred twenty-nine of the rates are nonzero. Most rates are positive (71), indicating change toward larger size, but there is a substantial number of negative rates (58) as well. The median rate is 0.000002 standard deviation per generation ($10^{-5.648}$), on a median scale of 199,000 generations ($10^{5.299}$). The median rate is substantially higher than that described earlier for *Microtus,* and the median interval is substantially longer as well.

All of the *Cosomys* rates taken together have an RML slope of -0.841, which falls between stasis (-1.0) and random (-0.5), but is well below the slope expected for a real process (0.0). The bootstrapped 95% confidence interval on this slope ranges from -0.630 to -1.196 (Figure 6.4). This interval includes -1.0, but not -0.5, which means

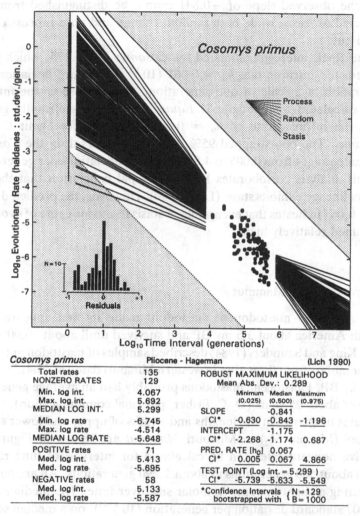

Cosomys primus		Pliocene - Hagerman			(Lich 1990)
Total rates	135	ROBUST MAXIMUM LIKELIHOOD			
NONZERO RATES	129	Mean Abs. Dev.: 0.289			
Min. log int.	4.067		Minimum	Median	Maximum
Max. log int.	5.692		(0.025)	(0.500)	(0.975)
MEDIAN LOG INT.	5.299	SLOPE		-0.841	
Min. log rate	-6.745	CI*	-0.630	-0.843	-1.196
Max. log rate	-4.514	INTERCEPT		-1.175	
MEDIAN LOG RATE	-5.648	CI*	-2.268	-1.174	0.687
POSITIVE rates	71	PRED. RATE [h₀]		0.067	
Med. log int.	5.413	CI*	0.005	0.067	4.866
Med. log rate	-5.695	TEST POINT (Log int. = 5.299)			
NEGATIVE rates	58	CI*	-5.739	-5.633	-5.549
Med. log int.	5.133	*Confidence Intervals		N = 129	
Med. log rate	-5.587	bootstrapped with		B = 1000	

Figure 6.4. LR-LI graph for length and width measurements of the lower first molar (M_1) in 10 samples of late Pleistocene *Cosomys primus* from Idaho. Solid circles are individual rates. Bootstrapped 95% confidence intervals for slope, intercept, and predicted intrinsic rate h_0 are based on 1,000 resamplings of original 129 nonzero rates. Confidence intervals on slopes range from −0.630 to −1.196, which includes stasis (−1.0) but excludes pure process change (0.0) and random change (−0.5). Predicted intrinsic rate h_0 is 0.067, and the 95% confidence interval on this value ranges from 0.005 to 4.866 standard deviations per generation on a time scale of one generation. Vertical scale bar represents $N = 10$ on inset histogram of residuals (mean absolute deviation is 0.289).

that the observed slope of -0.841 cannot be distinguished from that expected of stasis, while both random change and pure process can be ruled out.

The RML intercept estimated for *Cosomys* is -1.175, which yields a predicted intrinsic rate $h_0 = 0.067$ $(10^{-1.175})$ standard deviations per generation at a scale of one generation. This is close to the intrinsic rate calculated for Sandalja II *Equus*, and it is significantly greater than the intrinsic rate of $h_0 = 0.003$ calculated for Strait Canyon *Microtus*. The bootstrapped 95% confidence interval on the *Cosomys* estimate ranges from 0.005 to 4.866 standard deviations per generation.

This analysis corroborates Lich's principal conclusion that the *Cosomys* lineage exhibits stasis (Lich, 1990). However, the predicted value $h_0 = 0.067$ indicates that even while in stasis its intrinsic rate of evolution remained relatively high.

North American Mammut

The American mastodon or mastodont made its first appearance in North America about 3.5 my BP and survived until about 10,000 years ago. King and Saunders (1984) described samples of mastodon upper and lower molars from three sites in western Missouri dated at 45.4, 31.4, and 13.3 ky BP. Pleistocene mastodons probably had an average generation time of about 25 years (D. C. Fisher, personal communication).

Rates of change in the lengths and widths of upper and lower second molars (M^2 and M_2) in Missouri *Mammut* are shown in Figure 6.5. Twelve nonzero rates can be calculated for intervals of time ranging from about 560 generations to about 1,290 generations. All are positive, reflecting a slight increase in molar size over time. The median rate is 0.0007 standard deviation per generation $(10^{-3.145})$, on a median scale of 637 generations $(10^{2.804})$. The median rate is substantially higher than that described earlier for *Microtus*, but here the median interval is shorter.

All of the *Mammut* rates taken together have an RML slope of -0.004, which is almost exactly zero. The RML intercept estimated for *Mammut* is -3.133. This yields a predicted intrinsic rate $h_0 = 0.001$ $(10^{-3.133})$, which is at (or below) the lower limit to be expected for intrinsic rates. This example suggests that change in *Mammut* was due to a real process carried out over many generations and operating at a very low rate. However, both the number of rates and the interval range they

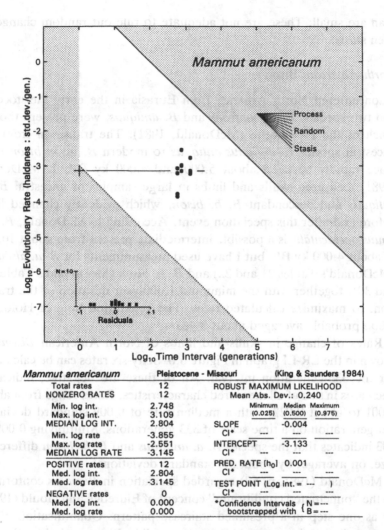

Mammut americanum	Pleistocene - Missouri	(King & Saunders 1984)
Total rates	12	ROBUST MAXIMUM LIKELIHOOD
NONZERO RATES	12	Mean Abs. Dev.: 0.240

		Minimum (0.025)	Median (0.500)	Maximum (0.975)
Min. log int.	2.748			
Max. log int.	3.109			
MEDIAN LOG INT.	2.804	SLOPE	-0.004	
Min. log rate	-3.855	CI*	---	---	---
Max. log rate	-2.551	INTERCEPT	-3.133	
MEDIAN LOG RATE	-3.145	CI*	---	---	---
POSITIVE rates	12	PRED. RATE [h₀]	0.001	
Med. log int.	2.804	CI*	---	---	---
Med. log rate	-3.145	TEST POINT (Log int. = ---)		
NEGATIVE rates	0	CI*	---	---	---
Med. log int.	0.000	*Confidence Intervals	N = ---
Med. log rate	0.000	bootstrapped with	B = ---

Figure 6.5. LR-LI graph for measurements of upper and lower molar (M² and M₂) lengths and widths in three samples of late Pleistocene *Mammut americanum* from western Missouri [sample statistics given by King and Saunders (1984)]. Solid circles are individual rates. Quantification of change yields 12 rates, all positive, ranging from 0.0001 to 0.003 haldane on a time scale of about 637 generations. This can be explained by process evolution at rates on the order of 0.001 standard deviation per generation sustained for the interval sampled, shown with the solid horizontal line. However, this sample of rates is too small and too closely spaced to rule out other interpretations, including random change or stasis at higher intrinsic rates.

span are small: These are not adequate to rule out random change or even stasis.

North American Bison

Bison entered North America from Eurasia in the early Pleistocene, and two species, *Bison latifrons* and *B. antiquus,* were present though much of the Pleistocene (McDonald, 1981). The transition from the ancestral species *B. antiquus antiquus* to modern *B. bison bison* took place rapidly between about 5.000 and 4.000 ky BP. J. McDonald (1981) measured skulls and limbs in large samples of ancestral *B. a. antiquus* and descendant *B. b. bison,* which evidently changed little before and after this speciation event. According to McDonald, *B. antiquus occidentalis* is a possible intermediate present from about 10.000 to about 4.000 ky BP, but I have used measurements for *B. a. antiquus* (McDonald's Tables 21 and 22) and *B. b. bison* (McDonald's Tables 29 and 30), together with the minimum 1,000-year duration of the transition, to maximize calculated rates. The generation time of Holocene *Bison* probably averaged about 3 years.

Rates of change in skulls and limbs of North American *Bison* are shown in the LR-LI graph in Figure 6.6. Sixty-six rates can be calculated for the 333-generation interval. All of these are negative, indicating decreases in size of all measured characteristics. Rates range from about 0.001 to about 0.022, with a median rate of 0.009 standard deviation per generation on a time scale of 333 generations. Multiplying 0.009 by 333 indicates that the species *B. a. antiquus* and *B. b. bison* differed in size, on average, by about 3.0 standard deviations.

McDonald (1981, p. 52) regarded speciation in *Bison* as conforming to the "punctuated equilibrium" concept of Eldredge and Gould (1972), or as one step in a presumed staircase pattern. Quantification of the best-studied and most rapid "punctuation" in *Bison* evolution, that between *B. a. antiquus* and *B. b. bison,* suggests that staircase evolution involves intrinsic rates ranging from as low as 0.001 to as high as 0.022 haldane, sustained over intervals ranging from 1 to about 333 generations (horizontal dashed line in Figure 6.6). The intrinsic rate may have been much higher if this distribution represents a lineage in stasis (oblique dashed line in Figure 6.6).

Isle of Jersey Cervus

The red deer *Cervus elaphus* has a 400,000-year history, and it is still widely distributed in Europe today. During the last interglacial, some

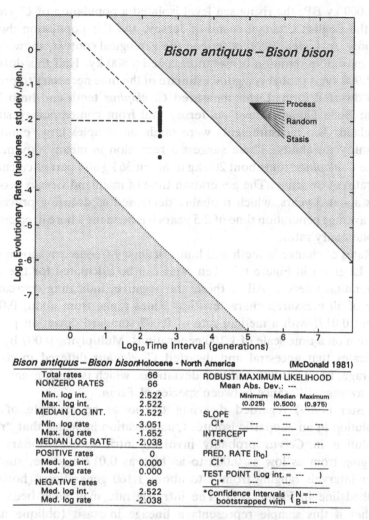

Bison antiquus – Bison bison		Holocene - North America		(McDonald 1981)

		ROBUST MAXIMUM LIKELIHOOD		
Total rates	66			
NONZERO RATES	66	Mean Abs. Dev.: ---		
Min. log int.	2.522		Minimum Median Maximum	
Max. log int.	2.522		(0.025) (0.500) (0.975)	
MEDIAN LOG INT.	2.522	SLOPE		
		CI* ---	--- ---	
Min. log rate	-3.051	INTERCEPT		
Max. log rate	-1.652	CI* ---	--- ---	
MEDIAN LOG RATE	-2.038			
POSITIVE rates	0	PRED. RATE [h₀]		
Med. log int.	0.000	CI* ---	--- ---	
Med. log rate	0.000			
NEGATIVE rates	66	TEST POINT (Log int. = ---)		
Med. log int.	2.522	CI* ---	--- ---	
Med. log rate	-2.038	*Confidence Intervals { N = ---		
		bootstrapped with { B = ---		

Figure 6.6. LR-LI graph for measurements of 66 dental, cranial, and postcranial characteristics in two samples of Pleistocene-Holocene *Bison antiquus antiquus* and Holocene *Bison bison bison* drawn from many sites in North America [sample statistics given by McDonald (1981)]. Solid circles are individual rates. Transition from one species to the other took place between 5.000 and 4.000 ky BP. Quantification of change across this 1,000-year interval yields 66 rates, all negative, ranging from 0.001 to 0.022 haldane on a time scale of about 333 generations. McDonald regards this transition as a "punctuation" event. It can be explained by process evolution at rates on the order of 0.001 to 0.022 standard deviation per generation sustained for 333 generations, shown with the horizontal dashed line. Samples of rates measured over a single interval are insufficient to permit calculation of slopes or projected intercepts, and this rate distribution could possibly represent stasis at a much higher intrinsic rate, as shown with the oblique dashed line.

120.000 ky BP, the rising sea level isolated a population of *C. elaphus* on the English Channel island of Jersey, and this population dwarfed rapidly. A. Lister (1989) described the geological context, showing that the maximum duration of isolation lasted 5.800 ky. Half this duration, or 2.900 ky, is probably a good estimate of the time necessary to produce this dwarfed form. Lister measured *C. elaphus* teeth and limb bones from Belle Hougue Cave on Jersey and from contemporary sites in England. Ten measurements were made on samples large enough to quantify variability. These suggest a reduction in mean body mass of male *C. elaphus* from about 200 kg to about 36 kg and permit calculation of rates of dwarfing. The generation time of mainland *Cervus* probably was about 3 years, which probably decreased as dwarfing proceeded. An average generation time of 2.5 years has been used here in calculating evolutionary rates.

Rates of change in teeth and limbs of Jersey *Cervus* are shown in the LR-LI graph in Figure 6.7. Ten rates can be calculated for the 1,160-generation interval. All of these are negative, indicating decreases in size of all measured characteristics. Rates range from about 0.003 to about 0.017, with a median rate of 0.007 standard deviation per generation on a time scale of 1,160 generations. Multiplying 0.007 by 1,160 indicates that ancestral and dwarfed *C. elaphus* differed in size, on average, by about 8.0 standard deviations, which is much greater than the average difference between species of *Bison*.

Lister (1989) regarded dwarfing *Cervus* as an example of rapid evolution in an allopatric isolate. Quantification indicates that "rapid" evolution in *Cervus* probably involved intrinsic evolutionary rates ranging from as low as 0.003 to as high as 0.017 haldane, sustained over intervals ranging from 1 to about 1,160 generations (horizontal dashed line in Figure 6.7). The intrinsic rate may have been much higher if this sample represents a lineage in stasis (oblique dashed line in Figure 6.7).

Discussion

Six case studies are too few to justify broad conclusions concerning evolutionary tempo and mode, especially when each of the individual studies has as many assumptions underlying temporal calibration as these do. However, each case analyzed here is an independent empirical study of population variation in samples of fossils in known stratigraphic context carried out with great effort and published to further understanding

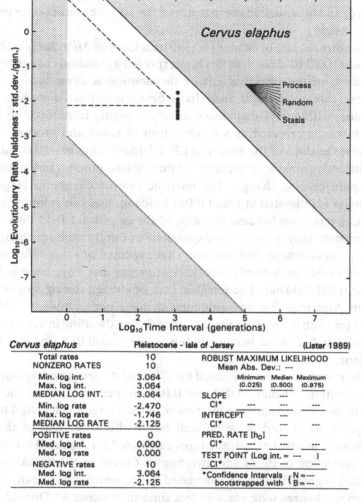

Cervus elaphus		Pleistocene - Isle of Jersey			(Lister 1989)
Total rates	10	ROBUST MAXIMUM LIKELIHOOD			
NONZERO RATES	10	Mean Abs. Dev.: ---			
Min. log int.	3.064		Minimum	Median	Maximum
Max. log int.	3.064		(0.025)	(0.500)	(0.975)
MEDIAN LOG INT.	3.064	SLOPE	---		
Min. log rate	-2.470	CI*	---	---	---
Max. log rate	-1.746	INTERCEPT		---	
MEDIAN LOG RATE	-2.125	CI*	---	---	---
POSITIVE rates	0	PRED. RATE [h₀]		---	
Med. log int.	0.000	CI*	---	---	---
Med. log rate	0.000	TEST POINT (Log int. = ---)			
NEGATIVE rates	10	CI*	---	---	---
Med. log int.	3.064	*Confidence Intervals {	N = ---		
Med. log rate	-2.125	bootstrapped with {	B = ---		

Figure 6.7. LR-LI graph for measurements of 10 dental and postcranial characteristics in two samples of late Pleistocene *Cervus elaphus*, one a British last-interglacial sample representing an ancestral stock, and the other a dwarfed descendant stock isolated on the Isle of Jersey in the English Channel [sample statistics given by Lister (1989)]. Solid circles are individual rates. Transition from one species to the other took place in a 2,900-year interval about 120,000 ky BP. Quantification of change across this interval yields 10 rates, all negative, ranging from 0.003 to 0.017 haldane on a time scale of about 1,160 generations. Lister regards this as an example of "rapid evolution in an allopatric isolate." It can be explained by process evolution at rates on the order of 0.003 to 0.017 standard deviation per generation sustained for 1,160 generations, shown with the horizontal dashed line. Samples of rates measured over a single interval are insufficient to permit calculation of slopes or projected intercepts, and this rate distribution could possibly represent stasis at a much higher intrinsic rate, as shown with the oblique dashed line.

of evolution. It would be less than scientific to ignore the empirical record, and quantitative comparison of the six case studies does provide some insight.

The intrinsic rate of change (h_0) in Strait Canyon *Microtus* is estimated at about 0.003 haldane, and this is interpreted as random change because the slope of the empirical LR-LI distribution is close to -0.5. The intrinsic rate of change in Sandalja II *Equus* is estimated at about 0.098 haldane, with a 95% confidence interval ranging from 0.058 to 0.202, and this is interpreted as a combination of stasis and process change because the slope of the empirical LR-LI distribution is -0.80, differing significantly from those expected of pure stasis, purely random change, and pure process change. The intrinsic rate of change in Hagerman *Cosomys* is estimated at about 0.067 haldane, and this is interpreted as evolutionary stasis because the slope of the empirical LR-LI distribution differs significantly from those expected of purely random change and pure process change, but not from that expected of stasis. The intrinsic rate of evolution in North American *Mammut* may have been as low as about 0.001 haldane. The intrinsic rate of change during speciation in North American *Bison* is estimated to have been between 0.001 and 0.022 (median 0.009) haldane, and that during dwarfing in Jersey *Cervus* is estimated to have ber ʲetween 0.003 and 0.017 (median 0.007) haldane.

The maximum rates calculated for *Bison* and *Cervus* undergoing rapid "punctuation" change, 0.022 and 0.017, respectively, lie at the lower end of the 95% confidence interval of intrinsic rates calculated for *Cosomys* in stasis, and these lie well below the lower end of the 95% confidence interval of intrinsic rates calculated for *Equus*. This suggests that speciation in *Bison* and dwarfing in *Cervus* may have taken place at rates higher than those calculated here, which may mean that the observed changes took place in less time than either McDonald (1981) or Lister (1989) considered in writing about them, but higher rates involve higher selection intensities (Gingerich, 1992), and species cannot survive when too small a proportion of individuals are selected in each generation.

Observation that rates of change during "punctuation" are no higher than intrinsic rates associated with stasis may also help to clarify the difference between "punctuation" and "stasis" in the punctuated equilibrium theory of Eldredge and Gould (1972). Eldredge and Gould's title proclaimed punctuated equilibrium as an alternative to phyletic gradualism, but sexual reproduction takes place generation by generation, step by step, which is why rates are appropriately calculated on a generational time scale. There is no alternative to gradualism because there is no alternative to generation-by-generation inheritance.

Process change ("punctuated" or not) and stasis differ in net rate but probably not in intrinsic rate, and thus the punctuation-or-stasis fate of lineages cannot, as Eldredge and Gould argued, be intrinsically determined (because what is intrinsic is observed not to be different). Further, when rates are examined in the appropriate quantitative LR-LI context, process rates (including punctuation in *Bison* and *Cervus*) and stasis rates (in *Cosomys*) are seen as the two extremes of an empirically fan-shaped spectrum of possible net rate distributions all rooted in the same narrow range of intrinsic rates (symbolized by the inset fan in the upper right quadrant of the LR-LI graphs). This fan-shaped spectrum includes random change in *Microtus* and a similar complex combination of process and stasis in *Equus*. Evolution and evolutionary change are manifestly more complicated than a simplistic punctuation-stasis dichotomy would indicate, and their full complexity can be characterized only by quantification.

The estimates of intrinsic rate (h_0) calculated here, spanning the two orders of magnitude from about 0.001 to about 0.100 haldane, appear to lie within the range of intrinsic rates a geneticist might suggest for change measured in phenotypic standard deviations per generation. My own analyses of intrinsic rates observed in artificial selection experiments indicate that these range from about 0.010 to 1.000 haldane, with a median h_0 of about 0.200 haldane. Economy and simplicity of design dictate that artificial selection experiments run at maximum rate, and it is reasonable that the range of rates calculated for the natural cases analyzed here is consistently about one order of magnitude less than the range observed experimentally.

In conclusion, stratigraphic sequences of fossils contain much information about the tempo of evolutionary change that can be extracted only by quantitative study of population variation and numerical calibration. When population variation and age are known to reasonable approximation, it is possible to characterize and compare changes in terms of rates. The six case studies reviewed here are exemplary in this regard and indicate that it is possible to relate change on a geological scale of time to change on a generational scale.

References

Barnosky, A. D. (1990). Evolution of dental traits since latest Pleistocene in meadow voles (*Microtus pennsylvanicus*) from Virginia. *Paleobiology*, 16:370–83.

Efron, B., and R. Tibshirani (1986). Bootstrap methods for standard errors, confidence intervals, and other measures of statistical accuracy. *Statistical Science*, 1:54–77.

Eisenberg, J. E. (1981). *The Mammalian Radiations: An Analysis of Trends in Evolution, Adaptation, and Behavior*. Chicago: University of Chicago Press.

Eldredge, N., and S. J. Gould (1972). Punctuated equilibria: an alternative to phyletic gradualism. In T. J. M. Schoph (ed.), *Models in Paleobiology* (pp. 82–115). San Francisco: Freeman.

Forsten, A. (1990). Dental size trends in an equid sample from the Sandalja II cave of northwestern Yugoslavia. *Paläontologische Zeitschrift*, 64:153–60.

Gingerich, P. D. (1983). Rates of evolution: effects of time and temporal scaling. *Science*, 222:159–61.

    (1992). Quantification and comparison of evolutionary rates. Unpublished manuscript.

Haldane, J. B. S. (1949). Suggestions as to quantitative measurement of rates of evolution. *Evolution*, 3:51–6.

King, J. E., and J. J. Saunders (1984). Environmental insularity and the extinction of the American mastodont. In P. S. Martin and R. G. Klein, (eds.), *Quaternary Extinctions: A Prehistoric Revolution* (pp. 315–39). Tucson: University of Arizona Press.

Kurtén, B. (1959). Rates of evolution in fossil mammals. *Cold Spring Harbor Symposia on Quantitative Biology*, 24:205–15.

Lewontin, R. C. (1966). On the measurement of relative variability. *Systematic Zoology*, 15:141–2.

Lich, D. K. (1990). *Cosomys primus*: a case for stasis. *Paleobiology*, 16:384–95.

Lister, A. M. (1989). Rapid dwarfing of red deer on Jersey in the last interglacial. *Nature*, 342:539–42.

McDonald, J. N. (1981). *North American Bison: Their Classification and Evolution*. Berkeley: University of California Press.

Simpson, G. G. (1944). *Tempo and Mode in Evolution*. New York: Columbia University Press.

7

Patterns of dental variation and evolution in prairie dogs, genus *Cynomys*

H. THOMAS GOODWIN

Prairie dogs are large, distinctive ground squirrels of the genus *Cynomys* that occur today on the Great Plains and on high plateaus and basins of the Rocky Mountains from southern Saskatchewan to northern Mexico. Two subgenera (*Cynomys* and *Leucocrossuromys*) and five extant species currently are recognized (Pizzimenti, 1975). The members of the two subgenera are referred to as black-tails and white-tails, respectively, designations that will be used in this chapter. The genus extends from the late Pliocene (late Blancan) to the Recent, but the record is best known for the late Pleistocene (late Irvingtonian and Rancholabrean) and Recent. The fossil record of the genus was recently reviewed (Goodwin, 1990).

The purpose of this chapter is to document patterns of morphological variation in several characters of fossil and Recent prairie dogs and to consider the tempo, mode, and causes of evolution as inferred from these patterns. The patterns that are treated include temporal variations in measures of dental size and "shape," in sample variability of measures of size, and in frequency of a nonmetric character, as well as latest Pleistocene (late Rancholabrean) geographic variations in measures of size and shape.

Materials and methods

Variables

This study was restricted to lower jaws and dentitions because relatively large sample sizes are available in the fossil record for these elements.

107

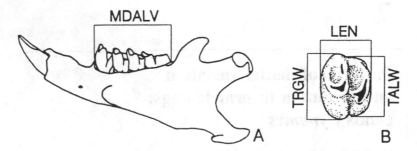

Figure 7.1. Illustrations of (A) MDALV (lower jaw viewed laterally to expose the posterior margin of the alveolar row) and (B) the three original measurements of P_4 (LEN, greatest length; TALW, talonid width; TRGW, trigonid width; tooth viewed occlusally).

Ten measurements originally were obtained: length of the lower alveolar row (MDALV); individual lengths of P_4–M_3; trigonid and talonid widths of P_4; and trigonid widths of M_1–M_3. Figure 7.1 illustrates MDALV and the three measurements of P_4; length and trigonid width measurements of M_1–M_3 were made as shown for P_4. Details of the protocol for these measurements are available elsewhere (Goodwin, 1990). In this study, I used the following as metric variables: MDALV and length × width of each tooth (LXWP4, etc.) as measures of size; length/width of each tooth (L/WP4, etc.) as measures of shape. For P_4, width was estimated as the average of trigonid and talonid measures of width. Ratios are not ideal variables for statistical comparisons (Sokal and Rohlf, 1981), but they offer the most direct way of estimating simple shape variation, as desired here. One problem is that ratio variables may depart from normality, but that was not found to be the case in this study, as discussed later. I also examined one nonmetric variable, the presence or absence of an ectostylid in the hypoflexid of M_3 (FRQEC).

Operational taxonomic units and sample selection

Table 7.1 lists the known fossil and Recent species-level taxa for prairie dogs (subspecies not given for Recent taxa), ordered by subgenus, and gives the time periods from which each taxon is known. I have revised the systematics of fossil prairie dogs elsewhere (Goodwin, 1990), and taxonomic details pertaining to the list in Table 7.1 can be obtained from that source.

Each taxon/time-period cell in Table 7.1 represents an operational

Table 7.1. *Species-level taxa of prairie dogs included in this study and their occurrence in the Plio-Pleistocene and Recent*

Taxon	Age							References
	SN	SP	CD	SD	ER	LR	RC	
Uncertain subgenus								
?*C. vetus*	X							Hibbard (1942)
C. hibbardi	X							Eshelman (1975)
Cynomys								
Cynomys new sp.A	X							Goodwin (1990)
Cynomys sp.		X						
Cynomys new sp.B				X	X			Goodwin (1990)
C. ludovicianus						X	X	
C. mexicanus						X?	X	
Leucocrossuromys								
C. gunnisoni			X			X	X	
C. n. niobrarius				X	X	X		Hay (1921)
C. n. churcherii						X		Burns and McGillivray (1989)
C. leucurus							X	
C. parvidens							X	

Note: SN, Senecan; SP, Sappan; CD, Cudahyan; ER, early Rancholabrean; LR, late Rancholabrean; RC, Recent.

taxonomic unit (OTU) in analyses of temporal variation, with one exception. *Cynomys vetus* is known only from cranial and upper dental material and thus was not included herein. Much of the focus in this chapter is on the following lineages: *Cynomys* sp. A–*Cynomys* sp. B–*C. ludovicianus* and *C. mexicanus*; *C. gunnisoni*; and *C. niobrarius* (divergence of two subspecies in the late Rancholabrean)–*C. leucurus* and *C. parvidens*. The southern subspecies of *C. niobrarius* (*n. niobrarius*) probably gave rise to the northern subspecies (*n. churcherii*) in the late Pleistocene and to *C. leucurus* and *C. parvidens* in the Recent.

The chronological scheme and terminology used in this chapter are those of Schultz et al. (1978), as modified by Lundelius et al. (1987). In addition, I subdivided the Rancholabrean into early and late portions. Prairie dog samples are known from the following chronological units: Senecan (late Blancan, ca. 2.5–1.8 my BP); Sappan (early Irvingtonian, ca. 1.8–0.75 my BP); Cudahyan (medial Irvingtonian, ca. 750–500 ky BP); Sheridanian (late Irvingtonian, or most of the "Illinoian," ca. 500–200 ky BP); early Rancholabrean (late "Illinoian" and Sangamon, ca. 200–100 ky BP); late Rancholabrean (Wisconsinan, ca. 100–10 ky BP); and Recent (10–0 ky BP). The absolute age estimates given generally follow Lundelius et al. (1987, Figure 7.3).

The OTUs for the study of late Rancholabrean geographic variation were samples of *C. ludovicianus* and *C. niobrarius* from the seven geographic areas outlined in Figure 7.2. *Cynomys niobrarius* was represented in areas 1–4 (two specimens also are known from the northern edge of area 5), and *C. ludovicianus* from areas 4–7. *Cynomys gunnisoni* was not included in this analysis because it is known only from area 6.

Many prairie dog localities could not be dated with confidence on the basis of criteria external to the prairie dogs themselves. Samples from these localities were not included in the study of temporal variation. However, a number of such localities, probably late Rancholabrean in age, were included in the study of geographic variation. Inclusion of these samples did not change the patterns of variation evident in the more restricted sample and served to increase sample sizes for this analysis.

For metric variables, an OTU sample represents the sum of all minimum numbers of individuals (MNI) across all localities included in that OTU. This was generally estimated by the greatest number of left or right specimens of a given element in a given locality or portion of a locality. MNI was used rather than total number of specimens to eliminate the effect of including right and left elements from the same individual in statistical comparisons. For two variables, LXWM3 and L/WM3, only left specimens were utilized, because preliminary analysis of the original data suggested a systematic difference in original measures of length on left and right M_3s. A similar systematic bias was not detected for original measurements of other teeth. For the nonmetric variable, FRQEC, an OTU sample represents the sum of all specimens, because left and right elements of a single individual may differ in this regard.

Analyses

I explored temporal and geographic patterns of variation, separately within each subgenus, using analysis of variance (ANOVA) and Scheffe's post-test comparisons of means (GLM routine of NCSS) (Hintze, 1990). Statistical comparisons were confined to better-sampled OTUs, but descriptive comparisons included all OTUs. Each variable was tested for normality within each sample with the Kolmogorov-Smirnov test, and for homogeneity of variances among samples used in each analysis with the F_{max} test. The assumption of homogeneity of variances was found to be valid in all cases, as was the assumption of normality in most cases (all but 5 of over 180 OTU/variable combinations), so the use of parametric statistics on untransformed variables seemed justified. Samples found to depart from normality were as fol-

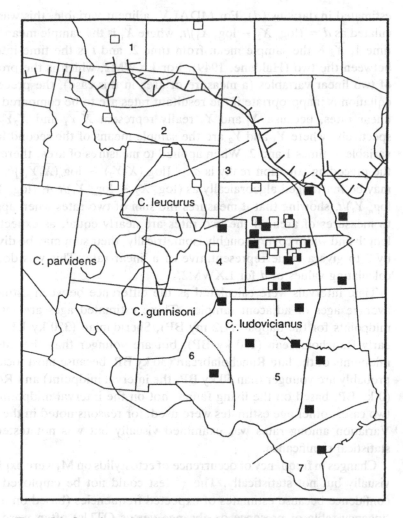

Figure 7.2. Locations of fossil sites used in study of late Rancholabrean geographic variation in *Cynomys niobrarius* and *C. ludovicianus*. Open squares represent white-tail localities, black squares black-tail localities, and half-black squares taxonomically mixed localities. Geographic areas are delimited by dotted lines. The base map gives ranges for four extant prairie dogs and the approximate boundary of maximum glacial advance during the Wisconsinan.

lows: Recent *C. gunnisoni* for MDALV; late Rancholabrean *C. gunnisoni* for LXWM1; early Rancholabrean *C. niobrarius* for L/WM3; geographic area 6 sample of *C. ludovicianus* for LXWM1; and geographic area 7 sample of *C. ludovicianus* for LXWM3.

For two measures of size (MDALV, LXWM2), rates of change were estimated in darwins (d). For MDALV, a linear variable, this was calculated as $d = (\log_n \overline{X}_2 - \log_n \overline{X}_1)/t$, where $\overline{X}_1$ is the sample mean from time 1, $\overline{X}_2$ is the sample mean from time 2, and t is the time interval between the two (Haldane, 1949). For LXWM2, which is the product of two linear variables (a measure of area in this case), the preceding equation is inappropriate if the resultant rates are to be compared with linear rates, because $\overline{X}_1$ and $\overline{X}_2$ really represent $\overline{X}_1\overline{Y}_1$ and $\overline{X}_2\overline{Y}_2$, respectively, where $\overline{Y}_1$ and $\overline{Y}_2$ are the sample means of the second linear variable at times 1 and 2. When applied to measures of area, therefore, the preceding equation really is $d = [\log_n(\overline{X}_2\overline{Y}_2) - \log_n(\overline{X}_1\overline{Y}_1)]/t$. This may be rearranged algebraically as $(\log_n \overline{X}_2 - \log_n \overline{X}_1)/t + (\log_n \overline{Y}_2 - \log_n \overline{Y}_1)/t$, showing that it measures the sum of two rates when applied to measures of area. If the two rates are nearly equal, as expected if length and width covary roughly isometrically, their sum may be divided by 2 to give a value representative of a linear rate. This was done in calculating values of d for LXWM2.

Time intervals were calculated as the difference between estimated average ages of adjacent samples. These estimated ages are interval midpoints for the Sappan (1.2 my BP), Sheridanian (350 ky BP), and early Rancholabrean (150 ky BP), but are younger than the interval midpoints of the late Rancholabrean (30 ky BP, because most localities probably are younger than 50 ky BP, the interval midpoint) and Recent (0 ky BP, based on the living fauna, not on the interval midpoint). In two cases, other age estimates were used, for reasons noted in the text. Variation among rates was examined visually but was not tested for statistical significance.

Changes in frequency of occurrence of ectostylids on M_3 were explored visually but not statistically. The χ^2 test could not be employed with confidence because estimates of expected frequencies (based on a contingency table of presence or absence versus OTUs) often were very low (<5).

Sample variability in all size variables of each well-represented OTU was estimated by its coefficient of variation: $CV = (SD/\overline{X})100$. Variation in CVs was explored visually but not statistically.

Patterns of morphological variation

Temporal variation

Tables 7.2 and 7.3 summarize descriptive statistics for all metric variables across uncertain-subgenus and black-tail OTUs and white-tail OTUs.

Table 7.2. *Descriptive statistics for uncertain-subgenus and black-tail OTUs* [$\bar{X}$ *(SE)n*]

Variable	SNHB	SPBT	CDBT	SDBT[a]	ERBT[a]	LRLD[a]	RCLD[a]	LRMX	RCMX[a]	F value
MDALV	13.65(−)1	13.80(−)1		14.26(0.12)17	14.43(0.27)4	15.27(0.09)43	15.20(0.10)28	14.13(0.28)3	13.71(0.10)10	27.42[b]
LXWP4	12.47(−)1	11.47(−)1	15.25(−)1	12.49(0.38)13	13.30(0.97)5	14.69(0.24)34	14.24(0.24)28	12.08(0.15)2	11.28(0.22)10	16.70[b]
LXWM1	11.06(−)1	11.29(−)1		12.53(0.34)13	13.47(0.62)4	14.95(0.21)33	14.83(0.19)28	12.24(−)1	12.19(0.29)10	21.05[b]
LXWM2	12.36(−)1	13.02(−)1		13.88(0.37)11	15.21(0.73)5	16.47(0.22)34	16.16(0.22)28	14.66(0.21)2	13.64(0.22)10	17.92[b]
LXWM3	16.81(−)1	15.87(−)1		20.20(1.2)4	20.67(0.93)4	23.71(0.42)18	23.04(0.63)12	20.70(−)1	19.66(0.46)5	7.13[b]
L/WP4	0.80(−)1	0.81(−)1	0.78(−)1	0.79(0.008)13	0.82(0.026)5	0.78(0.007)34	0.78(0.006)28	0.80(0.041)2	0.79(0.013)10	1.09
L/WM1	0.77(−)1	0.73(−)1		0.67(0.009)13	0.67(0.032)4	0.68(0.007)33	0.68(0.007)28	-0.67(−)1	0.69(0.012)10	0.18
L/WM2	0.76(−)1	0.74(−)1		0.69(0.012)11	0.67(0.018)5	0.69(0.007)34	0.68(0.006)28	0.69(0.021)2	0.69(0.008)10	0.32
L/WM3	1.00(−)1	1.05(−)1		1.00(0.018)4	1.04(0.021)4	1.06(0.009)18	01.04(0.008)12	1.03(−)1	1.00(0.024)5	3.44[c]

[a]These OTUs were included in the ANOVA producing the *F* statistic in the far right column. OTU abbreviations: CDBT, Cudahyan black-tail; ERBT, early Rancholabrean black-tail (*Cynomys* new sp. B); LRLD, late Rancholabrean *ludovicianus*; LRMX, late Rancholabrean *mexicanus*; RCLD, Recent *ludovicianus*; RCMX, Recent *mexicanus*; SDBT, Sheridanian black-tail (*Cynomys* new sp. B); SNHB, Senecan *hibbardi*; SPBT, Sappan black-tail (*Cynomys* new sp. A).
[b]Significant at 0.001 level.
[c]Significant at 0.05 level.

Table 7.3. Descriptive statistics for white-tail OTUs [$\bar{X}$ (SE)n]

Variable	CDGN	LRGN[a]	RCGN[a]	SDNB[a]	ERNB[a]	LRNB[a]	LRCH[a]	RCLC[a]	RCPV[a]	F value
					OTU					
MDALV		13.25(0.17)11	13.47(0.11)15	14.84(0.07)17	14.72(0.13)8	14.89(0.13)26	14.56(0.08)21	13.94(0.11)19	13.67(0.16)6	27.06[b]
LXWP4	10.89(–)1	10.63(0.23)9	10.76(0.18)15	13.78(0.21)16	14.06(0.28)10	14.04(0.24)30	13.45(0.23)21	11.36(0.25)19	10.80(0.40)6	33.39[b]
LXWM1	12.10(–)1	12.01(0.20)8	12.08(0.21)15	15.29(0.19)18	15.33(0.39)7	16.02(0.28)31	15.45(0.24)21	12.71(0.21)19	12.89(0.46)6	35.84[b]
LXWM2	13.47(–)1	13.38(0.42)6	13.83(0.18)14	16.78(0.27)17	17.02(0.30)10	18.02(0.31)30	17.19(0.24)21	15.05(0.27)19	15.06(0.47)6	26.63[b]
LXWM3	19.71(–)1	19.90(0.47)6	21.38(0.21)8	25.63(0.66)8	25.31(0.53)9	25.43(0.58)12	25.38(0.46)17	23.15(0.60)9	22.99(0.49)3	12.95[b]
L/WP4	0.72(–)1	0.74(.015)9	0.73(0.008)15	0.74(0.009)16	0.75(0.010)10	0.73(0.006)30	0.71(0.007)21	0.75(0.006)19	0.74(0.017)6	3.06[c]
L/WM1	0.63(–)1	0.63(0.005)8	0.63(0.008)15	0.63(0.007)18	0.64(0.012)7	0.62(0.006)31	0.58(0.007)21	0.63(0.006)19	0.63(0.015)6	7.61[b]
L/WM2	0.65(–)1	0.65(0.009)6	0.64(0.008)14	0.64(0.007)17	0.63(0.012)10	0.63(0.007)30	0.59(0.005)21	0.63(0.006)19	0.65(0.016)6	5.45[b]
L/WM3	0.98(–)1	0.99(0.007)6	1.02(0.016)8	1.00(0.017)8	0.98(0.010)9	0.96(0.013)12	0.95(0.009)17	0.97(0.023)9	0.94(0.022)3	2.84[d]

[a] These OTUs were included in the ANOVA producing the F statistic in the far right column. OTU abbreviations: CDGN, Cudahyan *gunnisoni*; ERNB, early Rancholabrean *niobrarius*; LRCH, late Rancholabrean *churcherii*; LRGN, late Rancholabrean *gunnisoni*; LRNB, late Rancholabrean *niobrarius*; RCGN, Recent *gunnisoni*; RCLC, Recent *leucurus*; RCPV, Recent *parvidens*; SDNB, Sheridanian *niobrarius*.
[b] Significant at 0.001 level.
[c] Significant at 0.01 level.
[d] Significant at 0.05 level.

Figures 7.3 and 7.4 depict temporal variations in two representative measures of size (MDALV and LXWM2) and dental shape (L/WM1 and L/WM3), respectively, with probable evolutionary pathways indicated.

Size. The early history of size evolution in prairie dogs is poorly documented (Figure 7.3). The earliest known fossil specimens (Senecan, ca. 2.2 my BP), of uncertain subgenus, already were large relative to all extant species of the ground squirrel subgenus *Spermophilus* (the putative sister group of *Cynomys*), but were smaller than most late Quaternary prairie dogs. There is little difference in size between Senecan and Sappan (ca. 1.8 my BP) species, but samples for this interval are limited. The size increase between Sappan and Sheridanian (ca. 350 ky BP) black-tails may be meaningful, especially in LXWM3 (Table 7.2), but that was not assessed statistically because of small sample size.

All measures of size exhibit highly significant ($p < 0.001$) variation among the better-sampled OTUs of black-tails (5 of 6 Sheridanian and younger samples, Table 7.2). The statistically significant aspects of size variation are a size increase between the Sheridanian (ca. 350 ky BP) and late Rancholabrean (ca. 30 ky BP) samples of the *Cynomys* new sp. B–*ludovicianus* lineage and size divergence between *C. ludovicianus* and *C. mexicanus;* both patterns are evident in Figure 7.3. There are no significant differences in size between late Rancholabrean and Recent *C. ludovicianus.*

Likewise, all measures of size exhibit highly significant ($p < 0.001$) variation among the better-sampled OTUs of white-tails (8 Sheridanian and younger samples, Table 7.3). The statistically significant aspects of size variation are visually evident in Figure 7.3 – size decrease between Pleistocene samples of *C. niobrarius* and Recent *C. leucurus/parvidens,* and the divergence between *C. gunnisoni* and *C. niobrarius.* Extant species of the *C. niobrarius–leucurus/parvidens* lineage converged in size toward *C. gunnisoni* and are not statistically different from the latter. There are no significant differences among Pleistocene samples of *C. niobrarius,* indicating long-term stasis in size from at least the Sheridanian to the end of the Pleistocene, a period of approximately 350 ky. Similarly, there was no significant change in size during the documented history of *C. gunnisoni.* A single Cudahyan specimen, indistinguishable from later samples of *C. gunnisoni,* and within its modern range (Hansen Bluff local fauna of south-central Colorado) (Rogers et al., 1985), suggests stasis for approximately 650 ky.

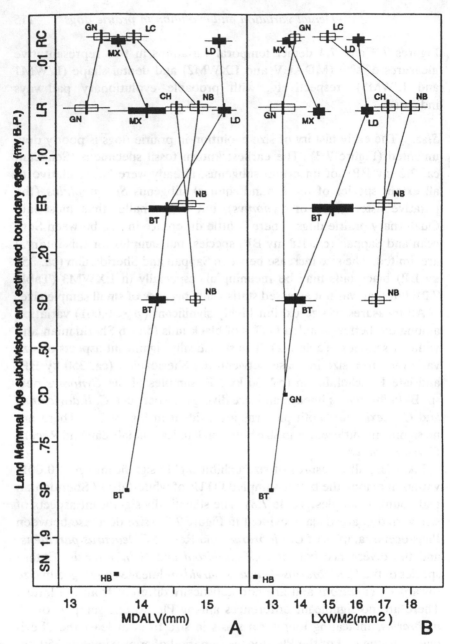

Figure 7.3. Temporal trends in (A) MDALV and (B) LXWM2 within several lineages of prairie dogs. Single specimens are plotted as small squares; samples are plotted as $\overline{X} \pm 1$ SE (boxes) and 2 SE (horizontal lines). Black-tail OTUs are represented by solid standard-error boxes, and white-tail OTUs by white standard-error boxes. Land-mammal-age abbreviations are given in Table 7.1. OTU abbreviations can be obtained by combining the land-mammal-age and taxonomic abbreviations; these are identified in Table 7.2 (black-tails) and Table 7.3 (white-tails). To minimize overlap of plotted samples within single time horizons, samples are slightly offset vertically. This is not intended to reflect differences in age.

Table 7.4. *Estimates of evolutionary rates (in darwins) for two measures of size (MDALV, LXWM2) in selected prairie dog lineage segments*

		Estimated time interval (my BP)					
Lineage	Variable	1.2–0.35	0.35–0.15	0.15–0.03	0.05–0.03	0.03–0	0.01–0
SPBT-RCLD[a]	MDALV	0.039	0.056	0.477		-0.153	
	LXWM2	0.038	0.229	0.332		-0.317	
LRLD-RCMX	MDALV				-3.22	-1.01	
	LXWM2				-2.00	-1.20	
LRGN-RCGN	MDALV					0.549	
	LXWM2					0.551	
SDNB-RCLC	MDALV			-0.041	0.096	-2.20	-6.59
	LXWM2			0.036	0.238	-3.00	-9.01

[a]Abbreviations as in Tables 7.2 and 7.3.

Table 7.4 gives estimates of evolutionary rates (d) in lineage segments of black-tails (*Cynomys* new sp. A–*ludovicianus; C. ludovicianus–mexicanus*) and white-tails (late Rancholabrean–Recent *C. gunnisoni; C. n. niobrarius–leucurus*) for MDALV and LXWM2. Rates were measured over time intervals ranging from 850 to 10 ky. As Gingerich (1983) has noted, rates measured over shorter intervals tend to be higher; thus comparisons of rates must take interval length into account. I confine most comparisons to intervals of 100–200-ky duration (Sheridanian–early Rancholabrean; early–late Rancholabrean) and 10–50-ky duration (within the late Rancholabrean; late Rancholabrean–Recent).

Several patterns are of interest. Both characters suggest low rates of change between the Sappan (ca. 1.2 my BP) and Sheridanian (ca. 350 ky BP) for black-tails (the longest interval in Table 7.4). However, the two variables suggest differing patterns of change in black-tails between the Sheridanian and late Rancholabrean (ca. 30 ky BP) – MDALV changed slowly, at a rate of 0.056 *d,* between the Sheridanian and early Rancholabrean (ca. 150 ky BP), and relatively rapidly (rate of 0.477 *d*) into the late Rancholabrean, whereas LXWM2 exhibited a more nearly constant rate of change throughout the interval. Compared with black-tails, white-tails exhibited low rates of change in both characters during both intervals (rates of −0.041 *d* and 0.096 *d* for MDALV, 0.036 *d* and 0.238 *d* for LXWM2), but LXWM2 appears to have changed more rapidly than MDALV between the early and late Rancholabrean.

Rates of change in MDALV and LXWM2 between the late Rancholabrean and Recent were relatively low for *C. ludovicianus* (−0.153

d and $-0.317\ d$), intermediate for *C. gunnisoni* ($0.549\ d$ and $0.551\ d$), and high for *C. niobrarius-leucurus* ($-2.20\ d$ and $-3.00\ d$). A right P_4 of *C. niobrarius* from the latest Pleistocene (in a layer dated between 14 and 10 ky BP) of north-central Kansas (Johnson, 1989; Stewart, 1989) is very large (LXWP4 = 16.37), indicating persistence of large size to the end of the Pleistocene. Thus, size reduction in this lineage was particularly rapid, occurring since the end of the Pleistocene (since 10 ky BP, Table 7.4).

The rates of change associated with the origin and subsequent history of *C. mexicanus* deserve comment. Genetic evidence has been interpreted to indicate that this taxon diverged from *C. ludovicianus* during the late Rancholabrean (McCullough and Chesser, 1987). A small sample from the mid-Wisconsinan (probably >29,000 yr BP) (Harris, 1987) Lost Valley local fauna of southeastern New Mexico resembles *C. mexicanus* in several respects (Goodwin, 1990) and may be within that lineage. Thus, I have estimated that the origin of *C. mexicanus* occurred between 50 and 30 ky BP (Table 7.4). In that event, the origin of *C. mexicanus* involved high rates of change in MDALV ($-3.22\ d$) and LXWM2 ($-2.00\ d$), followed by moderate rates of change ($-1.01\ d$ and $-1.20\ d$) into the Recent.

Most evidence suggests continuous increases in size of black-tails between the Sappan (ca. 1.2 my BP) and late Rancholabrean (ca. 30 ky BP). However, a single P_4 from the Cudahyan (ca. 650 ky BP) Hall Ash local fauna (Eshelman and Hager, 1984) is much larger than expected (Table 7.2). This single specimen may be intrusive, or it may represent the extreme of a variable population. Alternatively, black-tails may have been larger during that time. The available evidence does not distinguish among these possible interpretations. Additional materials are needed from this poorly known time interval.

Shape.   Compared with other ground squirrels, all prairie dogs have relatively wide teeth (indicated by low L/W ratios in Tables 7.2 and 7.3). However, the earliest prairie dogs had relatively narrower M_1s and M_2s than did later forms. Throughout their documented history, white-tails have had significantly wider P_4s through M_2s, and usually M_3s, than black-tails (Figure 7.4).

Black-tails exhibit no significant shape variation in P_4–M_2 (exemplified by L/WM1, Figure 7.4A), but do exhibit significant shape variation in M_3 (Figure 7.4B). The M_3 exhibits decreasing relative width between the Sheridanian (ca. 350 ky BP) and late Rancholabrean (ca. 30 ky BP),

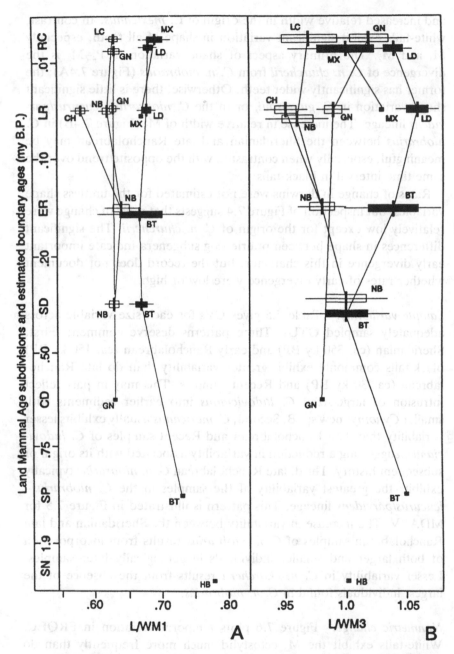

Figure 7.4. Temporal trends in (A) L/WM1 and (B) L/WM3 in several lineages of prairie dogs. For details of the symbols and abbreviations, see legend for Figure 7.3.

and increased relative width in the origin of *C. mexicanus*. In contrast, white-tails exhibit significant variation in shape of all teeth, especially M_1 and M_2. The primary aspect of shape variation in P_4–M_2 is the divergence of *C. n. churcherii* from *C. n. niobrarius* (Figure 7.4A); the former has significantly wider teeth. Otherwise, there is little significant shape variation in *C. gunnisoni,* or in the *C. niobrarius–leucurus/parvidens* lineage. The increase in relative width of M_3 (Figure 7.4B) in *C. niobrarius* between the Sheridanian and late Rancholabrean may be meaningful, especially when contrasted with the opposite trend over the same time interval in black-tails.

Rates of change in darwins were not estimated for the unitless shape variables, but inspection of Figure 7.4 suggests that rates of change were relatively low except for the origin of *C. n. churcherii*. The significant differences in shape between prairie dog subgenera indicate important early divergence in this character, but the record does not document whether rates of early divergence were low or high.

Sample variability.   Table 7.5 gives CVs for each size variable across adequately sampled OTUs. Three patterns deserve comment. First, Sheridanian (ca. 350 ky BP) and early Rancholabrean (ca. 150 ky BP) black-tails commonly exhibit greater variability than do late Rancholabrean (ca. 30 ky BP) and Recent samples. This may in part reflect intrusion of large-sized *C. ludovicianus* into earlier sediments with smaller *Cynomys* new sp. B. Second, *C. mexicanus* usually exhibits lesser variability than late Rancholabrean and Recent samples of *C. ludovicianus,* suggesting a reduction in variability associated with its origin or subsequent history. Third, late Rancholabrean *C. n. niobrarius* typically exhibits the greatest variability of the samples in the *C. niobrarius–leucurus/parvidens* lineage. This pattern is illustrated in Figure 7.5 for MDALV. The increase in variability between the Sheridanian and late Rancholabrean samples of *C. n. niobrarius* results from incorporation of both larger and smaller individuals in geologically later samples. Lesser variability in *C. n. churcherii* results from the absence of the largest individuals found in *C. n. niobrarius.*

Nonmetric change.   Figure 7.6 plots temporal variation in FRQEC. White-tails exhibit the M_3 ectostylid much more frequently than do black-tails. *Cynomys gunnisoni* exhibits considerable reduction in this character between the late Rancholabrean and Recent, a pattern that is probably meaningful. *Cynomys niobrarius–C. leucurus* exhibit consis-

Table 7.5. Coefficients of variation for size variables

| | | | OTU | | | | | | | | | | |
| | Black-tails | | | | | White-tails | | | | | | | |
Variable	SDBT[a]	ERBT	LRLD	RCLD	RCMX	LRGN	RCGN	SDNB	ERNB	LRNB	LRCH	RCLC	RCPV
MDALV	3.55	3.77	3.66	3.55	2.23	4.34	3.03	1.93	2.51	4.45	2.57	3.54	2.93
LXWP4	11.10	16.32	9.40	8.80	6.13	6.48	6.57	6.12	6.24	9.41	7.90	9.65	9.10
LXWM1	9.90	9.23	8.10	6.94	7.59	4.65	6.79	5.21	6.76	9.79	7.11	7.34	8.68
LXWM2	8.78	10.74	7.90	7.20	5.19	7.68	4.92	6.73	5.51	9.44	6.44	7.74	7.61
LXWM3	12.02	8.97	7.59	9.46	5.20	5.73	2.82	7.29	6.31	7.87	7.43	7.75	3.69

[a]Abbreviations for black-tails are given in Table 7.2, for white-tails in Table 7.3.

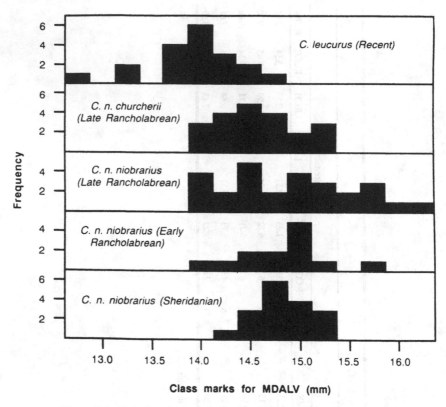

Figure 7.5. Size-class frequency distributions for OTUs in the *C. niobrarius–leucurus* lineage.

sistent, gradual increase in FRQEC through time, whereas black-tails show an apparent trend toward decreases in FRQEC.

Late Rancholabrean geographic variation

Table 7.6 summarizes descriptive statistics of all metric variables for geographic samples (Figure 7.2) of late Rancholabrean (ca. 30 ky BP) *C. niobrarius* and *C. ludovicianus*. *Cynomys niobrarius* exhibits significant geographic variation in one size variable (LXWM2), and *C. ludovicianus* does so in two size variables (LXWP4, LXWM2). Typical patterns of geographic size variation are illustrated by LXWM2 in Figure 7.7A. For *C. niobrarius,* the smallest individuals typically occur in the north (areas 1 and/or 2), the largest individuals typically occur in the center of its range (area 3), and intermediate or small-sized individuals

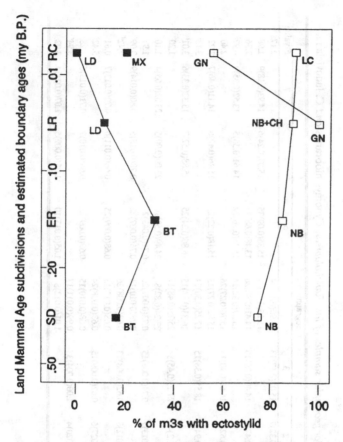

Figure 7.6. Temporal trends in the frequency of occurrence of M₃ ectostylids. Black squares represent black-tails, and white squares white-tails. For details of the abbreviations, see legend for Figure 7.3.

occur in the south (area 4, northern part of area 5). In contrast, black-tails usually exhibit increases in size from the south (area 7) and west (area 6) toward the center of their late Rancholabrean range (area 5), with subsequent decreases in size to the north (area 4).

White-tails exhibit significant geographic variation in all shape variables except L/WP4, whereas black-tails exhibit significant shape variation only in L/WP4. Typical patterns of geographic variation are shown for L/WM1 in Figure 7.7B. For *C. niobrarius,* the most significant feature of geographic variation is great relative tooth width in the northern sample (corresponding to *C. n. churcherii*); relative width commonly decreases southward toward areas 3 and 4. For black-tails, the only

Table 7.6. *Descriptive statistics [$\bar{X}$ (SE)n] for geographic samples of late Rancholabrean* Cynomys niobrarius *and* C. ludovicianus

Variable	Taxon	Geographic area[a]							F value
		1	2	3	4	5	6	7	
MDALV	niobrarius	14.56(0.08)21	14.79(0.32)7	15.01(0.17)13	14.90(0.12)24	15.36(0.09)35			2.17
	ludovicianus				15.16(0.12)17		15.07(0.24)6	14.93(0.18)9	1.98
LXWP4	niobrarius	13.48(0.22)22	13.30(0.47)7	14.40(0.29)15	14.41(0.32)26	13.48(0.68)2			2.36
	ludovicianus				15.07(0.32)14	15.25(0.26)22	14.61(0.36)5	13.52(0.35)9	4.97[b]
LXWM1	niobrarius	15.56(0.25)22	15.28(0.62)8	16.69(0.40)13	15.76(0.27)26	15.18(0.22)25			2.40
	ludovicianus				15.02(0.27)13		15.46(0.42)6	14.11(0.46)7	2.22
LXWM2	niobrarius	17.19(0.24)21	17.67(0.94)6	18.55(0.38)15	17.25(0.30)24	16.86(0.24)25			3.13[c]
	ludovicianus				16.83(0.34)15		15.88(0.52)7	15.32(0.33)6	3.01[c]
LXWM3	niobrarius	25.38(0.46)17	24.55(1.2)4	27.23(0.45)3	25.75(0.48)17	24.49(0.53)10			1.20
	ludovicianus				23.24(0.82)8		23.00(0.90)5	23.02(0.65)4	1.10
L/WP4	niobrarius	0.71(0.007)22	0.73(0.013)7	0.74(0.008)15	0.73(0.007)26	0.71(0.043)2			1.97
	ludovicianus				0.78(0.010)14	0.76(0.007)22	0.78(0.015)5	0.80(0.014)9	3.09[c]
L/WM1	niobrarius	0.58(0.007)22	0.61(0.012)8	0.62(0.008)13	0.63(0.008)26				7.42[d]
	ludovicianus				0.68(0.012)13	0.68(0.009)25	0.66(0.011)6	0.70(0.013)7	0.61
L/WM2	niobrarius	0.59(0.005)21	0.63(0.027)6	0.62(0.009)15	0.63(0.008)24				5.42[b]
	ludovicianus				0.68(0.010)15	0.69(0.009)25	0.68(0.010)7	0.71(0.013)6	1.28
L/WM3	niobrarius	0.95(0.009)17	0.93(0.020)4	1.00(0.004)3	0.99(0.011)17				3.94[c]
	ludovicianus				1.01(0.012)8	1.05(0.013)10	1.05(0.009)5	1.07(0.021)4	3.02

[a] Geographic areas are illustrated in Figure 7.2.
[b] Significant at 0.01 level.
[c] Significant at 0.05 level.
[d] Significant at 0.001 level.

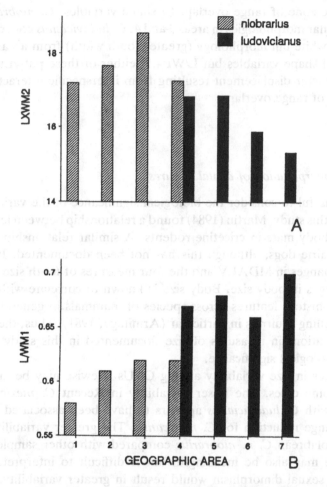

Figure 7.7. Histograms plotting mean values of (A) LXWM2 and (B) L/WM1 for geographic-area OTUs of late Rancholabrean *C. niobrarius* and *C. ludovicianus*. The geographic areas are identified in Figure 7.2.

consistent pattern is that the southernmost sample (area 7) always exhibits greatest relative tooth length; variation among the other geographic samples shows no clear pattern.

White-tails and black-tails overlapped broadly on the central Great Plains during the Pleistocene, primarily in area 4 (a few white-tail specimens have been recovered from the northern edge of area 5). The range overlap appears to have been real, not an artifact of a fluctuating zone of contact during the late Rancholabrean (Goodwin, 1990). In most size variables, both lineages exhibit size reductions between adjacent samples and the zone of range overlap. In shape variables, *C. niobrarius* exhibits similar morphologies in areas 3 and 4; *C. ludovicianus* converges toward the white-tail morphology (greater tooth width) from area 5 to area 4 in all shape variables but L/WP4. Neither of these patterns indicates character displacement resulting from interspecific interactions in the zone of range overlap.

Discussion

Biological interpretation of dental features

It may be useful to consider the biological significance of the variables included in this study. Martin (1984) found a relationship between length of M_1 and body mass in cricetine rodents. A similar relationship may hold for prairie dogs, although this has not been documented. If so, temporal changes in MDALV and the four measures of tooth size may reflect changes in body size. Body size is known to correlate with important life-history features across species of mammals in general and ground-dwelling squirrels in particular (Armitage, 1981). Thus, the significant variations in measures of size documented in this study may have had biological significance.

Differences in size variability among OTUs likewise may be meaningful in some cases. The lesser variability in Recent *C. mexicanus* compared with *C. ludovicianus* appears to have been associated with dramatic range reduction for *C. mexicanus*. The greater variability in late Rancholabrean *C. n. niobrarius* compared with other samples of this lineage may also be meaningful but is difficult to interpret. An increase in sexual dimorphism would result in greater variability, but should be reflected in increased bimodality of the frequency distribution. Such a pattern is not obvious in these data (Figure 7.5).

The marked differences in dental shape between the two subgenera of prairie dogs through time probably have adaptive importance given the critical role of teeth in mammalian food processing. However, the biological significance of these differences is unclear. The lesser shape differences among temporal and geographic samples within each subgenus may also have had adaptive value, although this is less certain. Patterns of late Rancholabrean geographic variation in most shape variables, as documented earlier, suggest that tooth shape may vary according to latitudinally controlled environmental factors – widest teeth tend to occur toward the north, and narrowest teeth toward the south.

Unfortunately, it is not yet possible to reconstruct with confidence late Rancholabrean environmental gradients from the southern to northern Great Plains that might have selected for differences in shape.

In contrast, the presence of ectostylids in the hypoflexid of M_3 seems to have had no obvious function. These structures appear to be too low to have been useful in mastication and may have been adaptively neutral. Other nonmetric characters of rodents have been interpreted in this manner (McLellan and Finnegan, 1990).

Evolutionary patterns and process

General patterns.   Morphological changes in prairie dogs through time illustrate several general evolutionary patterns. The early record of the genus suggests gradual increases in size and in relative tooth width, but the data points are so scattered in time that patterns of change cannot be discriminated.

The record is much better for the Sheridanian and younger intervals (from ca. 500 ky BP to the present). Significant examples of long-term stasis include size and shape in *C. gunnisoni* (possibly for 650 ky) and the shapes of P_4–M_2 in the *C. n. niobrarius–leucurus/parvidens* lineage (possibly for 350 ky). *Cynomys n. niobrarius–leucurus/parvidens* also exhibited stasis in size to the end of the Pleistocene, followed by abrupt change into the Recent. This may exemplify stairstep change (stasis, abrupt change, stasis) if size reduction was restricted to the early Holocene rather than extending throughout the Holocene. Some size variables (MDALV, LXWM3) also suggest stairstep change in the *Cynomys* new sp. B.–*ludovicianus* lineage – relative stasis between the Sheridanian (ca. 350 ky BP) and early Rancholabrean (ca. 150 ky BP), relatively rapid change into the late Rancholabrean (ca. 30 ky BP), and stasis into the Recent. However, other size variables (LXWP4, LXWM1, LXWM2) suggest more or less constant-rate phyletic change between the Sheridanian and late Rancholabrean, followed by stasis into the Recent. Rapid size change also is evident in the origin of *C. mexicanus* between 50 and 30 ky BP.

Other examples of constant-rate phyletic change may include change in M_3 shape between the Sheridanian (ca. 350 ky BP) and late Rancholabrean (ca. 30 ky BP) for *Cynomys* new sp. B–*ludovicianus* and for *C. niobrarius,* and change in ectostylid frequency in the same lineages from the Sheridanian to the Recent. In both cases, the two lineages exhibit opposite constant-rate trends.

The best examples of mosaic change are between size and shape variables, which appear to be evolving more or less independently of each other. Clearly, a change in tooth shape, described by a simple ratio of linear measurements, requires a change in size of one or both of the linear measurements, and in this sense shape and size cannot be independent. However, a directional change in tooth shape (e.g., an increase in relative width) need not be associated with a directional change in size (e.g., size increase), and in this sense size and shape do seem to be evolving independently in prairie dogs.

Tempo and mode of evolution.   These general patterns indicate that prairie dog evolution has been neither strictly punctuational nor strictly gradualistic. Barnosky (1987) recently reviewed the literature on Quaternary mammals as it bears on the models of punctuated equilibrium (Eldredge and Gould, 1972) and phyletic gradualism (Gingerich, 1976). He found that the Quaternary record provides examples supportive of both models, and my data support a similar conclusion within a single genus.

In addition to predictions about tempo, punctuated equilibrium theory predicts that most morphological change occurs during speciation, when small, isolated populations break away from the constraints of the parent species. Alternatively, phyletic gradualism suggests that most evolution occurs by phyletic transformation of the entire species over time. Several examples of significant temporal change, such as size increase in black-tails between the Sheridanian (ca. 350 ky BP) and late Rancholabrean (ca. 30 ky BP), appear to involve transformation of an entire large population, if not the whole species, thus fitting the picture of phyletic gradualism. The origin of *C. leucurus* and *C. parvidens* from ancestral *C. niobrarius* likewise appears to represent phyletic transformation. Size decrease probably occurred across the entire population, which subsequently diverged allopatrically to form the two small-sized modern species. The origin of *C. mexicanus* provides the only documented example of significant evolutionary change likely associated with speciation. Unfortunately, the mode of early subgeneric divergence in size and shape remains unknown. With the available evidence, it is clear that significant evolutionary change in prairie dogs has not been dependent on speciation.

Causes of evolution.   Allmon and Ross (1990) have suggested that the causes of evolution can usefully be classified as intrinsic versus extrinsic, and biotic versus abiotic. In this discussion I shall consider three causal hypotheses for patterns of evolution in prairie dogs – biotic interactions (extrinsic, biotic), climatically induced environmental selection (extrinsic, abiotic), and internal genetic mechanisms (intrinsic, biotic).

Biotic interactions that might influence evolution include predation and competition. Hoogland (1981) suggested that differential predation pressure has been responsible for divergence in coloniality between white-tails (moderately colonial) and black-tails (highly colonial), but this hypothesis is not obviously related to the character evolution discussed herein. Essentially nothing is known about predator-prey interactions for fossil prairie dogs, and thus the role of this factor is difficult to evaluate. Paleoindian culling might have resulted in selection for body size reduction, as seen in the *C. niobrarius–leucurus/parvidens* lineage at the Pleistocene/Holocene boundary. However, the same would be expected for *C. ludovicianus,* but has not been observed.

Prairie dogs differ from other ground squirrels that inhabit their range in having a much larger body size and also in regard to a number of ecological features; thus competition with other genera of ground squirrels is probably minimal. More intense competition would be expected with other prairie dog species in the limited areas of modern range overlap. The ranges of white-tails and black-tails apparently overlapped broadly on the central Great Plains at least from the Sheridanian (ca. 350 ky BP) to the end of the Pleistocene (Goodwin, 1990), and thus abundant opportunities for interspecific competition existed. Nonetheless, the patterns of geographic variation summarized here provide no evidence of character displacement in the zone of overlap. In addition, the increase in size of black-tails into the late Rancholabrean resulted in significant convergence in overall size toward that of white-tails, despite their continued broad sympatry on the central Great Plains. Finally, the reduction in size of the *C. niobrarius–leucurus/parvidens* lineage into the Recent resulted in divergence from black-tails at a time when the species' ranges overlapped less than they had throughout the late Pleistocene. None of these patterns of change appears to indicate an important role for interspecific competition.

Two lines of evidence suggest that climatically related environmental changes may have been important in prairie dog evolution. First, the patterns of geographic variation are generally consistent with predictions of a climatic model. The reverse Bergmann's response in fossil white-tails is consistent with a similar pattern in modern *C. leucurus,* which appears to reflect variation in environmental productivity (Pizzimenti,

1975, 1976). White-tails with the widest teeth occur farthest north, and black-tails with the narrowest teeth occur farthest south, again suggestive of climatic influence. Second, the size reduction in white-tails between the late Rancholabrean (ca. 30 ky BP) and the Recent occurs across a time of significant environmental change. Terminal Pleistocene size reduction also is known in several large mammal lineages (Guthrie, 1984), but is less well documented for small mammals. For adequately known ground squirrel lineages, the general pattern is of overall stasis [e.g., *Spermophilus richardsonii* complex (Neuner, 1975); *S. townsendii* ratio of M^3 length/width, but not size (Barnosky, 1987; Rensberger and Barnosky, Chapter 13, this volume); *Cynomys gunnisoni* and *C. ludovicianus* (this chapter)] or even size increase [*S. parryii* (Guthrie, 1984)] into the Holocene.

Although some of the evidence for *Cynomys* seems to be consistent with a climatic model of causation, other evidence is, at face value, contradictory. First, each of the known episodes of change was taxonomically selective. Why was *C. niobrarius* unaffected by climate change at the beginning of the Wisconsinan, and *C. gunnisoni* and *C. ludovicianus* at the end of the Wisconsinan? Interspecific differences in habitat and ecology might effect differential responses, but it is not always clear in what way. Nadler, Hoffmann, and Pizzimenti (1971) suggested that the degree of evolution in prairie dog lineages is inversely related to the vegetative and physiographic complexity of the lineage's habitat, and this might explain why *C. gunnisoni,* which has inhabited the complex habitats of the southern Rocky Mountains since the Cudahyan (ca. 650 ky BP), has changed little during this time.

Second, all prairie dog lineages exhibit stasis in one character or another over periods of time that presumably incorporate several glacial-interglacial transitions. If climate change resulted in morphological change at the end of the Wisconsinan, why not at other glacial-interglacial transitions? Unfortunately, we still know very little about pre-Wisconsinan environmental variation in continental North America, and thus the foregoing question is difficult to address. It is possible that previous environmental fluctuations were qualitatively or quantitatively different from that at the end of the Wisconsinan. Significantly, the prairie dog distributional history seems to tell a similar story. Pleistocene ranges remained relatively constant from the Sheridanian (ca. 350 ky BP) to the end of the Pleistocene (Goodwin, 1990).

Finally, intrinsic genetic mechanisms may play roles in constraining or even directing evolutionary change. If environments did fluctuate several times between the Cudahyan (ca. 650 ky BP) and late Rancholabrean (ca. 30 ky BP), as suggested by the Milankovitch-Croll astronomical theory of glaciation (Hays, Imbrie, and Shackleton, 1976), then the lack of significant morphological change may reflect internal constraints on variation. If FRQEC is adaptively neutral, as seems likely, changes in that variable through time might reflect genetic drift or a pleiotropic side effect of selection for another trait, in which case the proximal agent of change would be intrinsic.

Acknowledgments

This study originally formed part of my doctoral dissertation at the University of Kansas. I wish to thank the many people and institutions who generously loaned fossil and Recent materials that formed the basis for this study (they are listed in the acknowledgments in the dissertation). I especially wish to thank L. D. Martin, my graduate adviser, for initiating and encouraging my interest in fossil prairie dogs, and A. D. Barnosky, R. A. Martin, and an anonymous reviewer for helpful criticisms of an earlier draft of this chapter.

References

Allmon, W. D., and R. M. Ross (1990). Specifying causal factors in evolution: the paleontological contribution. In R. M. Ross and W. D. Allmon (eds.), *Causes of Evolution: A Paleontological Perspective* (pp. 1–17). Chicago: University of Chicago Press.

Armitage, K. B. (1981). Sociality as a life-history tactic of ground squirrels. *Oecologia*, 48:36–49.

Barnosky, A. D. (1987). Punctuated equilibrium and phyletic gradualism: some facts from the Quaternary mammalian record. *Current Mammalogy*, 1: 109–47.

Burns, J. A., and W. B. McGillivray (1989). A new prairie dog, *Cynomys churcherii*, from the late Pleistocene of southern Alberta. *Canadian Journal of Zoology*, 67:2633–9.

Eldredge, N., and S. J. Gould (1972). Punctuated equilibrium: an alternative to phyletic gradualism. In T. J. Schopf (ed.), *Models in Paleobiology* (pp. 82–115). San Francisco: Freeman.

Eshelman, R. E. (1975). Geology and paleontology of the early Pleistocene (late Blancan) White Rock fauna from north-central Kansas. In *Studies on Cenozoic Paleontology and Stratigraphy* (pp. 1–60). Claude W. Hibbard Memorial Volume 4. University of Michigan Papers on Paleontology, no. 13.

Eshelman, R. E., and M. W. Hager (1984). Two Irvingtonian (medial Pleistocene) vertebrate faunas from north-central Kansas. In H. H. Genoways and M. R. Dawson (eds.), *Contributions in Quaternary Vertebrate Paleontology: A Volume in Memorial to John E. Guilday* (pp. 384–404). Carnegie Museum of Natural History, Special Publication 8.

Gingerich, P. D. (1976). Paleontology and phylogeny: patterns of evolution at the species level in early Tertiary mammals. *American Journal of Science,* 276:1–28.

  (1983). Rates of evolution: effects of time and temporal scaling. *Science,* 222:159–61.

Goodwin, H. T. (1990). Systematics, biogeography, and evolution of fossil prairie dogs (genus *Cynomys*). Ph.D. dissertation, University of Kansas, Lawrence.

Guthrie, R. D. (1984). Alaskan megabucks, megabulls, and megarams: the issue of Pleistocene gigantism. In H. H. Genoways and M. R. Dawson (eds.), *Contributions in Quaternary Vertebrate Paleontology: A Volume in Memorial to John E. Guilday* (pp. 482–510). Carnegie Museum of Natural History, Special Publication 8.

Haldane, J. B. S. (1949). Suggestions as to quantitative measurement of rates of evolution. *Evolution,* 3:51–6.

Harris, A. H. (1987). Reconstruction of mid-Wisconsin environments in southern New Mexico. *National Geographic Research,* 3:142–217.

Hay, O. P. (1921). Description of Pleistocene Vertebrata, types or specimens of which are preserved in the United States National Museum. *Proceedings of the United States National Museum,* 59:617–38.

Hays, J. D., J. Imbrie, and N. J. Shackleton (1976). Variation in the earth's orbit: pacemaker of the ice ages. *Science,* 194:1121–32.

Hibbard, C. W. (1942). Pleistocene mammals from Kansas. *Bulletin of the Kansas Geological Survey,* 41:261–69.

Hintze, J. L. (1990). *Number Cruncher Statistical System Version 5.03: Installation and Reference Manual.* Kaysville, Utah: J. L. Hintze.

Hoogland, J. L. (1981). The evolution of coloniality in white-tailed and black-tailed prairie dogs (Sciuridae: *Cynomys leucurus* and *C. ludovicianus*). *Ecology,* 62:252–72.

Johnson, W. C. (1989). Stratigraphy and late Quaternary landscape evolution. In M. J. Adair (ed.), *Archaeological Investigations at the North Cove Site, Harlan County Lake, Harlan County, Nebraska* (pp. 22–52). Report submitted to the Kansas City District, U.S. Army Corps of Engineers; Kaw Valley Engineering and Development, Junction City, Kansas.

Lundelius, E. L., Jr., C. S. Churcher, T. Downs, C. R. Harrington, E. H. Lindsay, G. E. Schultz, H. A. Semken, S. D. Webb, and R. J. Zakrzewski (1987). The North American Quaternary sequence. In M. O. Woodburne (ed.), *Cenozoic Mammals of North America: Geochronology and Biostratigraphy* (pp. 211–35). Berkeley: University of California Press.

McCullough, D. A., and R. K. Chesser (1987). Genetic variation among populations of the Mexican prairie dog. *Journal of Mammalogy,* 68: 555–60.

McLellan, L. J., and M. Finnegan (1990). Geographic variation, asymmetry, and sexual dimorphism of nonmetric characters in the deer mouse (*Peromyscus maniculatus*). *Journal of Mammalogy,* 71:524–33.

Martin, R. A. (1984). The evolution of cotton rat body mass. In H. H. Genoways and M. R. Dawson (eds.), *Contributions in Quaternary Vertebrate Paleontology: A Volume in Memorial to John E. Guilday* (pp. 179–83). Carnegie Museum of Natural History, Special Publication 8.

Nadler, C. F., R. S. Hoffmann, and J. J. Pizzimenti (1971). Chromosomes and serum proteins of prairie dogs and a model of *Cynomys* evolution. *Journal of Mammalogy,* 52:545–55.

Neuner, A. M. (1975). Evolution and distribution of the *Spermophilus richardsonii* complex of ground squirrels in the middle and late Pleistocene: a multivariate analysis. M.A. thesis, University of Kansas, Lawrence.

Pizzimenti, J. J. (1975). Evolution of the prairie dog genus *Cynomys. Occasional Papers, Museum of Natural History, University of Kansas,* 19:1–73.

(1976). Genetic divergence and morphological convergence in the prairie dogs, *Cynomys gunnisoni* and *Cynomys leucurus*. I. Morphological and ecological analysis. *Evolution,* 30:345–66.

Rogers, K. L., et al. (1985). Middle Pleistocene (late Irvingtonian: Nebraskan) climatic changes in south-central Colorado. *National Geographic Research,* 1:535–65.

Schultz, C. B., L. D. Martin, L. G. Tanner, and R. G. Corner (1978). Provincial land mammal ages for the North American Quaternary. *Bulletin of the University of Nebraska State Museum,* 9:183–95.

Sokal, R. R., and F. J. Rohlf (1981). *Biometry.* San Francisco: Freeman.

Stewart, J. D. (1989). Paleontology and paleoecology of the 1987 excavation of the North Cove Site, 25HN164. In M. J. Adair (ed.), *Archaeological Investigations at the North Cove Site, Harlan County Lake, Harlan County, Nebraska* (pp. 63–80). Report submitted to the Kansas City District, U.S. Army Corps of Engineers. Kaw Valley Engineering and Development, Junction City, Kansas.

8

Quantitative and qualitative evolution in the giant armadillo *Holmesina* (Edentata: Pampatheriidae) in Florida

RICHARD C. HULBERT, JR., AND GARY S. MORGAN

The formation of the Panamanian isthmus in the late Pliocene permitted widespread faunal interchange between North and South America. Previously, edentates (or xenarthrans) had been limited to South America since the Paleogene and had there undergone a moderate radiation into three major clades: the shelled cingulates (three families), the anteaters (one family), and the sloths (a minimum of four families). Two sloth genera, *Thinobadistes* and *Pliometanastes,* appeared in North America in the late Miocene, heralding the forthcoming interchange. The emergence of Panama increased the probability of faunal interchange, and eight edentate families dispersed into North America during the late Pliocene or early Pleistocene with varying success. The Pampatheriidae is a distinct lineage of shelled edentates, now accorded familial rank (Engleman, 1985; Edmund, 1987) and separated from both the Dasypodidae (within which they were long regarded as a subfamily) and the Glyptodontidae. Edmund (1987, pp. 6–7) listed character states that distinguish pampatheres from the other two cingulate families.

The presence of pampatheres in North America was first recognized by Leidy (1889) on the basis of isolated osteoderms collected in Florida. During the ensuing century, numerous additional specimens of pampatheres were collected in North America, predominantly from Florida, and to a lesser extent from Texas (James, 1957; Edmund, 1987). Webb (1974) and Robertson (1976) were the first to describe pre-Rancholabrean pampatheres from Florida. They demonstrated that a very large size difference separated Blancan and Rancholabrean pampatheres in Florida, with some Irvingtonian populations being of intermediate size. Prior studies had necessarily been limited to

Rancholabrean specimens and thus could not have detected any significant evolutionary change. Robertson (1976) designated the Florida Blancan samples as a new species of the South American pampathere genus *Kraglievichia, K. floridanus,* and referred early Irvingtonian specimens to *Kraglievichia* sp. Martin (1974) and Robertson (1976) considered the late Irvingtonian Coleman 2A pampathere as conspecific with the typical Rancholabrean species, which they recognized as either *Chlamytherium septentrionalis* or *Pampatherium septentrionalis.* After extensive comparisons with South American pampatheres, Edmund (1985b, 1987) concluded that both Florida species, "*Glyptodon*" *septentrionalis* Leidy and *Kraglievichia floridanus* Robertson, were best regarded as members of the genus *Holmesina,* as were the Brazilian *Chlamydotherium humboldti* Lund and the Andean *Chlamytherium occidentale* Hoffstetter. *Holmesina* was named by Simpson (1930), with *H. septentrionalis* (Leidy) as its type species. According to Edmund (1987), the four species of *Holmesina* are distinguished from other South American pampathere genera (*Kraglievichia, Pampatherium* s.s., *Vassallia,* and *Plaina*) on the basis of osteoderm characters, but the exact relationships within the family remain unresolved. Pending a comprehensive phylogenetic analysis of the group, we follow Edmund (1987) in recognizing a single pampathere genus in Florida: *Holmesina.*

Edmund (1987) also quantified the size trends described by Robertson (1976). He found that osteoderm area (length multiplied by width) increased by about 225% from the Blancan to the Rancholabrean, and average limb length increased by 150% during the same period. Because $(1.5)^2 = 2.26$, the increase in limb length was isometric with the increase in osteoderm area. Edmund (1987) also graphed metapodial length versus time and compared measurements of tibiae, radii, and calcanea. He grouped the specimens into three time categories: Blancan, Irvingtonian, and Rancholabrean. At one point Edmund (1987, p. 14) characterized the size increase in Florida *Holmesina* as gradual. However, elsewhere in the same paper (p. 4) he noted that late Irvingtonian specimens were the same size as Rancholabrean individuals, which would argue for a pattern of episodic evolution and stasis.

Our study of evolution in *Holmesina* expands upon that of Edmund (1987) in several respects. First, a number of important samples of pre-Rancholabrean *Holmesina* that were unavailable to Edmund (1987) have been collected in Florida in the past decade, several within the past few years. Second, rather than grouping specimens into just three chronological samples, we have predominantly analyzed them on a site-by-site basis. This has been possible only because the recently improved bio-chronology of Blancan and Irvingtonian sites in Florida (Morgan and Hulbert, in press) allows the age of most faunas to be estimated within ± 100,000 years or less. This permits a much more detailed resolution of evolutionary patterns. Finally, we analyzed more measured characters on the limbs and osteoderms than were reported by Edmund (1987), and we added dental dimensions and qualitative skeletal characters.

The overall goal of the study was to elucidate the patterns of evolutionary change in *Holmesina* to the best resolution possible given the sampling limitations and imprecision of our biochronological methods. We were particularly interested in determining whether evolutionary rates varied or were more or less constant for the genus over its chronological range in Florida. We also wanted to compare the evolutionary rates and patterns between elements of the dentition, appendicular skeleton, and the osteoderms. That was intended to test Edmund's (1987) claim of equivalent rates of change in limb length and osteoderm area, with other parameters and dental rates added to the analysis.

Materials and methods

Specimens of *Holmesina* in the collections of the Florida Museum of Natural History (UF) and the Florida Geological Survey (FGS) formed the primary basis of this study (Table 8.1). These were augmented with a few others housed in the American Museum of Natural History (AMNH, McLeod Limerock Mine) and in private collections. Analyses were limited to specimens from Florida, thus eliminating the possible effects of geographic variation. This does not greatly diminish the available pool of specimens, as most are from Florida anyway, especially those older than Rancholabrean. Figure 8.1 shows the general locations of the major study sites.

Measurements were taken with calipers to the nearest 0.1 mm, with the exception of limb lengths in excess of 200 mm, which were measured to the nearest 0.5 mm. Three dental parameters, length, anterior width, and posterior width (Figure 8.2), were measured on the fourth through ninth lower teeth (t4-t9). Only length and maximum width were measured on t1 through t3. This system of tooth nomenclature is used because of the uncertainty as to homology between the tooth loci of edentates and epitherian mammals. These measurements were taken both at the occlusal surface of the tooth and internally for the alveolus. Alveolar measurements proved to have fewer missing values, as the

Table 8.1. *Fossil localities and* Holmesina *samples used in the study*

Locality	Age			Material
Santa Fe River 1 & 2	lBL	2.3	Ma	L, O
Haile 15A	lBL	2.3	Ma	L, M, O
Haile 12B	lBL	2.3	Ma	L
Macasphalt Shell Pit	lBL	2.3	Ma	L, O
Kissimmee River	lBL	2.3	Ma	O
Haile 7C	vlBL	2.0	Ma	A, L, M
De Soto Shell Pit	veIR	1.8	Ma	A, L, M
Inglis 1A	veIR	1.7	Ma	L, M, O
Leisey Shell Pit	leIR	1.3	Ma	L, M, O
Crystal River Power Plant	leIR	1.3	Ma	O
Haile 16A	emIR	0.8	Ma	L, O
McLeod Limerock Mine	lmIR	0.6	Ma	L, M
Coleman 2A	lIR	0.4	Ma	L, O
Coleman 3	veRA	0.3	Ma	L, O
Haile 7A	eRA	0.2	Ma	L, O
Branford 1A	lRA	12	ka	A, L
Ichetucknee River	lRA	12	ka	L
Santa Fe River 1 & 2	lRA	12	ka	L, O
Hornsby Springs	lRA	12	ka	L, M
Paynes Prairie	lRA	12	ka	L
Oklawaha River	lRA	12	ka	M
Silver Springs	lRA	12	ka	M
Waccasassa River	lRA	12	ka	L
Rock Springs	lRA	12	ka	L
Vero	lRA	12	ka	L, M
Tapir Hill	lRA	12	ka	A, L
Horse Creek	lRA	12	ka	M
Blue Ridge Waterway	lRA	12	ka	M

Note: Chronological abbreviations: BL, Blancan; IR, Irvingtonian; RA, Rancholabrean; e, early; m, middle; l, late; v, very. Abbreviations of fossil material: A,
associated osteoderms of a single individual; L, limb elements; M, mandible; O,
unassociated osteroderms. Biochronology follows Morgan and Hulbert (in press).

teeth often fall out of the mandible, break, or are otherwise not measurable. Isolated teeth were not measured because of the inability to
accurately assign them to a specific tooth locus. Length, proximal
breadth, and distal breadth were measured on humeri, radii, femora,
tibiae, astragali, calcanea, metacarpals 2–4, and metatarsals 2–4. Six
types of osteoderms were measured (Figure 8.3): the normal movable
osteoderms; those from the anterior row of the pelvic buckler; those
from the posterior row of the pectoral buckler (Edmund, 1985a, Figure
8); four-sided, rectangular buckler osteoderms; six-sided, regular-
hexagonal buckler osteoderms; and buckler osteoderms with five or six

Figure 8.1. Map of Florida showing the most important Pliocene and Pleisto-
cene localities for *Holmesina*. Late Blancan: 1, Santa Fe River 1 & 2, Columbia
County; 2, Haile 15A, Alachua County; 3, Macasphalt Shell Pit, Sarasota
County; 4, Kissimmee River, Okeechobee County. Latest Blancan: 5, Haile
7C, Alachua County. Earliest Irvingtonian: 6, Inglis 1A, Citrus County; 7,
De Soto Shell Pit, De Soto County. Late early Irvingtonian: 8, Crystal River
Power Plant, Citrus County; 9, Leisey Shell Pit 1 & 3, Hillsborough County;
10, Pool Branch, Polk County. Middle Irvingtonian: 11, Haile 16A, Alachua
County; 12, McLeod Limerock Mine, Levy County. Late Irvingtonian: 13,
Coleman 2A, Sumter County. Early Rancholabrean: 14, Haile 7A, Alachua
County; 15, Coleman 3, Sumter County; 16, Bradenton, Manatee County. Late
Rancholabrean: 17, Branford 1A, Suwannee County; 18, Santa Fe River 1 &
2, Columbia County; 19, Hornsby Springs, Alachua County; 20, Oklawaha
River, Marion County; 21, Vero, Indian River County; 22, Peace River &
Horse Creek, De Soto County; 23, Tapir Hill, Charlotte County.

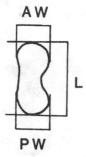

Figure 8.2. Occlusal outline of a tooth of *Holmesina* showing the three dental measurements taken in this study: L, length; AW, anterior width; PW, posterior width.

sides, with one pair of sides much longer than the others. The latter variety are especially common in *Holmesina*. We did not measure irregularly shaped buckler osteoderms, nor the small, wide, movable osteoderms from near the margin of the carapace, nor all the caudal, cephalic, and limb osteoderms. Three measurements were taken on each osteoderm: overall maximum length, width (perpendicular to length, taken in the middle of the osteoderm), and thickness (along the edge, at the point where the width was taken). All data were stored and analyzed on an AT-class microcomputer using the Quattro spreadsheet program.

Biochronology

We present a brief summary of the biochronology of the late Pliocene and Pleistocene vertebrate sites from Florida that contain samples of *Holmesina* (Table 8.1). It is based on a more detailed biochronological analysis of Florida Plio-Pleistocene sites and over 120 mammalian species by Morgan and Hulbert (in press). For the purposes of calculating evolutionary rates in darwins, we had to assign each fauna a numeric age. For each we chose our best estimate to the nearest 0.1 Ma (except for late Rancholabrean sites, which were assigned an age of 12 ka). It must be emphasized that the "true" numeric age for most of the faunas is known only to within a range of ±100,000 years. The Pliocene/Pleistocene boundary is now placed at 1.64 Ma, just above the top of the Olduvai Normal Subchron (Berggren et al., 1985; Harland et al., 1990). The boundary between the Blancan and Irvingtonian is transitional in

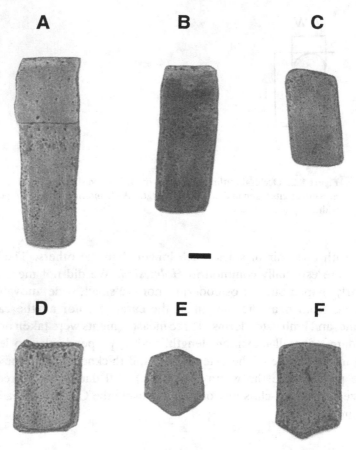

Figure 8.3. Dorsal views of representatives of the six types of *Holmesina* osteoderms measured in this study. The specimens belong to an associated carapace and skeleton, UF 125227, Haile 7C, Alachua County. (A) Normal movable osteoderm. (B) Anterior row of posterior buckler. (C) Posterior row of pectoral buckler. (D) Rectangular immovable osteoderm. (E) Hexagonal immovable osteoderm. (F) Five- and six-sided immovable osteoderms, with lateral sides much longer than anterior or posterior sides. Length of scale bar is 10 mm.

nature, but is usually placed at about 1.9 Ma (Lundelius et al., 1987; Repenning, 1987). Therefore, earliest Irvingtonian faunas between 1.9 and 1.64 Ma are latest Pliocene in age, not early Pleistocene as they have generally been regarded (e.g., Kurtén and Anderson, 1980; Lundelius et al., 1987). The Irvingtonian/Rancholabrean boundary is traditionally defined by the appearance of *Bison* in North America at about 0.3 Ma.

Late Blancan faunas in Florida are recognized by the co-occurrence of the horses *Nannippus peninsulatus* and *Equus* (*Dolichohippus*) sp., the gomphothere *Rhynchotherium*, and seven genera of Neotropical immigrants including *Dasypus, Holmesina, Glyptotherium, Glossotherium*, and *Neochoerus* (Morgan and Ridgway, 1987). These faunas are correlated with well-dated localities in Arizona and Texas (Johnson, Opdyke, and Lindsay, 1975; Galusha et al., 1984; Lundelius et al., 1987) that are associated with numeric dates of 2.5–2.0 Ma. Our analysis includes samples of *Holmesina* from four sites of this age (Table 8.1): Haile 15A, Santa Fe River, Kissimmee River, and Macasphalt Shell Pit. Haile 15A is the type locality of *H. floridanus* (Robertson, 1976).

Haile 7C has a limited mammalian fauna, which hinders precise age estimation. The presence of Neotropical immigrants (including *Holmesina, Eremotherium*, and *Erethizon*) implies a maximum age of late Blancan. Abundant remains of two turtles, *Trachemys platymarginata* and *Chelydra* n.sp., are identical with specimens from the late Blancan Haile 15A site. However, the stages of evolution for several species are more suggestive of an earliest Irvingtonian age (Hulbert, Morgan, and Poyer, 1989). We tentatively regard Haile 7C as intermediate in age between the typical late Blancan faunas listed earlier and the earliest Irvingtonian sites, with an age of about 2.0 Ma.

Earliest Irvingtonian (1.9–1.6 Ma) faunas in Florida contain *Chasmaporthetes, Canis edwardii, Ondatra idahoensis, Sigmodon curtisi*, and *Geomys propinetis. Nannippus* and *Mammuthus* are conspicuously absent. A primary correlative fauna is Curtis Ranch in Arizona. Two Florida localities with abundant *Holmesina* fall within this interval, Inglis 1A and De Soto Shell Pit. The two are not exactly contemporaneous, as De Soto was deposited during a high sea-level stand, and Inglis during a low sea-level stand.

Late early Irvingtonian (1.6–1.0 Ma) faunas in Florida are characterized by the association of *Mammuthus, Megalonyx wheatleyi, Nothrotheriops texanus, Canis armbrusteri, C. edwardii, Tapirus haysii, Platygonus vetus, Geomys pinetis*, and *Sigmodon libitinus*. Primary correlative faunas are Gilliland (Texas), Holloman (Oklahoma), and Sappa (Nebraska). The Leisey Shell Pit local fauna (l.f.) is the largest and best-known assemblage of this age in Florida (Hulbert and Morgan, 1989). The Haile 16A l.f. contains several species in common with Leisey, including *M. wheatleyi, T. haysii*, and *S. libitinus*. On the basis of the absence of definitive early Irvingtonian indicators (such as *C. edwardii* or *Nothrotheriops*) and the presence of *Atopomys salvelinus* (Winkler and Grady, 1990), Haile 16A is tentatively assigned a slightly younger age than Leisey, about 0.8 Ma, or early middle Irvingtonian. The ages in this interval in particular will be subject to change as new sites are discovered and taxonomic analyses are completed. In any event, the age of Haile 16A is unlikely to be as young as indicated by Lundelius et al. (1987, Figure 7.3), who provisionally assigned it to the late Irvingtonian.

The other Florida site here included in the middle Irvingtonian (1.0–0.6 Ma) is McLeod Limerock Mine. Its fauna includes a mix of species that are absent after the middle Irvingtonian (*Smilodon gracilis, T. haysii*) and the muskrat *Neofiber leonardi,* otherwise a late Irvingtonian species (Frazier, 1977). Accordingly, we assign a latest middle Irvingtonian age to the site: about 0.6 Ma.

Coleman 2A is the only Florida fauna securely assigned to the late Irvingtonian (0.6–0.3 Ma). This age assignment is based on the co-occurrence of *Arctodus pristinus, C. armbrusteri, Tapirus veroensis, Platygonus cumberlandensis, Didelphis virginiana,* and *Neofiber alleni* (Martin, 1974). The Coleman 2A sample of *Holmesina* is very limited, but is of interest because of the large size of the specimens. However, a much more numerous sample of *Holmesina* is known from the nearby Coleman 3 site. Geographic proximity to Coleman 2A does not necessarily imply a similar age. For example, geologically similar fissure-fill deposits located adjacent to one another in the Haile Quarry Complex range in age from early Hemphillian to late Rancholabrean. Dating the Coleman 3 fauna is extremely difficult because of its very limited diversity. The presence of *T. veroensis* and *N. alleni* suggests a late Irvingtonian or Rancholabrean age. *Bison* and all other typical indicators of the Rancholabrean are absent, however, so we tentatively regard this site as falling near the Irvingtonian/Rancholabrean boundary, about 0.3 Ma.

Rancholabrean faunas can be divided into two groups in Florida. Early Rancholabrean (0.3–0.13 Ma) faunas are characterized by *Bison latifrons, Sigmodon bakeri* (a late Irvingtonian holdover), and *Pitymys hibbardi.* Late Rancholabrean (130–10 ka) sites instead include *Bison antiquus, Panthera atrox, Sigmodon hispidus, Pitymys pinetorum,* and *Ondatra zibethicus.* The best sample of early Rancholabrean *Holmesina* comes from Bradenton (Simpson, 1930), although records also exist at Haile 7A, Haile 8A, and Daytona Beach. Late Rancholabrean records of *Holmesina* are very common throughout Florida, although it is never a very abundant taxon at any particular site. The age of the type specimen of *H. septentrionalis* from the Peace River is uncertain, but probably late Rancholabrean, based on the common occurrence of *Bison antiquus* in deposits along this river.

Results

Osteoderm variation

Standard univariate sample statistics were computed for the three osteoderm parameters (Table 8.2). Specimens from two groups of sites were combined to produce larger sample sizes. Late Blancan specimens from Haile 15A, Santa Fe River, Macasphalt Shell Pit, and Kissimmee River formed one of these lumped samples; late Rancholabrean specimens from Branford 1A, Tapir Hill, and the Santa Fe River formed the other. There has been no published information regarding quantitative variation in pampathere osteoderms. Ideally, comparisons between different individuals would involve osteoderms from the same position on the carapace, but this is impossible in practice. Even in sets of associated osteoderms, such as the Branford 1A (UF 9336) and Haile 7C (UF 125226/125227) specimens, the exact position on the carapace generally cannot be determined. Therefore, we have selected six generalized types of osteoderms to analyze (Figure 8.2); although comparisons between exactly the same osteoderms are not being made, comparisons among the same groups are. It is well known that coefficients of variation (CVs) for univariate, linear dimensions of mammalian skeletal elements usually range between 4 and 10 if the measured sample represents a random sample of a single population (Simpson, Roe, and Lewontin, 1960; Yablokov, 1974). Prior to using osteoderms to calculate evolutionary rates, we examined their CVs to determine if they were of similar magnitudes. Our a priori assumption was that their variation would be somewhat greater than those found on single limb elements or teeth, because of the lack of direct comparison between identical elements.

Coefficients of variation for osteoderms are summarized in Table 8.3. Only large samples ($N > 10$) are included. Overall, osteoderm CVs were less than 10.0 in about 90% of the samples. Large CVs were most common in the thickness measurement (8 of 13 cases, where CV > 10.0). All 13 of these cases are from the immovable buckler osteoderms.

Table 8.2. *Sample statistics for* Holmesina *osteoderms from Florida*

Locality	Length	Width	Thickness
Normal movable osteoderms			
Blancan	67.0, 4.09, 36	23.3, 1.81, 50	7.2, 0.62, 50
Haile 7C	74.6, 5.29, 36	25.0, 1.89, 39	7.0, 0.55, 39
De Soto Shell Pit	—[a]	28.4, 2.64, 12	8.5, 0.36, 13
Inglis 1A	83.3, 5.81, 18	27.4, 1.42, 41	7.9, 0.69, 41
Leisey Shell Pit	79.7, 6.01, 26	25.9, 1.55, 58	6.6, 0.36, 58
Crystal River	90.2, 5.82, 3	27.7, 2.39, 3	7.2, 0.26, 3
Haile 16A	90.1, 10.19, 5	27.4, 2.08, 9	8.7, 0.79, 9
Coleman 3	103.6, 5.93, 13	34.5, 2.71, 30	8.8, 0.59, 28
1 Rancholabrean	107.7, 5.62, 12	37.1, 3.15, 28	9.5, 0.81, 38
Anterior row, pelvic buckler			
Blancan	56.1, 3.33, 12	23.2, 2.47, 16	7.8, 0.53, 16
Haile 7C	60.3, 4.11, 7	24.3, 0.77, 7	7.9, 0.32, 7
De Soto Shell Pit	75.1, 1.68, 4	30.3, 1.96, 13	11.1, 0.37, 13
Inglis 1A	66.3, 3.24, 14	26.9, 1.25, 22	9.7, 0.40, 22
Leisey Shell Pit	63.0, 2.32, 13	26.6, 1.71, 17	8.3, 0.58, 17
Haile 16A	69.7, 2.72, 4	27.0, 1.39, 4	10.2, 0.26, 4
Coleman 3	75.3, 6.33, 10	37.7, 3.60, 11	11.1, 0.75, 11
1 Rancholabrean	90.7, 2.91, 3	36.3, 2.22, 9	12.2, 0.73, 9
Posterior row, pectoral buckler			
Blancan	34.3, 2.74, 14	21.5, 2.07, 14	5.6, 0.52, 14
Haile 7C	42.9, 3.07, 15	23.4, 2.01, 15	6.2, 0.34, 15
De Soto Shell Pit	47.6, 2.12, 13	27.6, 1.19, 16	7.1, 0.52, 17
Inglis 1A	46.9, 3.89, 19	27.5, 1.79, 19	6.6, 0.40, 19
Leisey Shell Pit	40.3, 2.72, 15	23.3, 1.99, 15	5.6, 0.33, 15
Crystal River	48.3, — , 1	26.6, 1.33, 2	5.6, 0.14, 2
Haile 16A	47.6, 5.27, 9	27.3, 2.88, 9	6.8, 0.68, 9
Coleman 3	52.8, 3.63, 8	32.9, 2.26, 8	8.6, 0.62, 8
1 Rancholabrean	57.9, 4.55, 9	33.9, 2.92, 9	8.7, 0.97, 9
Rectangular immovable osteoderms			
Blancan	29.3, 2.69, 8	23.2, 2.75, 8	7.1, 0.40, 8
Haile 7C	33.7, 3.13, 9	23.6, 3.50, 9	7.7, 0.73, 9
De Soto Shell Pit	34.1, 3.37, 12	28.0, 1.92, 12	9.4, 0.51, 12
Inglis 1A	37.7, 3.80, 25	28.1, 2.80, 25	8.4, 0.89, 25
Leisey Shell Pit	30.6, 4.25, 11	23.7, 1.92, 11	7.0, 1.01, 11
Haile 16A	37.5, 3.05, 5	27.4, 1.90, 5	8.7, 0.58, 5
Coleman 3	41.4, 7.84, 3	28.8, 0.76, 3	10.7, 1.22, 3
1 Rancholabrean	42.2, 4.10, 49	34.8, 2.75, 49	11.9, 1.12, 49

Table 8.2 (*cont.*)

Locality	Length	Width	Thickness
Hexagonal immovable osteoderms			
Blancan	28.4, 2.07, 20	23.2, 1.40, 20	7.7, 0.89, 20
Haile 7C	31.1, 2.26, 27	25.3, 2.08, 27	8.5, 0.43, 27
De Soto Shell Pit	33.2, 2.41, 12	27.0, 1.90, 12	9.2, 0.55, 12
Inglis 1A	32.3, 2.32, 26	26.7, 1.97, 26	9.2, 0.98, 26
Leisey Shell Pit	30.4, 2.10, 39	25.7, 2.35, 39	7.6, 0.77, 39
Haile 16A	33.0, 2.56, 11	27.2, 2.42, 11	8.6, 0.50, 11
Coleman 3	39.5, 1.36, 6	32.6, 1.56, 6	11.1, 1.08, 6
l Rancholabrean	44.0, 3.11, 29	36.8, 2.25, 29	11.9, 1.54, 29
Five/six-sided immovable osteoderms			
Blancan	32.9, 3.21, 61	23.0, 2.25, 61	7.4, 0.68, 61
Haile 7C	33.4, 2.73, 76	23.8, 1.92, 76	7.9, 0.69, 76
De Soto Shell Pit	40.3, 2.91, 29	27.6, 1.99, 29	8.9, 0.42, 29
Inglis 1A	39.6, 3.02, 65	26.9, 2.17, 65	8.0, 0.62, 65
Leisey Shell Pit	34.8, 3.72, 84	25.2, 2.20, 84	7.4, 0.81, 84
Crystal River	42.2, 2.46, 3	31.0, 2.53, 3	6.8, 0.83, 3
Haile 16A	40.0, 2.16, 13	27.8, 2.13, 13	8.4, 0.85, 13
Coleman 2A	40.4, 2.78, 4	28.3, 2.86, 4	7.3, 0.63, 4
Coleman 3	45.2, 3.70, 28	32.2, 3.00, 28	10.4, 0.76, 28
Haile 7A	41.1, 0.89, 2	34.2, 0.13, 2	11.2, 0.33, 2
l Rancholabrean	49.4, 4.63, 118	36.2, 2.63, 118	11.4, 0.96, 118

Note: Each listing gives the mean (in mm), standard deviation, and sample size. Blancan samples include specimens from Haile 15A, Santa Fe River, Macasphalt Shell Pit, and Kissimmee River. Late Rancholabrean samples include specimens from Branford 1A, Santa Fe River, and Tapir Hill.
[a]Not measured.

Variations in osteoderm length and width, as measured by CVs, are of magnitudes similar to those found for skeletal dimensions in typical mammalian populations, as is thickness in the movable osteoderms.

Table 8.3 also compares variations in three types of data sets: samples from a single individual, samples from a single locality, and combined samples from multiple localities of similar ages. As expected, CVs were least for the associated osteoderms from single individuals, with no values exceeding 10.0. Values for samples composed of multiple individuals are not much greater. Comparison of CVs between movable and immovable osteoderms revealed that values for "movables" are, on average, lower, with most samples having values less than 8.0. The majority of samples of immovable scutes have CVs greater than 8.0. The reason for this is that movable scutes come from a very limited

Table 8.3. *Distribution of coefficients of variation (CVs) for samples of* Holmesina *osteoderms from Florida*

	Individuals			Single sites			Multiple sites		
	<7	7-10	>10	<7	7-10	>10	<7	7-10	>10
Length	2	9	0	6	8	3	3	6	0
Width	4	9	0	5	12	0	2	5	2
Thickness	7	6	0	5	6	6	1	6	2

Note: Minimum sample size is 10. Three types of osteoderm samples were analyzed: those from a single, associated carapace; those from multiple individuals from a single locality; and those combined from several localities of the same geologic age. Reported values are the number of samples observed for each different range of CV.

region of the shell and form a functional unit, whereas immovable osteoderms come from more varied regions on the shell and have no functional constraints. Thus, even small samples of movable osteoderms will, on average, provide adequate estimates for population mean values of length, width, and thickness, but very large samples should be used for immovable buckler osteoderms. This assumes that samples come from multiple individuals. A sample of osteoderms from a single individual provides adequate estimates for mean values for the osteoderms of that individual, but naturally provides more limited information for the whole population.

Evolutionary change in Holmesina osteoderms

Osteoderm evolution in *Holmesina* was analyzed graphically by plotting sample means versus geologic age (Figure 8.4) and by computing rates of evolution in darwins (Table 8.4). In general, all of the osteoderm parameters showed the same trend. The smallest specimens belonged to the oldest samples (late Blancan). The Haile 7C individual was consistently larger, intermediate in size between the late Blancan and earliest Irvingtonian populations (Table 8.2). Values for the latter (specimens from Inglis 1A and De Soto Shell Pit) are not significantly different from those for the much younger Haile 16A site. A notable exception is in the length of the movable osteoderms (Table 8.2), where the Inglis 1A specimens are small. The best sample of *Holmesina* intermediate in age between Inglis 1A and Haile 16A is from Leisey Shell

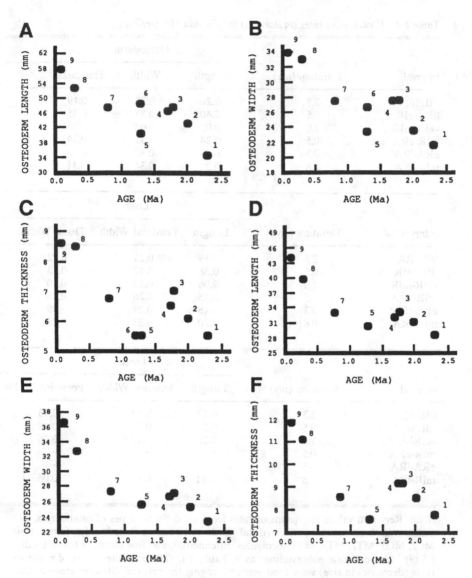

Figure 8.4. Graphs of *Holmesina* osteoderm size versus geological age. Each point represents a mean value for a sample of certain age (values listed in Table 8.2). Osteoderm sizes measured in millimeters. (A-C) From posterior row of pectoral buckler (Figure 8.3C). (D-F) Hexagonal immovable osteoderms (Figure 8.3E). (A, D) Length. (B, E) Width. (C, F) Thickness. Key to sites: 1, late Blancan (combined sample); 2, Haile 7C; 3, De Soto Shell Pit; 4, Inglis 1A; 5, Leisey Shell Pit; 6, Crystal River Power Plant; 7, Haile 16A; 8, Coleman 3; 9, late Rancholabrean (combined sample).

Table 8.4. *Evolutionary rates (in darwins) for Florida* Holmesina

Interval	Duration (my)	Osteoderm		
		Length	Width	Thickness
lBL-lRA	2.3	0.20	0.20	0.19
lBL-veIR	0.5	0.40	0.35	0.35
veIR-mIR	1.0	0.03	0.02	0.03
mIR-eRA	0.5	0.24	0.37	0.34
eRA-lRA	0.3	0.31	0.32	0.26
mIR-lRA	0.8	0.27	0.34	0.31

Interval	Duration (my)	Limb		
		Length	Proximal Width	Distal Width
lBL-lRA	2.3	0.19	0.21	0.22
lBL-veIR	0.5	0.38	0.42	0.43
veIR-mIR	1.0	0.06	0.13	0.10
mIR-eRA	0.5	0.15	0.26	0.26
eRA-lRA	0.3	0.28	0.24	0.29
mIR-lRA	0.8	0.20	0.23	0.28

Interval	Duration (my)	Teeth		
		Length	Anterior Width	Posterior Width
lBL-lRA	2.3	0.17	0.12	0.13
lBL-veIR	0.5	0.28	0.33	0.48
veIR-mIR	1.0	0.20	0.17	0.25
mIR-eRA	0.5	–	–	–
eRA-lRA	0.3	–	–	–
mIR-lRA	0.8	0.11	0.11	0.06

Note: Reported values are means of rates from six different types of osteoderms, 11 different limb elements [humerus (distal width only), radius (length only), femur, tibia, MC2, MC3, MT2, MT3, MT4, astragalus, calcaneum], and five different lower teeth (t5-t9). Chronological abbreviations as in Table 8.1. The absolute values of negative rates (decreases in size) were used when averaging the rates of different elements, to produce an overall measure of change regardless of direction. Late early Irvingtonian specimens were substituted for middle Irvingtonian individuals if the latter are not known (with an adjusted duration of 0.50 my).
[a]Data not available.

Pit. For most osteoderm parameters, Leisey specimens are more similar to those from Blancan sites than they are to other Irvingtonian samples (Figure 8.4, Table 8.2), especially by being extremely thin. The sample sizes for Leisey osteoderms are large (Table 8.2), so their small mean size is not an artifact of inadequate sampling. Osteoderms are known from another late early Irvingtonian site, Crystal River Power Plant, but with much smaller sample sizes. The Crystal River osteoderms resemble those from Leisey in their thinness (Figure 8.4C), but are more like those of Inglis and Haile 16A in length and width (Figures 8.4A and 8.4B). Until further late early Irvingtonian samples are collected, it cannot be determined whether the small osteoderm size of the Leisey *Holmesina* is a local or widespread phenomenon.

There is no adequate sample of osteoderms chronologically between the middle Irvingtonian Haile 16A site and the early Rancholabrean Bradenton and Coleman 3 localities. For most osteoderm parameters there is a dramatic increase in size between the Haile 16A and Coleman 3 samples (movable osteoderm thickness is an exception), with much less of a difference between the latter and the late Rancholabrean specimens (Figure 8.4). These are clearly the largest specimens, although in some cases not significantly so.

Evolutionary rates were calculated for the six types of osteoderms and the three measured parameters. The rates for each of the six types were averaged to produce mean rates of change in osteoderm length, width, and thickness (Table 8.4). The overall rates, from the Blancan to the late Rancholabrean, were very similar for each of the three parameters, about 0.20 darwin (*d*). Evolutionary rates for osteoderms were not constant over the chronological range of *Holmesina*. The entire range was divided into four intervals, with the boundaries chosen at points where rates appeared to have changed, and for which adequate samples of *Holmesina* were available. These intervals were as follows: late Blancan–very early Irvingtonian (2.3–1.8 Ma); very early Irvingtonian–middle Irvingtonian (1.8–0.8 Ma); middle Irvingtonian–early Rancholabrean (0.8–0.3 Ma); and early Rancholabrean–late Rancholabrean (0.3 Ma–12 ka). As early Rancholabrean samples were lacking for some characters of the dentition and limbs, the rate from the middle Irvingtonian to the late Rancholabrean was also calculated for comparative purposes. For osteoderms, evolutionary rates were about 10 times greater during the Blancan and the Rancholabrean than they were during the early to middle Irvingtonian.

Evolution of quantitative skeletal characters

Changes in *Holmesina* limb proportions were analyzed similarly as the osteoderms (Appendix 8.1, Figure 8.5, Table 8.4), except that individual specimens were plotted, rather than sample means, because of the small sample sizes. The evolutionary patterns observed in the limb elements are essentially the same as those seen in the osteoderms. There was a period of rapid increase in size during the late Blancan (at rates of about 0.40 *d*), followed by an interval during which size increased at a much slower rate (0.06–0.13 *d*). During the late Irvingtonian through the Rancholabrean the rate of increase in limb length of articular breadth elevated to nearly Blancan levels (about 0.30 *d*). The overall, average rate of increase, from the late Blancan to latest Rancholabrean, was 0.20 d, the same as in the osteoderms.

One interesting difference from the osteoderms is that limb elements from Leisey Shell Pit are of typical Irvingtonian size, and not unusually small as was the case with the osteoderms. With this exception, the changes in sizes of osteoderms and limb elements were similar in both magnitude and timing (Table 8.4), as was previously observed by Edmund (1987).

Evolution of dental characters

Analysis of the teeth and interpretation of the results are especially limited by sample size considerations (Appendix 8.2, Figure 8.6). Only the late Rancholabrean sample is represented by more than three individuals, and it demonstrates that there is considerable individual variation in tooth length. UF/FGS 336, a mandible from Vero, is from a well-dated late Rancholabrean site, but its posterior teeth (t6-t9) are relatively small, comparable to Irvingtonian specimens. A second Vero mandible, UF/FGS 337, is of "normal" Rancholabrean size. It is possible that UF/FGS 336 represents an immature individual. Another apparently anomalous specimen is UF 129051 from Leisey Shell Pit. The sizes of its more anterior teeth (t4-t5) are similar to those of other early Irvingtonian specimens, but its posterior teeth are very large and narrow, like those of late Irvingtonian and Rancholabrean individuals (Figure 8.6).

Overall, the rate of size increase in the teeth is less than in the limbs and osteoderms, 0.17 d for average length and about 0.12 d for tooth width (Table 8.4). Because length increased more than width, the teeth

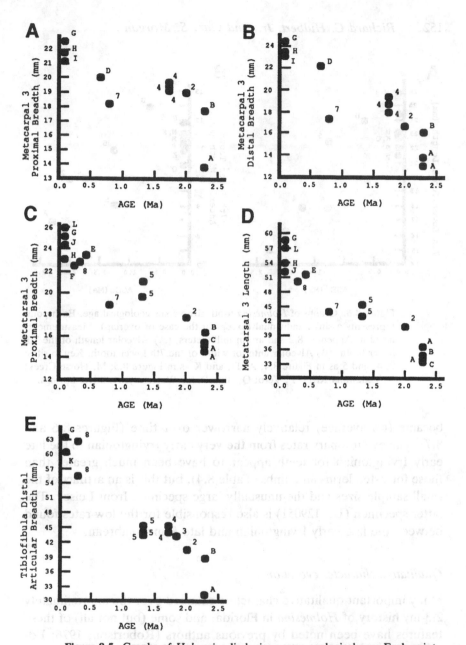

Figure 8.5. Graphs of *Holmesina* limb size versus geological age. Each point represents a single individual (except in the case of overlap). Measurements are listed in Appendix 8.1 and are in millimeters. (A) Metacarpal 3 proximal breadth. (B) Metacarpal 3 distal breadth. (C) Metatarsal 3 proximal breadth. (D) Metatarsal 3 length. (E) Distal articular breadth of tibia. Key to sites: 2–8 in Figure 8.4; A, Haile 15A; B, Santa Fe River (Blancan); C, El Jobean; D, McLeod Limerock Mine; E, Coleman 2A; F, Haile 7A; G, Branford 1A; H, Santa Fe River (Rancholabrean); I, Withlacoochee River; J, Tapir Hill; K, Hornsby; L, Myakkahatchee Waterway.

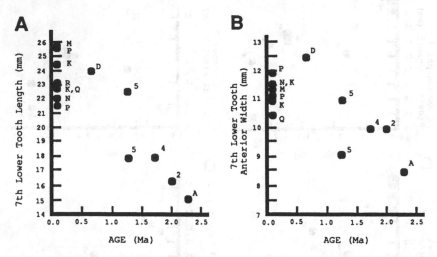

Figure 8.6. Graphs of *Holmesina* tooth size versus geological age. Each point represents a single individual (except in the case of overlap). Measurements listed in Appendix 8.2 and are in millimeters. (A) Alveolar length of the 7th lower tooth. (B) Alveolar anterior width of the 7th lower tooth. Key to sites: 2, 4, and 5 as in Figure 8.4; A, D, and K as in Figure 8.5; M, Horse Creek; N, Oklawaha River; P, Vero; Q, Blue Ridge Waterway; R, Silver Springs.

became (on average) relatively narrower over time (Figures 8.6 and 8.7). The evolutionary rates from the very early Irvingtonian to the late early Irvingtonian for teeth appear to have been much greater than those for osteoderms and limbs (Table 8.4), but this is an artifact of the small sample sizes and the unusually large specimen from Leisey. The latter specimen (UF 129051) is also responsible for the low rates found between the late early Irvingtonian and late Rancholabrean.

Qualitative character evolution

Many important qualitative characters changed over the approximately 2.3-my history of *Holmesina* in Florida, and some (but not all) of these features have been noted by previous authors (Robertson, 1976; Edmund, 1985b, 1987). Our discussion stresses, for each character, the time period during which the morphological transition(s) occurred, so that we may compare these data with the quantitative changes (Table 8.5). In this section we use the adjective "older" to collectively refer to late Blancan and early Irvingtonian specimens, and "younger" to indicate individuals of Rancholabrean age.

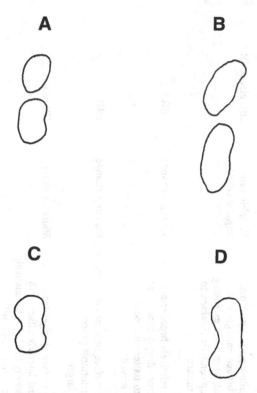

Figure 8.7. Occlusal outlines of *Holmesina* teeth, all approximately natural size. (A) UF 121742, right T3 and T4; Haile 7C, latest Blancan. (B) Uf 889, right T3 and T4; Hornsby Springs, late Rancholabrean. (C) UF 125226, left t6; Haile 7C. (D) UF 890, right t6 (reversed); Hornsby Springs.

Edmund (1985b) noted that the third and fourth upper teeth (T3 and T4) of *Holmesina* tended to become more reniform (= "molariform") through time. T3 is elliptical in all older specimens (Figure 8.7A), whereas this tooth is more elongated and at least incipiently, if not strongly, reniform in younger individuals (Figure 8.7B). T4 is generally slightly reniform in older specimens, but in younger specimens is always more strongly reniform (Figures 8.7A and 8.7B). The most obvious change in the teeth is that the posterior teeth (sixth through eighth) tended to become more narrow and elongated over time (Figures 8.7C and 8.7D). Unfortunately, there are no mandibles or maxillae of *Holmesina* known between the late early Irvingtonian Leisey Shell Pit and the late middle Irvingtonian McLeod Limerock Mine. A mandible from McLeod (AMNH 99233) has large, elongated teeth that are ac-

Table 8.5. *Summary of qualitative character changes in* Holmesina

Elements	Character	Morphological change (Blancan to Rancholabrean)	Mode of Morphological change	Time of maximum Morphological change
T3	Occlusal shape	Elliptical to reniform	Insufficient data	eIR-eRA
T4	Occlusal shape	Slightly reniform to strongly reniform	Insufficient data	eIR-eRA
Posterior teeth (T6-8)	Occlusal shape	Short and broad to long and narrow	Stairstep change	mIR
Humerus (distal)	Size and depth of entepicondylar fossa	Small, shallow pit just proximal articular surface enlarges into very broad, deep fossa extending from entepicondylar foramen to articular surface	Variable-rate phyletic change	lBL-eIR & IR
MC2	Shape of proximal articular surface	Becomes relatively broader and changes from deeply concave or notched to flatter and more smoothly curved	Stairstep change	mIR
MC3	Shape of proximal articular surface	Articular surface becomes concave, and posteriormost process becomes larger	Stairstep change	mIR
MT2	Shape of proximal articular surface	Articular surface changes from narrow and laterally offset, with deep concavity for MT1, to relatively broader and not offset, with concavity absent	Stairstep change	lBL-eIR

MT3	Shape of proximal articular surface	Posterior portion becomes broader, and medial notch is reduced	Stairstep change	mIR-IIR
MT4	Shape of proximal articular surface	Articular surface becomes relatively broader	Stairstep change	eIR-IIR
Astragalus	Overall shape	Breadth and anteroposterior length about equal (squarish); changes to much broader (rectangular)	Stairstep change	eIR-eRA
Astragalus	Dorsal surface of astragalar neck	Shallow, narrow depression; deep, broad concavity	Stairstep change	eIR-eRA
Astragalus	Anterior border of dorsal articulation	Depth of articular surface becomes reduced, with surface development of deep medial indentation on anterior border	Stairstep change	eIR-eRA
Astragalus & calcaneum	Separation of articular facets	Two articular facets change from connected to slightly separated to widely separated	Variable-rate phyletic change	lBL-eIR & eIR-eRA
Astragalus & calcaneum	Relative size of articular facets	Two facets change from nearly equal in size, with the lateral becoming much larger	Variable-rate phyletic change	eIR-eRA
Calcaneum	Robustness of calcanear tuber	Tuber becomes more robust, broader, and distally expanded	Variable-rate phyletic change	mIR-IIR

Note: All elements increased in size during the Plio-Pleistocene (Figure 8.5). Chronological abbreviations as in Table 8.1.

tually larger than those in many Rancholabrean specimens (Appendix 8.2). It appears that the primary morphological transition in the dentition occurred sometime during the middle Irvingtonian.

All specimens of *Holmesina* have a prominent entepicondylar foramen on the distal end of the humerus (Figure 8.8). Three distal humeri from the Blancan have a small, very shallow pit distal to the entepicondylar foramen and proximal to the medial edge of the articular surface (Figure 8.8A). This pit becomes progressively larger and deeper in humeri from the latest Blancan and early Irvingtonian, although the pit is always separated from the entepicondylar foramen by a convexity (Figure 8.8B). In Rancholabrean humeri this pit is greatly expanded in size to form a broad, deep fossa that extends from the entepicondylar foramen to the proximal edge of the articular surface (Figure 8.8C). A humerus from McLeod (AMNH 99227) is somewhat intermediate between these two morphologies, but is more similar to humeri of Rancholabrean individuals. The most dramatic morphological change in the distal humerus of *Holmesina* occurred during the middle Irvingtonian, although there was gradual change between the Blancan and middle Irvingtonian as well.

The proximal articular surface of the second metacarpal (MC2), composed principally of the trapezoid facet, is relatively broader in Rancholabrean specimens than in older individuals (Figure 8.9). In proximal view, the anterolateral portion of the articular surface is flat or slightly concave in younger specimens, but is sharply notched or concave in older individuals. In lateral view, the trapezoid facet is sharply V-shaped in older specimens, but broader and more smoothly curved in younger individuals. The transition between the two morphologies of the MC2 occurs in the middle Irvingtonian. A MC2 from Haile 16A (UF 24963) resembles specimens from the Blancan and early Irvingtonian, whereas a specimen from McLeod (AMNH 99225) agrees with MC2s from Rancholabrean faunas.

The facet for the magnum on the proximal end of the MC3 is much deeper anteroposteriorly in Rancholabrean *Holmesina*, and the posteriormost process is sharply upturned dorsally (Figure 8.10). In older individuals, the proximal articular surface of MC3 does not extend as far posteriorly, and the posterior process is flatter. The major morphological break in the MC3 occurs between specimens from Haile 16A and McLeod (during the middle Irvingtonian).

The proximal articular surface of the second metatarsal (MT2) is narrower and appears to be laterally offset relative to the shaft in Blan-

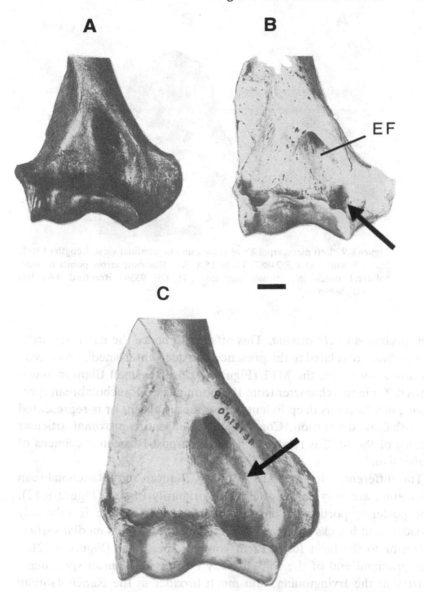

Figure 8.8. Distal humeri of *Holmesina* from Florida, anterior view. EF, en-tepicondylar foramen. Length of scale bar is 10 millimeters. (A) UF 10432, Santa Fe River 1, late Blancan. (B) UF 125226, Haile 7C, latest Blancan; arrow points to deep pit distal to EF. (C) UF 15140, Coleman 3, early Rancholabrean; arrow points to enlarged entepicondylar fossa formed by merged EF and distal pit.

A   **B**

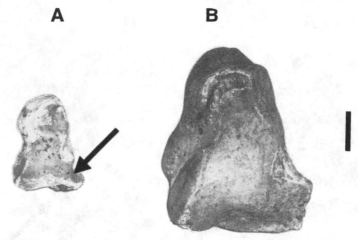

Figure 8.9. Left metacarpal 2's of *Holmesina* in proximal view. Length of scale bar is 5 mm. (A) UF 24887, Haile 15A, late Blancan; arrow points to anterolateral notch in articular surface. (B) UF 9336, Branford 1A, late Rancholabrean.

can specimens of *Holmesina*. This offset and hence the narrower articular surface are related to the presence of a deep anteromedial concavity for articulation with the MT1 (Figure 8.11). The small Blancan specimens differ in this character from Irvingtonian and Rancholabrean specimens, in which this deep indentation either is absent or is represented by a shallow depression. Consequently, the entire proximal articular surface of the MT2 is relatively broader in post-Blancan specimens of *Holmesina*.

The differences in the MT3 between Blancan and Rancholabrean *Holmesina* are very subtle and relate primarily to size (Figure 8.12). The posterior portion of the proximal articular surface is relatively broader, and it lacks or has a very reduced notch on its medial surface posterior to the facet for MT2 in younger specimens (Figure 8.12B). The proximal end of the MT4 is very narrow in Blancan specimens, narrow in the Irvingtonian, and much broader in the Rancholabrean (Figure 8.13). The most noticeable morphological break occurs after Inglis 1A; however, no specimens are known from Leisey, Haile 16A, or McLeod.

The astragalus of *Holmesina* was not discussed by previous authors (Robertson, 1976; Edmund, 1985b, 1987) because there were no known Blancan specimens. We have recently identified a Blancan astragalus of *Holmesina* from the Santa Fe River (UF 125610) (Figures 8.14A and

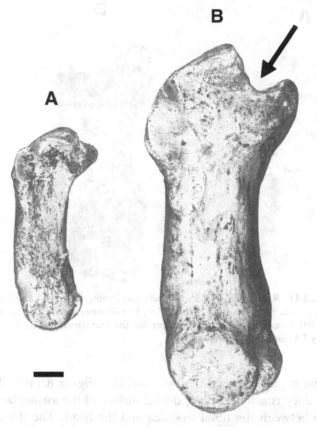

Figure 8.10. Left metacarpal 3's of *Holmesina* in lateral view. Length of scale bar is 5 mm. (A) UF 24871, Haile 15A, late Blancan. (B) UF 9336, Branford 1A, late Rancholabrean; arrow points to deeper depression in the proximal articular surface and sharply upturned posteriormost process.

8.15A) and have compared it with astragali from Inglis 1A, Leisey, and various Florida Rancholabrean faunas. The astragali from Santa Fe, Inglis, and Leisey are similar except for a slight increase in size, but there are several major morphological differences between these specimens and astragali from the Rancholabrean (Figures 8.14 and 8.15). The Rancholabrean astragali are much broader than those of the older specimens, but are only slightly deeper in the anteroposterior dimension. In general, the older astragali are more squarish, and younger ones are more rectangular.

The dorsal articular surface on the astragalus of Rancholabrean *Holmesina* has a deep indentation about midway along its anterior border

A   **B**

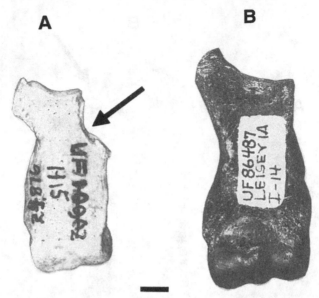

Figure 8.11. Right metatarsal 2's of *Holmesina* in anterior view. Length of scale bar is 5 mm. (A) UF 24876, Haile 15A, late Blancan; arrow points to antero-medial concavity with articular facet for the metatarsal 1. (B) UF 86487, Leisey 1A, late early Irvingtonian.

in the concave region between the two trochlea (Figure 8.14C). There is also a very deep concavity on the dorsal surface of the astragalar neck in the region between the tibial trochlea and the head. The dorsal articular area of the astragalus in older individuals is much deeper antero-posteriorly, the deep medial indentation is lacking, and there is only a shallow and narrow depression on the dorsal surface of the neck (Figures 8.14A and 8.14B).

The two articular facets on the ventral surface of the astragalus are broadly confluent along most of their length in the Blancan specimen (Figure 8.15A). These two facets are distinctly separated in the Inglis and Leisey astragali, although the separation is very narrow in most specimens (Figure 8.15B). In all Rancholabrean astragali these two facets are widely separated by a broad, deep groove (Figure 8.15C). There is also a difference between the older and younger specimens in the relative sizes of the ventral articular facets. The facets are more nearly equal in size in older individuals, with the medial facet the slightly smaller of the two. In younger individuals, the lateral or astragaloical-canear facet is much larger owing to a more prominent anterolateral process. In dorsal view, this projection is readily visible in all Ran-

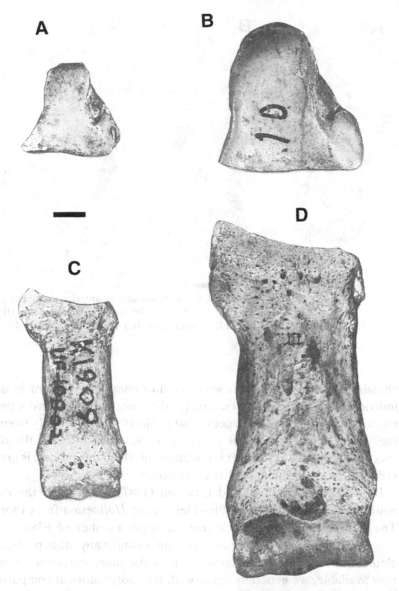

Figure 8.12. Right metatarsal 3's of *Holmesina* in proximal (A, B) and anterior (C, D) views. Length of scale bar is 5 mm. (A, C) UF 24875, Haile 15A, late Blancan. (B, D) UF 9336, Branford 1A, late Rancholabrean.

A   **B**   **C**

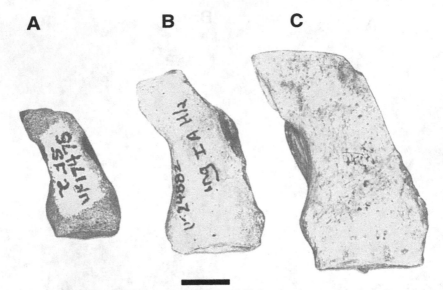

Figure 8.13. Right metatarsal 4's of *Holmesina* in anterior view. Length of scale bar is 10 mm. (A) UF 17475, Santa Fe River 2, late Blancan. (B) UF 24882, Inglis 1A, very early Irvingtonian. (C) UF 9336, Branford 1A, late Rancholabrean.

cholabrean specimens, but is very small or completely absent in older individuals (Figure 8.14). Conversely, the older astragali have a prominent, rounded, knob-like process that projects ventromedially from the medial edge. This process is visible in dorsal view in all of the older specimens, but not in younger individuals, in which this process is greatly reduced, particularly its medial extension.

Both Robertson (1976) and Edmund (1987) emphasized the calcaneum in their discussions of Plio-Pleistocene *Holmesina* from Florida. The calcaneum is represented from a larger number of Blancan, Irvingtonian, and Rancholabrean sites than is almost any other postcranial element. Based upon an examination of the more extensive material now available, we generally agree with the morphological comparisons of these authors. Robertson (1976) pointed out that the two facets on the calcaneum for articulation with the astragalus, the medial sustentacular facet and lateral astragalocalcanear facet, are connected in Blancan *Holmesina* to form a bilobed facet (Figure 8.16A). The two astragalar facets are fused or confluent in the Haile 15A calcaneum, whereas the facets are contiguous but have a definite separation between them in a Blancan specimen from Santa Fe. The two astragalar facets

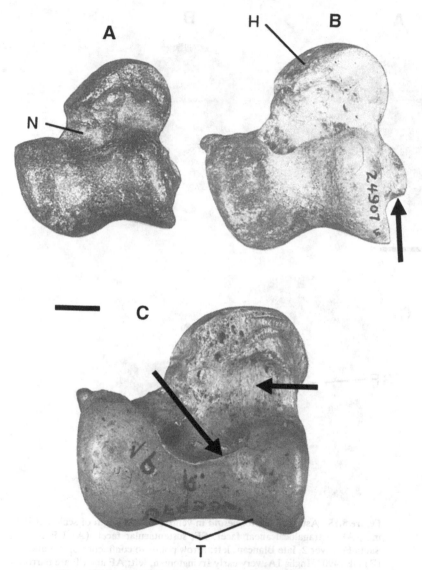

Figure 8.14. Astragali of *Holmesina* in dorsal view. Length of scale bar is 10 mm; N, neck; H, head; T, tibial trochlea. (A) UF 125610, Santa Fe River 2, late Blancan, left. (B) UF 24907, Inglis 1A, very early Irvingtonian, left; arrow points to enlarged medial projection absent in Rancholabrean individuals. (C) UF 9336, Branford 1A, late Rancholabrean, right (reversed); right arrow points to deep concavity on neck; left arrow points to strong indentation on anterior border of the trochlea.

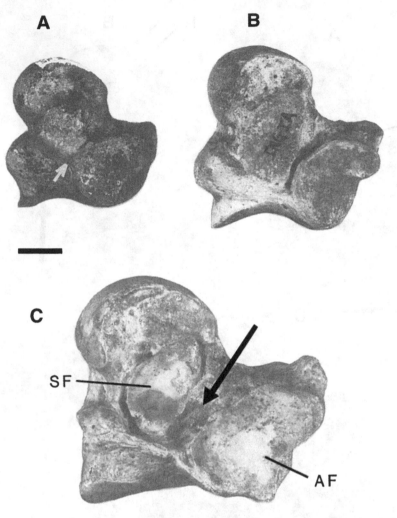

Figure 8.15. Astragali of *Holmesina* in ventral view. Length of scale bar is 10 mm; AF, astragalocalcanear facet; SF, sustentacular facet. (A) UF 125610, Santa Fe River 2, late Blancan, left; arrow points to confluence of AF and SF. (B) UF 24907, Inglis 1A, very early Irvingtonian, left; AF and SF are narrowly separated. (C) UF 9336, Branford 1A, late Rancholabrean, right (reversed); arrow points to widely separated AF and SF.

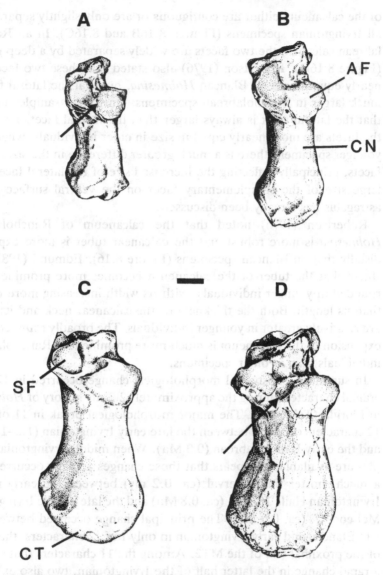

Figure 8.16. Right calcanea of *Holmesina* in dorsal view. Length of scale bar is 10 mm; AF, astragalocalcanear facet; CN, calcanear neck; CT, calcanear tuber; SF, sustentacular facet. (A) UF 125223, Haile 15A, late Blancan; arrow points to confluence of AF and SF. (B) UF 20946, Inglis 1A, very early Irvingtonian. (C) UF 20954, Haile 16A, middle Irvingtonian (left, reversed). (D) UF 24653 (cast of RHB 3151), Tapir Hill, late Rancholabrean; arrow points to widely separated AF and SF. Modified from Edmund (1987).

of the calcaneum either are contiguous or are only slightly separated in all Irvingtonian specimens (Figures 8.16B and 8.16C). In all Rancholabrean calcanea, the two facets are widely separated by a deep groove (Figure 8.16D). Robertson (1976) also stated that these two facets are nearly equal in size in Blancan *Holmesina,* but that the lateral facet is much larger in Rancholabrean specimens. Our larger samples indicate that the lateral facet is always larger than the medial facet. However, the facets are more nearly equal in size in older individuals, whereas in younger specimens there is a much greater difference in the sizes of the facets, principally reflecting the increased size of the lateral facet. The large size of the complementary facet on the ventral surface of the astragalus has already been discussed.

Robertson (1976) noted that the calcaneum of Rancholabrean *Holmesina* is more robust and the calcanear tuber is more expanded distally than in Blancan specimens (Figure 8.16). Edmund (1987) also stated that the tuber of the calcaneum becomes more prominent and rounded in younger individuals, with its width increasing more rapidly than its length. Both the thickness of the calcanear neck and its depth are relatively greater in younger individuals. The broadly rounded distal expansion of the calcaneum is much more prominent in Rancholabrean individuals than in older specimens.

In summary, substantial morphological change occurred in 12 postcranial characters during the approximately 2.3-my history of *Holmesina* in Florida (Table 8.5). The major morphological break in 11 of these 12 characters occurred between the late early Irvingtonian (1.2–1.0 Ma) and the early Rancholabrean (0.3 Ma). When middle Irvingtonian samples are available, it appears that these changes actually occurred over a much shorter time interval (ca. 0.2 my) between the early middle Irvingtonian Haile 16A l.f. (ca. 0.8 Ma) and the late middle Irvingtonian McLeod l.f. (ca. 0.6 Ma). The principal change occurred between the late Blancan and early Irvingtonian in only 1 of 12 characters, the shape of the proximal end of the MT2. Among the 11 characters that showed a rapid change in the latter half of the Irvingtonian, two also exhibited gradual evolution from the late Blancan through the middle Irvingtonian. In the remaining nine postcranial characters we examined, between the Blancan and early middle Irvingtonian there was an increase in size of the particular element, but no change in morphological shape until the late middle Irvingtonian. The predominant mode of evolutionary change in postcranial morphology is stairstep change (Martin and Barnosky, Chapter 1, this volume).

Many of the morphological changes that occurred in the postcranial skeleton of *Holmesina* could be related to the size increase that animals in this lineage underwent during the middle Irvingtonian and early Rancholabrean. Most of these changes were in features such as greater breadth or depth of articular surfaces or robustness of the shaft that probably were related to the weight-bearing capacity of the bone. In particular, the distal expansion of the calcanear tuber and the comparatively greater breadth of the astragalus in Rancholabrean *Holmesina* likely were related to the increased weight-bearing responsibilities of these two elements in this species.

Discussion

Taxonomic ramifications

Paleontologists disagree on how to divide unbranching chronoclines into discrete taxonomic species. One extreme position, exemplified by the stratophentic approach, arbitrarily divides up lineages to keep intraspecific variation within the limits of modern species (Gingerich, 1974, 1985; Rose and Bown, 1986). At the opposite end of the spectrum, some would group all the individuals in an unbranching chronocline into a single species, no matter how different the two end members might be (Wiley, 1978, 1979). Intermediate positions also exist, wherein short intervals of very rapid (geologically instantaneous) change are selected to demarcate species boundaries. It is not our intention to debate the strengths and weaknesses of these philosophies, but rather to discuss possible schemes for classifying species of *Holmesina* with respect to our data (Table 8.2, Appendixes 8.1 and 8.2).

There is no clear evidence for any branching in the genus *Holmesina* in Florida (or the rest of North America for that matter). The genus also appears to have been a continual inhabitant of the state from the late Blancan through the end of the Rancholabrean. Nor is there any evidence that the morphological change observed in successive Florida faunas resulted from dispersal from other regions; rather, *in situ* evolution is the most likely explanation. At least three reasonable schemes have been suggested for species-level systematics of *Holmesina*, termed the one-, two-, and three-species hypotheses, respectively.

According to the one-species hypothesis [in effect following the logic of Wiley (1978)], all specimens are referred to *H. septentrionalis*, the oldest available name for North American pampatheres. This hypothesis is followed if only branching in a lineage (cladogenetic speciation) is used to delineate species boundaries. The result is an extremely heterogeneous species, with end members of vastly different sizes, although that would be the case in many other lineages if this method were widely applied. One advantage of this scheme is its unambiguity with respect to extinction. There was no true extinction within this lineage, but if more than one species is recognized, then it must be emphasized that only pseudoextinction took place. Use of informal lineage segments (Krishtalka and Stucky, 1985) or chronological subspecies with this scheme would at least allow some retention of biostratigraphic information.

The two-species hypothesis would recognize both *H. septentrionalis* and *H. floridanus*. The question then becomes where to set the boundary between the two, an issue avoided by Edmund (1987). Two intervals are the most obvious candidates: between the late Blancan and the earliest Irvingtonian, or within the middle Irvingtonian. The strong point in favor of the former is that it coincided with the period of most rapid size increase (Table 8.4) and was followed by a long interval of near stasis. The drawbacks are that there would be almost no morphological differences (other than size) that would separate the two species, with the Haile 7C sample almost perfectly intermediate between the two. Its assignment to either would be fairly arbitrary. Placing the boundary in the middle Irvingtonian (chronologically between Haile 16A and McLeod, ca. 0.7 Ma) is the other option. There would then be at least 11 qualitative character differences between the two species in the postcranial skeleton alone, plus dental differences and a moderate difference in size. For many characters that appears to have been a time of very abrupt size change, although this could be an artifact of the interval's admittedly poor fossil record. A final point in favor of the two-species hypothesis is that it recognizes both species widely regarded as valid in the paleontologic literature (i.e., it is taxonomically conservative and does not require naming a new species). This is the arrangement we presently favor, using the middle Irvingtonian boundary.

The final alternative is to recognize three species and use both of the boundaries discussed in the preceding paragraph. This would recognize the large size difference (certainly significant in the statistical sense) that existed between late Blancan and earliest Irvingtonian populations, which was of similar or greater magnitude than the size differences regarded as specifically diagnostic for many congeneric pairs of species. This scheme would recognize a late Blancan *H. floridanus,* a presently undescribed species for the early and early middle Irvingtonian, and *H. septentrionalis* extending from the late middle Irvingtonian through the Rancholabrean. The assignment of the Haile 7C sample would be problematic.

Evolutionary patterns and rates in Holmesina

Most of the quantitative characters observed in *Holmesina* are examples of episodic, or variable-rate, phyletic change (as defined by Martin and Barnosky, Chapter 1, this volume). Rates were more rapid, by a factor of 3 to 10 times, during some intervals than during others. The patterns shown by the size variables of the osteoderms and limbs closely track one another (Table 8.4), suggesting that both were responding equally to increased body size. The rates of increase in proximal and distal breadths were twice as fast as that for limb length during the Irvingtonian. It is possible that during this interval the genus passed through some body mass threshold that required increased relative massiveness in the limbs (graviportal adaptations). Recall that this is also the interval when all but one of the qualitative changes in limb morphology took place. The qualitative changes in limb morphology also appear to have been adaptations for supporting a more massive body, and they mostly exemplify the stairstep change pattern of evolution (Table 8.5).

The measured dental characters show variable-rate phyletic change, but at generally slower rates than those observed in the osteoderms and limbs. Notably, tooth width increased at only about 60% to 70% of the rates observed in other characters, leading to the evolution of relatively narrow teeth in younger populations. There is some variation in this, and larger sample sizes are needed to confirm this pattern.

How do the calculated evolutionary rates for *Holmesina* (Table 8.4) compare with those of other vertebrates? Gingerich (1983) summarized the data regarding evolutionary rates in 228 fossil vertebrates. For an average duration of 1.6 my, the mean rate of change was 0.08 *d*. Obviously, because most investigators do not analyze or report samples in which very little or no change has occurred, this is not a random sample of overall vertebrate evolution. Using Gingerich's (1983) data, "mean" expected rates for an interval of 0.5–1.0 my should fall in the range of 0.18–0.30 *d*. Rates substantially greater than this can be considered examples of fairly rapid evolution, and lower values represent slow evolution. For the intervals middle Irvingtonian–early Rancholabrean and early Rancholabrean–late Rancholabrean, the rates of increase in most osteoderm and limb parameters fall within this "normal" range. The rates during the late Blancan-very early Irvingtonian interval were slightly greater, and for the interval very early Irvingtonian-middle Irvingtonian, they are substantially less than "normal."

For an interval of 2.3 my, Gingerich's (1983) data suggest that "normal" rates of evolution are about 0.06–0.07 d. For the period late Blancan to late Rancholabrean, which had this duration, the rates for *Holmesina* were two to three times greater than this, implying fairly rapid evolution. How can the rates for the subdivisions be approximately normal (or even below normal in one case), but their overall rate be much greater than normal? Gingerich (1983, p. 160) noted that "the shorter the interval of measurement, the more likely one is to observe high rates. The longer the interval, the more stasis and evolutionary reversal are likely to be averaged in the result." Apparently what was unusual about the evolution of *Holmesina* was not the rate at which size increased, but that it was sustained almost continuously for such a long interval.

Acknowledgments

We thank Bob Martin and Tony Barnosky for their helpful reviews of the manuscript. Richard Tedford and John Alexander (AMNH) loaned critical specimens on short notice. The following allowed us to study important specimens in their private collections: Roy Burgess, Bill and Lelia Brayfield, Barbara Fite, and James Pendergraft. Frank Garcia and Mike Stallings donated specimens to UF. The study was supported by a Georgia Southern University Faculty Research Grant to the senior author and Fred Rich. Ann Pratt and Bruce MacFadden aided in production of the graphics. This is University of Florida Contribution to Paleobiology No. 394.

References

Berggren, W. A., D. V. Kent, J. J. Flynn, and J. A. Van Couvering (1985). Cenozoic geochronology. *Geological Society of America Bulletin*, 96: 1407–18.

Edmund, A. G. (1985a). The armor of fossil giant armadillos (Pampatheriidae, Xenarthra, Mammalia). *Pearce-Sellards Series, Texas Memorial Museum*, 40:1–20.

   (1985b). The fossil giant armadillos of North America (Pampatheriinae, Xenarthra = Edentata). In G. G. Montgomery (ed.), *The Evolution and*

Ecology of Armadillos, Sloths, and Vermilinguas (pp. 83–93). Washington, D.C.: Smithsonian Institution Press.

(1987). Evolution of the genus *Holmesina* (Pampatheriidae, Mammalia) in Florida, with remarks on taxonomy and distribution. *Pearce-Sellards Series, Texas Memorial Museum,* 45:1–20.

Engleman, G. F. (1985). The phylogeny of the Xenarthra. In G. G. Montgomery (ed.), *The Evolution and Ecology of Armadillos, Sloths, and Vermilinguas* (pp. 51–64). Washington, D.C.: Smithsonian Institution Press.

Frazier, M. K. (1977). New records of *Neofiber leonardi* (Rodentia: Cricetidae) and the paleoecology of the genus. *Journal of Mammalogy,* 58:368–73.

Galusha, T., N. M. Johnson, E. H. Lindsay, N. D. Opdyke, and R. H. Tedford (1984). Biostratigraphy and magnetostratigraphy, late Pliocene rocks, 111 Ranch, Arizona. *Geological Society of America Bulletin,* 95:714–22.

Gingerich, P. D. (1974). Size variability of the teeth in living mammals and the diagnosis of closely related sympatric fossil species. *Journal of Paleontology,* 48:895–903.

(1983). Rates of evolution: effects of time and temporal scaling. *Science,* 222:159–61.

(1985). Species in the fossil record: concepts, trends, and transitions. *Paleobiology,* 11:27–41.

Harland, W. B., R. L. Armstrong, A. V. Cox, L. E. Craig, A. G. Smith, and D. G. Smith (1990). *A Geologic Time Scale: 1989.* Cambridge University Press.

Hulbert, R. C., and G. S. Morgan (1989). Stratigraphy, paleoecology, and vertebrate fauna of the Leisey Shell Pit local fauna, early Pleistocene (Irvingtonian) of southwestern Florida. *Papers in Florida Paleontology,* 2: 1–19.

Hulbert, R. C., G. S. Morgan, and A. R. Poyer (1989). Associated skeletons of megathere, pampathere, tapir, and turtles from the latest Pliocene or earliest Pleistocene of north-central Florida. *Journal of Vertebrate Paleontology,* 9:26A.

James, G. T. (1957). An edentate from the Pleistocene of Texas. *Journal of Paleontology,* 31:796–808.

Johnson, N. M., N. D. Opdyke, and E. H. Lindsay (1975). Magnetic polarity stratigraphy of Pliocene-Pleistocene terrestrial deposits and vertebrate faunas, San Pedro Valley, Arizona. *Geological Society of America Bulletin,* 86:5–12.

Krishtalka, L., and R. K. Stucky (1985). Revision of the Wind River faunas, early Eocene of central Wyoming. Part 7. Revision of *Diacodexis* (Mammalia, Artiodactyla). *Annals of the Carnegie Museum,* 54:413–86.

Kurtén, B., and E. Anderson (1980). *Pleistocene Mammals of North America.* New York: Columbia University Press.

Leidy, J. (1889). Fossil vertebrates from Florida. *Proceedings of the Academy of Natural Sciences of Philadelphia,* 1889:96–7.

Lundelius, E. L., C. S. Churcher, T. Downs, C. R. Harrington, E. H. Lindsay, G. E. Schultz, H. A. Semken, S. D. Webb, and R. J. Zakrzewski (1987). The North American Quaternary sequence. In M. O. Woodburne, (ed.),

Cenozoic Mammals of North America: Geochronology and Biostratigraphy (pp. 211–35). Berkeley: University of California Press.

Martin, R. A. (1974). Fossil mammals from the Coleman 2A fauna, Sumter County. In S. D. Webb (ed.). *Pleistocene Mammals of Florida* (pp. 35–99). Gainesville: University Presses of Florida.

Morgan, G. S., and R. C. Hulbert (in press). Overview of the geology and vertebrate biochronology of the Leisey Shell Pit local fauna, Hillsborough County, Florida. *Bulletin of the Florida Museum of Natural History.*

Morgan, G. S., and R. B. Ridgway (1987). Late Pliocene (late Blancan) vertebrates from the St. Petersburg Times site, Pinellas County, Florida, with a brief review of Florida Blancan faunas. *Papers in Florida Paleontology,* 1:1–22.

Repenning, C. A. (1987). Biochronology of the microtine rodents of the United States. In M. O. Woodburne (ed.), *Cenozoic Mammals of North America: Geochronology and Biostratigraphy* (pp. 236–68). Berkeley: University of California Press.

Robertson, J. S. (1976). Latest Pliocene mammals from Haile 15A, Alachua County, Florida. *Bulletin of the Florida State Museum, Biological Sciences,* 20:111–86.

Rose, K. D., and T. M. Bown (1986). Gradual evolution and species discrimination in the fossil record. In K. M. Flanagan and J. A. Lilligraven (eds.), *Vertebrates, Phylogeny, and Philosophy* (pp. 119–30). University of Wyoming Contributions to Geology, Special Paper 3.

Simpson, G. G. (1930). *Holmesina septentrionalis,* extinct giant armadillo of Florida. *American Museum Novitates,* 442:1–10.

Simpson, G. G., A. Roe and R. C. Lewontin (1960). *Quantitative Zoology.* New York: Harcourt, Brace.

Webb, S. D. (1974). Chronology of Florida Pleistocene mammals. In S. D. Webb (ed.), *Pleistocene Mammals of Florida* (pp. 5–31). Gainesville: University Presses of Florida.

Wiley, E. O. (1978). The evolutionary species concept reconsidered. *Systematic Zoology,* 27:17–26.

(1979). Ancestors, species, and cladograms. In J. Cracraft, and N. Eldredge, (eds.), *Phylogenetic Analysis and Paleontology* (pp. 211–25). New York: Columbia University Press.

Winkler, A. J., and F. Grady (1990). The middle Pleistocene rodent *Atopomys* (Cricetidae: Arvicolinae) from the eastern and south-central United States. *Journal of Vertebrate Paleontology,* 10:484–90.

Yablokov, A. V. (1974). *Variability of Mammals.* New Delhi: Amerind Publishing Corp.

Appendix 8.1. *Measurements (in mm) of appendicular skeletal elements of* Holmesina *from Florida used in the study*

Specimen	Site	Element	Length	Proximal breadth	Distal breadth
UF 24919	Haile 15A	Hum		37.0	
UF 9354	Sante Fe	Hum			54.4
UF 10432	Sante Fe	Hum			61.3
UF 125226	Haile 7C	Hum	153.0	45.9	60.3
UF 97318	Inglis	Hum		50.4	
UF 86638	Lelsey 1A	Hum			62.0
UF 21006	Haile 16A	Hum			71.0
AM 99227	McLeod	Hum	190.5	52.8	74.5
UF 15140	Coleman 3	Hum	216.0	58.9	80.5
UF 10435	Sante Fe	Hum			84.8
UF 16371	Waccasassa	Hum			85.6
UF 24920	Haile 15A	Rad	89.5	22.6	23.6
UF 10830	Santa Fe	Rad	94.3	24.6	25.9
UF 125226	Haile 7C	Rad	98.4	29.1	28.3
BF	De Sota	Rad		26.6	
UF 86470	Leisey 1A	Rad		23.8	
AM 99226	McLeod	Rad	129.5	35.5	36.3
AM 99229	McLeod	Rad			34.1
UF 10462	Santa Fe	Rad	135.0	34.2	37.3
UF 125400	Leisey 2	Rad	135.0	39.5	37.6
UF 24887	Haile 15A	MC2	32.9	11.1	12.3
UF 10722	Sante Fe	MC2	34.6	12.8	13.2
UF 125226	Haile 7C	MC2	37.5	13.6	14.5
BF	De Sota	MC2	38.6	13.9	16.2
UF 24883	Inglis 1A	MC2	40.6	15.5	15.7
UF 24963	Haile 16A	MC2	42.5	16.9	17.4
AM 99225	McLeod	MC2	42.4	17.8	20.2
FGS 2016	Vero	MC2	42.8	16.1	19.2
UF 9336	Branford	MC2	51.5	19.2	23.1
UF 24871	Haile 15A	MC3	34.3	13.8	13.0
UF 24879	Haile 15A	MC3	36.3	3.8	3.8
UF 24896	Santa Fe	MC3	39.9	17.8	16.1
UF 125226	Haile 7C	MC3	42.0	19.0	16.6
UF 24870	Inglis 1A	MC3			18.6
UF 24884	Inglis 1A	MC3	45.3	19.1	17.8
UF 24885	Inglis 1A	MC3	45.5	19.2	19.3
UF 24886	Inglis 1A	MC3	43.9	19.4	17.9
UF 24908	Inglis 1A	MC3	45.9	19.4	19.3
UF 24962	Haile 16A	MC3	43.6	18.3	17.3
AM 99225	McLeod	MC3		20.1	22.2
UF 9336	Branford	MC3	59.5	22.5	24.4
UF 10436	Santa Fe	MC3	55.5	21.8	23.5
UF 129052	With. Riv.	MC3	55.1	21.1	23.2
UF 24918	Haile 15A	Fem	192.0	59.6	48.6
UF 69756	Macasphalt	Fem			50.8

Appendix 8.1 (*cont.*)

Specimen	Site	Element	Length	Proximal breadth	Distal breadth
UF 17476	Haile 12B	Fem			55.3
UF 121748	Haile 7C	Fem	213.0	68.7	53.8
BF	De Soto	Fem	249.0	78.9	
UF 97322	Inglis 1A	Fem			63.7
UF 84749	Leisey 1A	Fem	248.5	74.2	56.5
UF 15145	Coleman 3	Fem			79.1
UF 9336	Branford	Fem	335.0	106.8	81.2
UF 3994	Hornsby	Fem			83.8
UF 4007	Hornsby	Fem	311.5		85.6
UF 4008	Hornsby	Fem	312.0		78.8
UF 4011	Hornsby	Fem			98.9
UF 24916	Haile 15A	Tib	122.6		30.8
UF 10434	Sante Fe	Tib	151.9	59.3	38.8
UF 125227	Haile 7C	Tib	142.4	60.2	40.3
BF	De Soto	Tib			42.4
UF 97323	Inglis 1A	Tib			45.1
UF 97324	Inglis 1A	Tib			44.0
UF 65900	Leisey 1A	Tib			43.8
UF 86469	Leisey 1A	Tib			43.8
UF 86535	Leisey 1A	Tib	176.1	65.8	45.1
UF 15141A	Coleman 3	Tib	186.5	76.6	55.3
UF 15141B	Coleman 3	Tib		62.9	
UF 15144	Coleman 3	Tib		83.9	
UF 9336	Branford	Tib	217.0	91.6	63.5
UF 4016	Hornsby	Tib	191.0	84.7	60.3
UF 4017	Hornsby	Tib	188.5	83.5	
UF 125610	Santa Fe	Ast	35.9	31.2	21.8
BF	De Soto	Ast	38.3	39.7	24.9
UF 20947	Inglis 1A	Ast	38.8	37.0	23.3
UF 20950	Inglis 1A	Ast	42.6	38.1	26.5
UF 24905	Inglis 1A	Ast	38.3	34.2	24.4
UF 24906	Inglis 1A	Ast	43.7	39.1	24.8
UF 24907	Inglis 1A	Ast	41.8	37.3	24.0
UF 83226	Leisey 1A	Ast	39.8	35.8	23.1
UF 15138	Coleman 3	Ast	44.5	44.8	27.6
UF 15199	Coleman 3	Ast	48.1	50.4	30.9
UF 9336	Branford	Ast	51.3	50.1	32.1
RHB 3151	Tapir Hill	Ast	48.2	47.1	29.2
UF 125947	Rock Springs	Ast	51.9	53.4	30.2
UF 125228	Haile 15A	Cal	64.6	28.4	35.5
UF 20953	Santa Fe	Cal	62.9	24.9	33.7
UF 125226	Haile 7C	Cal	71.9	31.3	38.3
UF 20946	Inglis 1A	Cal	80.7	33.2	41.9
UF 20951	Inglis 1A	Cal	80.9	41.2	
UF 86646	Leisey 1A	Cal		31.3	43.1
UF 20954	Haile 16A	Cal	89.5	34.6	43.5

Appendix 8.1 (*cont.*)

Specimen	Site	Element	Length	Proximal breadth	Distal breadth
UF 15138	Coleman 3	Cal	94.5	36.9	50.0
UF 9336	Branford	Cal	106.1	42.8	54.5
RHB 3151	Tapir Hill	Cal	103.1	50.3	58.8
UF 24876	Haile 15A	MT2	33.4	11.8	12.7
UF 24877	Haile 15A	MT2	33.1	12.5	12.9
UF 24878	Haile 15A	MT2	30.9	11.5	11.7
BF	De Soto	MT2	39.3	16.3	18.2
UF 86487	Leisey 1A	MT2	41.6	17.3	19.1
AM 99228	McLeod	MT2	46.1	19.2	22.5
UF 125949	Coleman 2A	MT2	47.5	22.0	22.5
UF 9336	Branford	MT2	53.4	23.9	25.6
RHB 3151	Tapir Hill	MT2	49.4	24.1	24.6
UF 9338	Paynes Pra.	MT2	48.8	20.9	22.2
UF 24875	Haile 15A	MT3	34.2	15.3	14.1
UF 24880	Haile 15A	MT3	36.5	14.6	13.7
UF 17472	Santa Fe	MT3	34.4	16.7	13.2
B 8475	El Jobean	MT3	33.3	15.6	17.4
UF 125226	Haile 7C	MT3	40.7	17.9	15.3
UF 83583	Leisey 1A	MT3	42.8	19.7	17.9
UF 124554	Leisey 3	MT3	45.6	21.1	21.0
UF 24964	Haile 16A	MT3	43.7	18.8	17.7
UF 13189	Coleman 2A	MT3	51.7	23.6	22.1
UF 24910	Coleman 3	MT3	50.3	22.7	22.2
UF 125932	Haile 7A	MT3		22.5	
UF 9336	Branford	MT3	58.5	25.0	24.7
UF 126332	Santa Fe	MT3	53.7	23.1	23.0
RHB 3151	Tapir Hill	MT3	52.7	24.5	24.0
B 2193	Myakkahatch.	MT3	57.0	25.8	25.9
UF 17475	Santa Fe	MT4	27.7	11.1	15.2
UF 125227	Haile 7C	MT4	31.5	15.0	18.4
BF	De Soto	MT4	37.5	16.7	20.5
UF 21005	Inglis 1A	MT4	34.7	15.3	17.4
UF 24882	Inglis 1A	MT4	36.1	15.1	20.4
UF 24888	Inglis 1A	MT4	35.7	14.8	18.8
UF 86124	Leisey 1A	MT4	36.1	14.1	19.7
UF 24911	Coleman 3	MT4	38.3	17.7	22.2
UF 125948	Ichetucknee	MT4	41.8	19.1	24.7
UF 9336	Branford	MT4	43.0	19.3	25.7

Note: Abbreviations for specimen numbers: BF, private collection of Barbara Fite; RHB, private collection of Roy H. Burgess; UF, University of Florida; AM, American Museum of Natural History. Abbreviations for elements: Hum, humerus; Rad, radius; MC, metacarpal; Fem, femur; Tib, tibia; MT, metatarsal; Ast, astragalus; Cal, calcaneum. All proximal breadths were measured across the articular surface, except for the radius and femur, which are maximum, not articular, dimensions. Likewise, all distal measurements are articular dimensions except in the case of the humerus.

Appendix 8.2. *Measurements (in mm) of lower teeth of* Holmesina *from Florida used in the study*

Specimen	Site	Tooth	Length	Anterior width	Posterior width
UF 125227	Haile 7C	t1	5.4	4.7	
B 9034	Horse Creek	t1	10.1	8.6	
UF 125227	Haile 7C	t2	7.4	5.7	
B 9034	Horse Creek	t2	13.5	8.3	
UF 95780	Oklawaha	t2	13.3	10.4	
UF 125227	Haile 7C	t3	8.7	6.1	
B 9034	Horse Creek	t3	14.2	9.1	
UF 95780	Oklawaha	t3	14.1	10.7	
UF 125227	Haile 7C	t4	11.9	8.0	8.2
UF 20948	Inglis 1A	t4	13.8	6.8	6.8
UF 66422	Leisey 1A	t4	12.1	17.0	7.3
UF 129051	Leisey 1A	t4	11.6	7.3	7.7
UF 115964	Leisey 3	t4	13.9	7.0	7.2
B 9034	Horse Creek	t4	17.7	9.3	9.2
UF 95780	Oklawaha	t4	15.3		
FGS 336	Vero	t4	15.0	8.0	8.1
UF 125227	Haile 7C	t5	16.1	9.1	9.3
UF 20948	Inglis 1A	t5	17.5	7.4	7.2
UF 66422	Leisey 1A	t5	18.5	9.0	10.1
UF 129051	Leisey 1A	t5	18.6	8.6	9.9
UF 115964	Leisey 3	t5	19.1	8.4	9.4
B 9034	Horse Creek	t5	23.8	12.2	12.5
UF 95780	Oklawaha	t5	23.2		
FGS 336	Vero	t5	20.5	9.5	10.1
RHB 3955	Blue Ridge	t5	20.6	9.1	9.6
UF 125227	Haile 7C	t6	17.4	9.6	10.0
UF20948	Inglis 1A	t6	18.5	9.6	10.2
UF 129051	Leisey 1A	t6	22.4	10.4	11.8
UF 66422	Leisey 1A	t6	19.2	10.0	10.0
B 9034	Horse Creek	t6	25.4	12.9	12.6
UF 95780	Oklawaha	t6	23.7	11.6	12.3
UF 4019	Hornsby	t6	24.4		
UF 890	Hornsby	t6	23.8	12.4	12.7
FGS 336	Vero	t6	22.9	10.9	11.4
RHB 3955	Blue Ridge	t6	23.3	9.8	10.8
UF 24915	Haile 15A	t7	15.1	8.6	8.1
UF 125227	Haile 7C	t7	16.3	9.8	9.4
UF 20948	Inglis 1A	t7	17.8	9.8	10.0
UF 129051	Leisey 1A	t7	22.5	10.9	10.9
UF 66422	Leisey 1A	t7	17.9	9.0	9.2
AM 99233	McLeod	t7	23.9	12.3	11.7
B 9034	Horse Creek	t7	25.7	11.3	11.2
UF 95780	Oklawaha	t7	22.0	11.5	11.2
UF 4019	Hornsby	t7	22.9		
UF 47219	Hornsby	t7	22.8	10.8	11.1
UF 890	Hornsby	t7	24.4	11.5	11.6

Appendix 8.2 (*cont.*)

Specimen	Site	Tooth	Length	Anterior width	Posterior width
FGS 336	Vero	t7	21.5	11.2	10.5
FGS 337	Vero	t7	25.6	11.8	11.7
RHB 3955	Blue Ridge	t7	22.9	10.4	10.7
UF 9329	Silver Spr.	t7	23.1		
UF 24915	Haile 15A	t8	13.8	7.2	6.6
UF 125227	Haile 7C	t8	14.7	9.1	8.6
UF 20948	Inglis 1A	t8	16.6	8.7	9.2
UF 129051	Leisey 1A	t8	9.7		
UF 66422	Leisey 1A	t8	15.3	8.7	8.0
AM 99233	McLeod	t8	22.0	10.7	10.9
B 9034	Horse Creek	t8	23.5	10.7	10.5
UF 95780	Oklawaha	t8	20.4	10.4	9.6
UF 47219	Hornsby	t8	21.5		
UF 890	Hornsby	t8	23.7	11.7	11.7
FGS 336	Vero	t8	18.7	9.6	9.1
FGS 337	Vero	t8	21.9	10.2	10.0
RHB 3955	Blue Ridge	t8	20.3	9.5	9.1
UF 9329	Silver Spr.	t8	21.4		
UF 24915	Haile 15A	t9	10.3	6.0	4.5
UF 24934	Haile 15A	t9	10.1	5.9	5.6
UF 125227	Haile 7C	t9	9.8	6.7	6.2
UF 20948	Inglis 1A	t9	12.6	7.8	7.0
UF 66422	Leisey 1A	t9	12.5	7.5	7.1
AM 99233	McLeod	t9	15.4	9.5	9.2
B 9034	Horse Creek	t9		8.6	
UF 95780	Oklawaha	t9	13.5	9.3	8.3
UF 47219	Hornsby	t9	14.5		
UF 890	Hornsby	t9	17.3		
FGS 336	Vero	t9	12.1	8.3	6.5
FGS 337	Vero	t9	13.8	8.8	7.0
RHB 3955	Blue Ridge	t9	13.6		
UF 9329	Silver Spr.	t9	13.2		

Note: Values reported here are actually internal measurements of the alveolus, which can be taken whether or not the tooth is present. Only a single, maximum width was taken on the anteriormost three teeth (t1-t3).

9

Evolution of mammoths and moose: the Holarctic perspective

ADRIAN M. LISTER

Mammalian species and genera that during the Pleistocene became dispersed in both Eurasia and North America often displayed interesting patterns and processes of evolution, illustrating phenomena such as speciation and parallel evolution between the two continents. Two lineages of Holarctic large mammals whose Pleistocene histories have several features in common are the mammoth (*Mammuthus* spp.) and moose (*Alces* spp.).

Although the ancient origins of these taxa probably lay elsewhere, both the mammoth and moose lineages underwent their early Pleistocene evolution in Eurasia. Each then dispersed, at some time during the early to middle Pleistocene, across the Bering land connection into North America. There they underwent apparently independent evolution from the Eurasian stock, producing endemic North American species. Meanwhile, evolution proceeded in Eurasia, resulting in specialized late Pleistocene forms that, in a second wave of emigration, entered North America. Understanding the Pleistocene evolution of these lineages therefore requires a Holarctic perspective, and in particular the Eurasian background is important to understanding the North American immigrants and their subsequent development.

Following Harland et al. (1990), the Pleistocene is here divided into three parts: early (ca. 1.6 my–790 ky BP), middle (ca. 790–128 ky BP), and late (ca. 128–10 ky BP).

Moose (*Alces* spp.)

The Eurasian record

The moose tribe (Alcini) is ultimately most closely related to the neo-cervine deer, currently distributed mainly in North America (Bubenik, 1990), but the origins of the present-day species [*Alces alces* (L.)] and its Pleistocene relatives are found in the Plio-Pleistocene of Europe. The earliest representative is *Alces gallicus* Azzaroli. This species already possessed many of the skeletal characteristics of modern moose: the peculiarities of dental and postcranial morphology that make moose bones and teeth unmistakable among cervids, the unique horizontal departure of the antler-bearing pedicles from the side of the skull, and the basic antler structure, comprising a horizontal beam devoid of tines, terminating in a palmation (Figure 9.1). *A. gallicus* differed from modern *A. alces* in its primitive skull morphology, with a low facial region and long nasal bones contacting the premaxillaries (Figure 9.2), the great length of its antler beams (Figures 9.1 and 9.3), and its small body size (Figure 9.4). The type skeleton of *A. gallicus,* the largest known individual of the species, has a shoulder height of approximately 140 cm (Azzaroli, 1952), compared with the typical modern European male *A. alces* shoulder height of 180 cm (Flerov, 1952).

Alces gallicus was founded on the basis of two skeletons from Senèze, France (Azzaroli, 1952), in deposits now dated to about 2.0–1.6 my BP (Thouveny and Bonifay, 1984; Lister, in press-a). The largest sample of antlers and teeth (Azzaroli, 1953) comes from East Runton, England, in deposits of Pastonian age, about 1.7 my BP (Gibbard et al., 1990; Lister, in press-a). Several other sites of late Pliocene to early Pleistocene age have yielded more fragmentary remains; all currently known localities are in Europe and western Asia (H.-D. Kahlke, 1990) (Figure 9.5). The time range of the species appears to have extended from about 2.5–2.0 my to about 1.2 my BP (Lister, in press-a).

In the early middle Pleistocene, broad-antlered moose were again found in Europe, and they are referred to the species *Alces latifrons* (Johnson). In dental and postcranial anatomy and basic antler plan, *A. latifrons* resembled *A. gallicus.* Its skull structure was also of essentially similar, primitive form (Azzaroli, 1981; Sher, 1987) (Figure 9.2). The most obvious difference between the two species is the much larger size of *A. latifrons.* Limb bones of *A. latifrons* are, on average, about 15% longer than those of modern European *A. alces* (Figure 9.4), suggesting

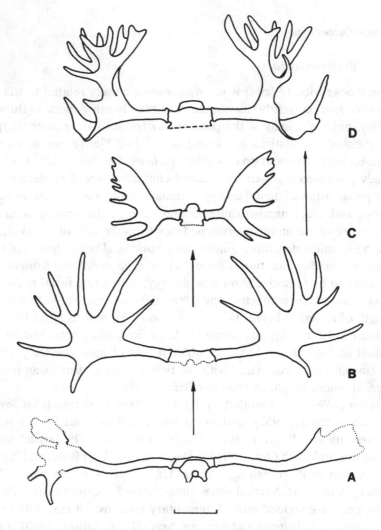

Figure 9.1. Evolution of antler structure in Quaternary moose. A–C show beam length reduction in the Eurasian lineage. (A) *A. gallicus*, early Pleistocene, Senèze, France, in posterior view. From Azzaroli (1952). (B) *A. latifrons*, early middle Pleistocene, Goldhofer Sands, Germany, in anterior view (from a slide, courtesy of Prof. W. von Koenigswald). (C) *A. alces*, modern, Sweden, in anterodorsal view (original, from a specimen in the University Museum of Zoology, Cambridge). (D) *A. scotti*, late Pleistocene, Mt. Hermon, New Jersey, in anterodorsal view. From Scott (1885). Scale bar 25 cm.

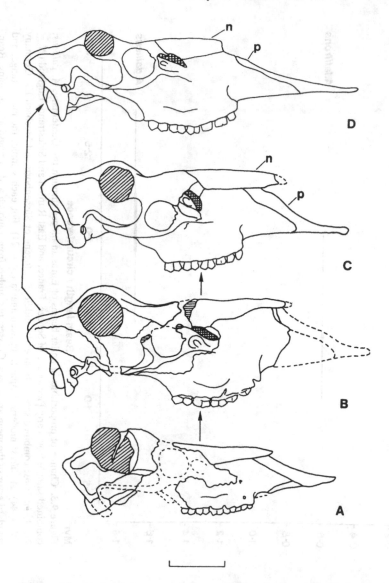

Figure 9.2. Skulls of Quaternary moose. Adapted from Sher (1987). (A) *A. gallicus*, early Pleistocene, Senèze, France. (B) *A. latifrons*, early middle Pleistocene, Süssenborn, Germany. (C) *A. scotti*, Mt. Hermon, New Jersey. (D) *A. alces*, modern; n, nasal; p, premaxillary. Scale bar 10 cm.

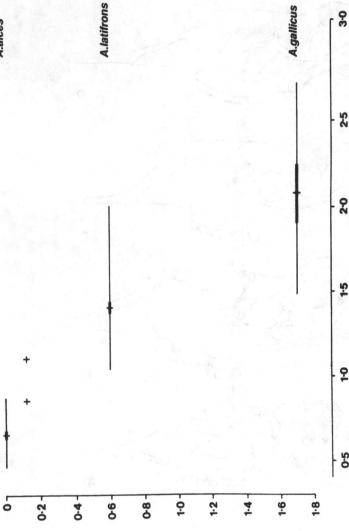

Myr

Beam length / circumference

Figure 9.3. Change of proportions in antler beams of Eurasian moose through the Quaternary. *A. gallicus*: combined sample from early Pleistocene of Senèze, France, and East Runton and Sidestrand, England (N = 7). *A. latifrons*: combined sample from early middle Pleistocene of Süssenborn and Mosbach III, Germany (N = 44). *A. alces*: modern sample from Scandinavia (N = 24). For each sample, the mean, range, and standard error of the mean are shown. Crosses: two antlers from Taubach and Ehringsdorf, Germany. Bear length measured from the burr to the beginning of the palmation (defined as the point at which beam circumference has increased by 50%). Beam circumference measured directly above the burr. Juvenile antlers (beam circumference <150 cm for *A. gallicus* and *A. alces*, <175 cm for *A. latifrons*) have been excluded. All data original; see Table 9.1 for sources and dating of samples.

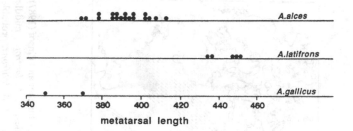

Figure 9.4. Changes in body size of the Eurasian moose lineage, illustrated by metatarsal length (mm). *A. gallicus*: holotype and cotype skeletons from early Pleistocene of Senèze, France. Data from Azzaroli (1952). *A. latifrons*: sample from early middle Pleistocene of Mosbach III, Germany. Data original. *A. alces*: sample from the Holocene of Europe. Data from Chaix and Desse (1981). Each point represents one specimen; see Table 9.1 for sources and dating of samples.

a shoulder height of about 210 cm. Compared with *A. gallicus* at 140 cm, this represents a 50% increase. Despite this size increase, the overall antler spans of the two species are similar, at about 2 m in adult individuals, with beam lengths typically 30–50 cm. There had thus been a significant *relative* reduction in antler beam length from *A. gallicus* to *A. latifrons* (Figure 9.1).

Alces latifrons remains are known in abundance from early middle Pleistocene sites in central Europe such as Mosbach and Süssenborn, Germany (H.-D. Kahlke, 1956, 1960), but the species also expanded its range across Asia to eastern Siberia (Figure 9.5). The best-known European samples date from about 700–500 ky BP (Lister, in press-a); some of the Asian material is probably later (Vangengeim and Flerov, 1965).

Well-preserved remains attributable to the living species *A. alces* are known in Europe and Asia from the early Weichselian (last) cold stage (ca. 100 ky BP) up to the present day. In addition to a decrease in size, they show two marked differences from *A. latifrons*. First, the antler beam has become very substantially shortened, to typically 15 cm, or about one-third of its length in the middle Pleistocene species (Figure 9.1). Second, there is a change in skull architecture, the facial region having deepened and the nasals shortened, so that they no longer contact the premaxillaries (Figure 9.2).

As a working hypothesis, it is assumed that the three species *A. gallicus, A. latifrons,* and *A. alces* represent an approximate line of descent. First, the various alcine peculiarities of teeth, skull, antlers,

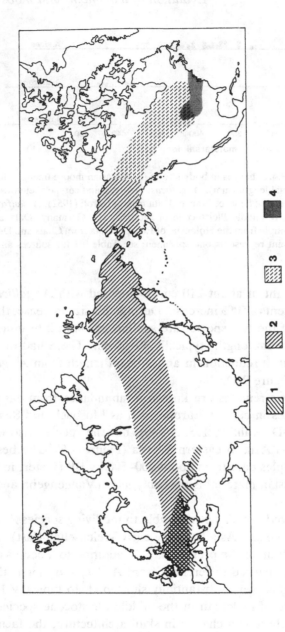

Figure 9.5. Total known ranges of extinct Quaternary moose species. Adapted from Churcher and Pinsof (1987) and H.-D. Kahlke (1990): 1, *A. gallicus*, early Pleistocene, based on 12 localities; 2, *A. latifrons*, middle Pleistocene, based on 49 localities in Europe, 31 in Siberia, and 5 in Alaska/Yukon; 3, *A. latifrons*, probable extension of range in North America, based on incomplete antlers from 4 localities; 4, *A. scotti*, late Pleistocene, based on definitely identifiable antlers from 4 localities.

and postcrania are present throughout. Second, the three species replace each other in time and geographic distribution. Third, no other presumptive ancestors are currently known. In the following analysis, the samples will be treated as an ancestor-descendant lineage; if this were to prove incorrect, the observations are still valid as comparisons between chronologically replacing species.

In Figure 9.3, an index of antler beam proportions is plotted for dated samples of the three species. The index is formed as the ratio between the length of the beam and its circumference, and Figure 9.3 clearly shows a directional trend in reduction of this ratio. This corresponds to a reduction in beam length relative to body size, because beam circumference is known to reflect body size, including both individual and ontogenetic variations (Lister, 1990). Some of the variation *within* each species is due to age- and size-related allometry of length/circumference ratio, but the clear displacement between the species samples (each of which spans individuals of different ages) indicates that the trend is a genuine change, not due to age or size bias of the different samples. Full data illustrating this point are given by Lister (1981).

The pattern of change in Figure 9.3 is consistent with an approximately constant-rate phyletic change. However, the three main samples are spaced about 1.0 and 0.5 my apart from each other, so we do not really have enough information to draw a straight line through the points and describe the changes as "gradual." Theoretically, there is enough time between them to have included rapid transitions. Unfortunately, the fossil record of moose in Eurasia between these points is poor. Some fossils intermediate in age between *A. gallicus* and *A. latifrons* appear intermediate in size (Heintz and Poplin, 1981), but there are no antler remains. Two antlers of Eemian age (ca. 120 ky BP) from the German travertines (H.-D. Kahlke, 1975, 1976) are plotted in Figure 9.3, and, appropriately, their beam proportions are intermediate between those of typical *A. latifrons* and *A. alces*. The sample size is small here, but the position of the two specimens on the plot suggests that we may provisionally speak of four samples in chronological order showing a unidirectional trend.

Possible adaptive reasons for the reduction in relative antler length can be briefly discussed. In the first step, from *A. gallicus* to *A. latifrons*, the explanation may lie in the great increase in body size, which occurred for unknown reasons. Linear dimensions increased by 40–50%, so if antler size had simply increased isometrically from the ~2-m span in *A. gallicus*, spans of ~3 m would have resulted in *A. latifrons*, or up to

4 m if the positive allometry between antler size and body size among cervids (Gould, 1974) is assumed. This probably would have caused mechanical problems and strains on the skull, so absolute antler span remained the same, at ~2 m, representing a relative reduction in beam length.

In the second step, from *A. latifrons* to *A. alces,* body size became somewhat reduced, but not nearly enough to account for the drastic reduction in antler beam length. Here a change of habitat may have been responsible. *A. latifrons,* to judge from associated flora and fauna, lived in a relatively open, steppe-like habitat (Sher, 1974, p. 198). At some point in the transition to *A. alces,* the lineage moved into the coniferous forest habitat that the moose occupies today. An antler span of 2 m or more would be a fine display organ in an open habitat, but in forest it could impede movement. Modern moose do sometimes have trouble moving rapidly through forest because of snagging their antlers (R. D. Guthrie, personal communication).

It is therefore possible that the unidirectional morphological trend in antler shape may have been linked to different adaptive pressures at different times during the sequence.

The North American record

Broad-antlered Pleistocene moose remains have been found in several parts of North America and are the subject of an excellent review by Churcher and Pinsof (1987). Numerous moose fossils from Alaska and the Yukon include long antler beams of sizes and proportions similar to those of European *A. latifrons* (Harington, 1977; Kurtén and Anderson, 1980), although Churcher and Pinsof caution that no specimen preserves the palmation for definitive species identification. In the eastern United States, the endemic form *Alces* (or *Cervalces,* discussed later) *scotti* (Lydekker) occurs. The four known specimens with sufficiently complete antlers (from New Jersey, Kansas, and South Dakota) indicate a form clearly different from *A. latifrons* (Figure 9.1), with a beam of intermediate length, between typical *A. latifrons* and *A. alces,* and terminating in two or three narrow palmations (Churcher and Pinsof, 1987). Other specimens of broad-antlered *Alces* from eastern and central North America lack the palmation, leading Churcher and Pinsof (1987) to prudently regard their specific status as uncertain. Kurtén and Anderson (1980) listed these as *A. scotti,* but H.-D. Kahlke (1990), following

Harington (1977), has tentatively suggested that those with relatively long beams represent *A. latifrons,* so that the latter species extended its range eastward as shown in Figure 9.5. The distribution map for *C. scotti* (Figure 9.5) is based on the four sufficiently complete specimens mentioned. Some additional material listed by Churcher (1991) as *C. scotti* may extend the distribution of the species farther to the west, but many of these specimens do not include the antler palmations required for definite specific diagnosis.

It is likely that *A. scotti* was an endemic offshoot from *A. latifrons* in North America (Azzaroli, 1981; H.-D. Kahlke, 1990). In addition to the long antler beams, its skull morphology is of particular significance in this respect, as it is of the "primitive" type similar to *A. gallicus* and *A. latifrons,* with long nasals contacting the premaxillaries (Azzaroli, 1981; Sher, 1987) (Figure 9.2). The alternative suggestion of Bubenik (1990, Figure 34), on the basis of antler branching pattern, that *A. scotti* is derived from *A. alces* is unlikely, as it would imply a reversal of skull structure.

The entry of *A. latifrons* into North America probably occurred at some time in the middle Pleistocene; a few specimens of broad-antlered moose are of Irvingtonian or even Kansan age (Kurtén and Anderson, 1980; Churcher and Pinsof, 1987). Most well-dated remains in the New World are, however, of Wisconsinan age, indicating a later survival there than in Eurasia. All dated remains of definite *C. scotti* are of Wisconsinan age (Churcher and Pinsof, 1987).

The morphological stage in the European chronocline at which *Alces* was "seeded" into North America is relevant to understanding the entry time and relationships of the North American forms. The transition from *A. latifrons* to *A. alces* occurred in Eurasia at some time in the late middle or early late Pleistocene; the entry of *A. latifrons* into the New World presumably would have occurred before this transition, unless populations of *A. latifrons* persisted in Siberia after the origin of *A. alces.* Some of the critical Siberian remains (the small form *A. latifrons postremus*) may be later than European *A. latifrons,* but they are poorly dated (Vangengeim and Flerov, 1965; Sher, 1974), so this remains a possibility. However, a middle Pleistocene migration into the New World would be consistent with the evidence from both continents.

If the distributions of *A. latifrons* and *A. scotti* implied by Kahlke's (1990) map (Figure 9.5) are correct, they are consistent with a model of allopatric speciation. Genuine speciation is one possibility considered by Churcher and Pinsof (1987), although they suggest that in view of the paucity of evidence from central North America, the possibility of a morphocline within a single species cannot be excluded.

Following its origin from *A. latifrons* in Eurasia at some time in the late middle or early late Pleistocene, *A. alces* entered North America across Beringia. This appears to have occurred in the Wisconsinan stage (Kurtén and Anderson, 1980) and was followed by the extinction of *A. latifrons* and *A. scotti*.

Taxonomy

On the basis of the evidence discussed earlier, the phylogeny for Pleistocene moose presented in Figure 9.6 seems likely and is essentially similar to those given by Azzaroli (1981) and H.-D. Kahlke (1990). The three alternative diagrams indicate identical relationships between the species, but differ in the generic allocation of the taxa. This question has been debated at some length; see Azzaroli (1981), Sher (1987), and H.-D. Kahlke (1990) for details. Early authors retained *Cervalces* for the North American *scotti*, and *Alces* for all the Eurasian forms (*gallicus, latifrons,* and *alces*). Azzaroli (1981) correctly noted the similar skull morphology for *gallicus, latifrons,* and *scotti* and placed these together in *Cervalces* (the earliest available name, which had been founded for *scotti*); living moose, with specialized skull morphology, remained as *A. alces*. This policy was supported by Sher (1987). Other authors, such as Geraads (1983), Geist (1987), and H.-D. Kahlke (1990), have placed all species in the single genus *Alces*. While recognizing the significance of skull form as an important morphological transition within this group, I have followed the single-genus policy for taxonomic purposes. There are three reasons for this: First, the important samples intermediate between *latifrons* and *alces*, such as those from the German travertines discussed previously, would be of uncertain generic status under the scheme of Figure 9.6A, because their skulls are unknown. Second, some Holocene moose from the Caucasus show a skull morphology tending toward the "*Cervalces*" condition (Heptner, Nasimovich, and Bannikov, 1988, p. 360). Third, there is the technical point that the taxonomies of Figures 9.6A and 9.6B leave *Cervalces* and *Alces* respectively as paraphyletic groups. Therefore, all known Pleistocene moose are here conventionally included in *Alces*.

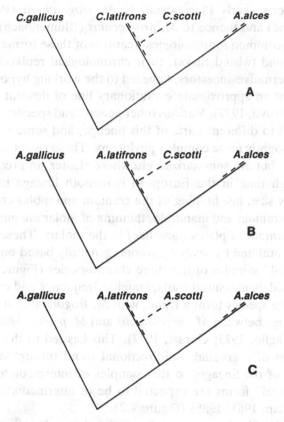

Figure 9.6. Probable relationships of Quaternary moose species expressed as a cladogram, and alternative taxonomies. The three diagrams show the same phylogeny, but different naming schemes; *A. = Alces*; *C = Cervalces*. Solid lines indicate occurrence in Eurasia, and broken lines North America.

Mammoths (*Mammuthus* spp.)

The Eurasian record

The mammoth lineage, like other elephant groups, appears to have originated in Africa (Maglio, 1973). Around 2.7 my BP, early mammoths migrated north, probably via the Middle East, and their earliest remains outside of Africa are found in deposits of this age in Europe (Aguirre and Morales, 1990). Through the Pleistocene in Europe, a sequence of forms has been established, leading from *Mammuthus meridionalis*

(Nesti) (late Pliocene–early Pleistocene) to *M. trogontherii* (Pohlig) (middle Pleistocene) and thence to *M. primigenius* (Blumenbach) (late Pleistocene). The common morphological features of these forms (such as the high skull and twisted tusks), their chronological replacement, and the lack of alternative ancestors, have led to the working hypothesis that they represent an approximate evolutionary line of descent (e.g., Maglio, 1973; Dubrovo, 1977). Various other generic and specific names have been applied to different parts of this lineage, and some authors have suggested a slightly more complex phylogeny. These issues will not be discussed here, but are summarized elsewhere (Lister, in press-b).

Changes through time in the European mammoth lineage include reductions in body size, heightening of the cranium and molar crowns, shortening of the cranium and mandible, thinning of molar enamel, and increases in the number of plates (lamellae) in the molars. These transitions have been outlined by several authors, generally based on comparisons of "typical" samples of the three chronospecies (Figure 9.7). In the two most widely measured traits, lamellar frequency and enamel thickness, the three species form a trend, with *M. trogontherii* of intermediate morphology between *M. meridionalis* and *M. primigenius* (e.g. Aguirre, 1969; Maglio, 1973; Garutt, 1977). This has led to the widespread assumption of a gradual, unidirectional trend throughout the ~.2.7-my history of the lineage, so that samples of intermediate ages between the "typical" forms are expected to be of intermediate morphology (e.g., Adam 1961, 1988) (Figure 9.2).

The recent increase in the number of well-dated samples allows this hypothesis to be put to the test. Some steps in this direction have been taken by, for example, Dubrovo (1977). In Figure 9.8, lamellar frequency has been plotted for 10 chronologically ordered samples of mammoth teeth. Data on the trend in enamel thickness show very similar patterns of change and will be published elsewhere (Lister, in press-b).

Figure 9.8 includes the best-known samples of the three "typical" stages: *M. meridionalis* from Valdarno, Italy (the type site of the species) (Azzaroli, 1977), *M. trogontherii* from Süssenborn, Germany (the type site of the species) (R.-D. Kahlke, 1990), and *M. primigenius* from Predmostí, Czechoslovakia, the most intensively studied sample of late European mammoth (Musil, 1968). In addition, earlier, later, and intermediate-aged samples are plotted.

In the lower part of the sequence, between about 2.0 my and 400 ky BP, a gradual pattern of change is evident, corresponding to approximately constant-rate phyletic change. The five sampling points are rather

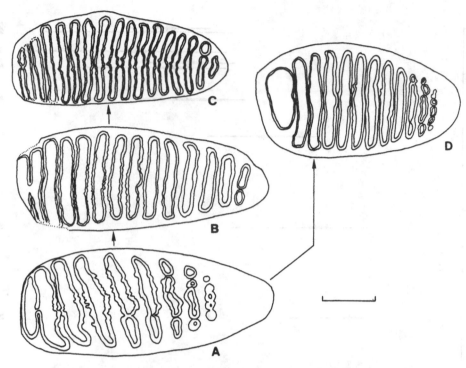

Figure 9.7. Right upper third molars of Quaternary *Mammuthus* species, illustrating changes in lamellar frequency and enamel thickness. Scale bar 5 cm. (A) *M. meridionalis*, early Pleistocene, Siniaya Balka, Russia (Paleontological Institute, Moscow, no. 1249/234). (B) *M. trogontherii*, early middle Pleistocene, Mosbach III, Germany (Natural History Museum, Mainz). (C) *M. primigenius*, late Pleistocene, Siberia (Trofimov's skeleton, Paleontological Institute, Moscow). (D) *M. columbi*, late Pleistocene, Hot Springs, South Dakota (Mammoth Site, Hot Springs, no. HS203).

widely spaced, with approximately 500 ky between successive stages. Although this might appear to leave open the possibility of unobserved "jumps" in morphology between the sampling points, this seems precluded by the smallness of the changes in mean morphology between the existing samples, allowing little morphological space for substantial jumps.

In the upper part of the sequence, from about 400 to 14 ky BP, there is no clear continuation of the directional trend. Instead, samples across this time period appear to indicate a pattern of stasis, with fluctuations around a mean value, but no net change. Additional dated samples in the later part of this period (ca. 150–14 ky BP) corroborate this pattern

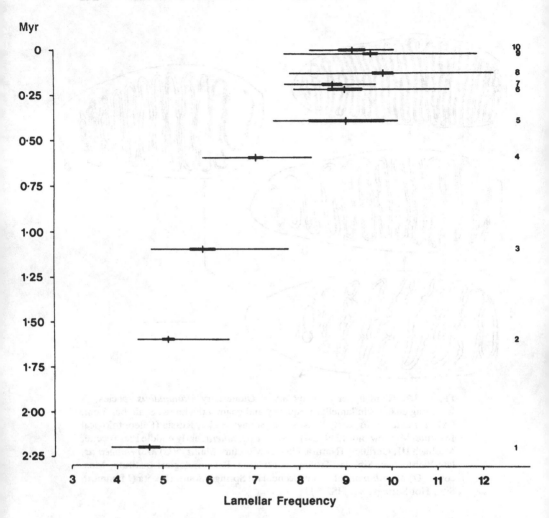

Figure 9.8. Changes in lamellar frequency of European mammoth molars through the Quaternary. Lamellar frequency indicates the number of lamellae (plates) in a typical 10-cm length of tooth and is measured on last upper molars according to the method of Maglio (1973). Samples 1–3 are conventionally named *M. meridionalis*, sample 4 *M. trogontherii*, and samples 5–10 *M. primigenius*. For each sample, the mean, range, and standard error of the mean are shown. All data original; see Table 9.1 for sources and dating of samples. 1, Khapry/Liventsovka, Russia (*N* = 7); 2, Upper Valdarno, Italy (*N* = 19); 3, Taman, Moldavia (*N* = 12); 4, Mosbach, Germany (*N* = 18); 5, Homersfield, England (*N* = 3); 6, Tourville, France (*N* = 8); 7, Ilford, England (*N* = 11); 8, Balderton, England (*N* = 23); 9, Predmosti, Czechoslovakia (*N* = 42); 10, Kostenki I, Russia (*N* = 6).

(A. M. Lister, unpublished data). On the other hand, it is possible that in the earlier part of this period (ca. 400–200 ky BP), dental morphology was on average slightly more primitive than in the later part (Figure 9.8) (Lister and Joysey, in press). Sample sizes are small, but the mean lamellar frequency of samples 5, 6, and 7 combined is significantly lower than that of samples 8, 9, and 10 combined (one-tailed t test, $p < 0.01$). More material will be required to test this possibility further. Nonetheless, it seems clear that dentally fully advanced mammoths already existed, at least in part of their range, by the late middle Pleistocene (ca. 140 ky BP), with no further advancement between then and extinction. Note, however, that a continuation of the slow rate of change seen in the lower-to-middle Pleistocene would not be expected to produce very noticeable change in a 125-ky interval.

Some terminal Siberian mammoths, dated to 12–10 ky BP and therefore later than any of the European samples, have been thought to represent a more "advanced" dental stage than the latter, but this has been shown to be an artifact of their small dental size (Lister and Joysey, in press).

A further important evolutionary transition in this lineage is the increase in crown height of the molars. In Figure 9.9, hypsodonty index is plotted for those sites providing sufficient sample sizes. Again, a unidirectional trend is seen in the lower part of the sequence, but the trend reaches its maximum extent by the typical *M. trogontherii* stage (ca. 600–500 ky BP), somewhat earlier than the trends in lamellar frequency and enamel thickness.

The dental changes in the Eurasian mammoth lineage reflect a change of habitat and feeding adaptation. *M. meridionalis* was a temperate woodland feeder, to judge both from its low-crowned, low-lamellar-frequency dentition and from its associated flora and fauna (e.g., Stuart, 1981). As the Quaternary Ice Ages proceeded, mammoths came to inhabit progressively more open habitats, and their dentition became adapted to a coarse, grassy diet, developing high crowns and a much higher lamellar frequency. Other changes included reductions in ear size and tail length and development of thick fur (Kubiak, 1982; Lister, in press-b).

The North American record

Mammoths appeared in the New World in the early Pleistocene, and like their Eurasian relatives, they persisted until nearly the end of the

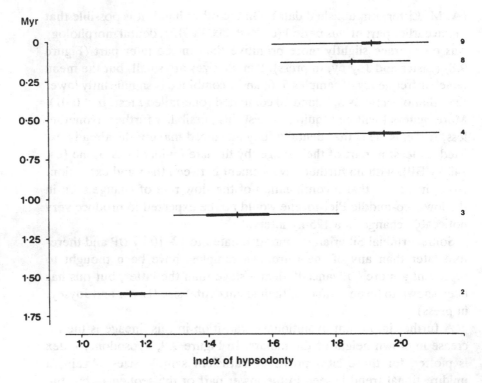

Figure 9.9. Changes in hypsodonty (crown height) of European mammoth molars through the Quaternary. Index of hypsodonty is calculated on third upper molars as the height of unworn plates in the region of maximum crown height (approximately the middle third of the tooth) divided by the maximum width of the molar. Sample codes as in Figure 9.8. Numbers of specimens: sample 2, $N = 8$; sample 3, $N = 5$; sample 4, $N = 10$; sample 8, $N = 6$; sample 9, $N = 20$. All data original.

last cold stage. The earliest records in temperate North America date to approximately 1.3–1.6 my BP (Kurtén and Anderson, 1980; Webb, 1993). They may already have been present in Alaska and northwest Canada by about 1.8 my BP, to judge, for example, by their presence in the Wellsch Valley fauna, believed to date from the Olduvai subchron (Lundelius et al., 1987). The immigration across the Bering Strait might then have occurred during Repenning's (1987) "Event 6," (ca. 1.9 my BP), when various microtines entered North America. Enigmatically, early Pleistocene mammoth remains from northeast Siberia are extremely sparse (Dubrovo, 1964).

The taxonomy of North American mammoths is confused and will not be treated here; see Maglio (1973), Madden (1981), and Graham (1986) for discussion. Following Maglio (1973), the earliest form can be referred to the Eurasian species *M. meridionalis,* although some authors distinguish it as *M. hayi* Barbour (Madden 1981; Webb, 1993). As with the moose lineage, the timing of the Eurasian evolutionary sequence is relevant to interpreting the earliest North American records. By 1.8– 1.3 my BP the progression from the most primitive forms was already under way in Europe (Figure 9.8), so that unless this process lagged behind in Siberia, one might expect to find a stage similar to that of Valdarno (ca. 1.6 my BP) in the earliest American mammoths. Maglio (1973) stated that some early records (dating from 1.36 my BP and younger) were of morphology equivalent to the "Bacton" stage of *M. meridionalis* (approximately between Siniaya Balka [Taman] and Valdarno in Figure 9.8), corresponding well enough to European morphology at that time. Churcher (1986) found that the most "primitive" teeth from the Old Crow Basin, Yukon, corresponded to the earlier "Montevarchi" stage of the European sequence (represented by the Khapry sample in Figure 9.1), and suggested that North American mammoths might therefore have an antiquity back to 2.5 my BP. Madden (1981) also suggested that some early North American mammoths had a lamellar count closer to early European *M. meridionalis* (or "*M. gromovi*" Alexeeva and Garutt) (i.e., sample 8 in Figure 9.8, ca. 2.0– 2.5 Ma). This question needs further investigation, bearing in mind particularly the wide spread seen in individual populations (Figure 9.8), and the need for sufficient sample sizes in comparing evolutionary stages.

In dental features, mammoths through the North American Pleistocene underwent morphological changes in some respects paralleling those in Europe. Middle Pleistocene fossils, conventionally named *M. imperator* (Leidy) by most current authors, are said to show advancement in lamellar frequency and enamel thickness beyond *M. meridionalis,* but to a lesser extent than in late Pleistocene *M. columbi* (Falconer) (Maglio, 1973; Kurtén and Anderson, 1980). Some data illustrating this transition have been given by Schultz and Martin (1970) and Schultz, Tanner, and Martin (1972), but the trend has not been quantified to the same detail as for the European lineage, and it is the subject of current research.

Most late Pleistocene mammoth samples from central North America (*M. columbi*) show a dental morphology clearly less "advanced" than the stage reached by *M. primigenius* at the same time (e.g., Agenbroad et al., in press). They are of a stage approximately intermediate between

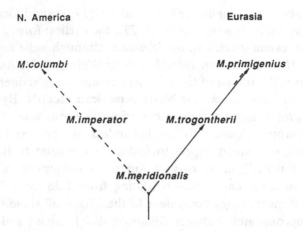

Figure 9.10. Schematic evolutionary tree of Holarctic Quaternary *Mammuthus* species, illustrating the cladogenetic origin of two lineages. Solid line indicates occurrence in Eurasia, and broken line North America.

European *M. trogontherii* and *M. primigenius* (Figure 9.7). Adaptively, this could correspond to the more southerly distribution of *M. columbi* compared with *M. primigenius*. On the other hand, some samples south of the *M. primigenius* range, and apparently of the *M. columbi* lineage, display advanced dental features rivaling *M. primigenius*. According to Maglio (1973), Roth (1982, p. 32), and Saunders (1970, in press), this material, termed *M. jeffersonii* (Osborn) by some authors, was of very large size and high lamellar frequency. The parallel development of two lines of mammoths, one in North America and the other in Eurasia, represents a speciation process from a common ancestral form (*M. meridionalis*) that became divided between the two continents. This process (Figure 9.10) is not easily accommodated in a cladistic scheme, unless the earliest North American form is given a name separate from *M. meridionalis*, on the basis not of any morphological difference but by virtue of having crossed Beringia. Further, the belief that the developments of the Eurasian and American branches represent parallel (and, by implication, entirely separate) processes depends on the assumption that there was no further genetic exchange across the Bering Strait after the initial colonization, a contention that would be difficult to test. Like the late Pleistocene immigration of *A. alces*, Eurasian *M. primigenius* came across the Bering Strait to join its American cousins. Most records of this species are of Wisconsinan age, although Kurtén and Anderson (1980) place its immigration as early as Illinoian. The woolly mammoth

Table 9.1. *Sites yielding mammoth or moose samples incorporated in the biometric study*

Site	Approximate age (my BP)	Dating methods	Collections	References
Kostenki I, Russia	~0.014	i, l, r	ZIN	Klein (1969)
Predmostí, Czechoslovakia	~0.026	i, r	AIB	Musil (1968)
Ehringsdorf & Taubach, Germany	~0.12	f, l	IQW	Kahlke (1975, 1976)
Balderton, England	~0.14	e, f, i, l	UMZC	Brandon & Sumbler (1991), Lister & Brandon (1991)
Ilford, England	~0.2	a, f, l	NHM, BGS	Stuart (1976), Sutcliffe & Kowalski (1976), Bowen et al. (1989)
Tourville, France	0.3–0.2	f, l, t	GCE	Lautridou et al. (1984), Descombes & Carpentier (1987), G. Carpentier (personal communication)
Homersfield, England	0.5–0.3	l	NCM	Coxon (1982), Stuart (1982), Berggren et al. (1980)
Mosbach III, Germany	0.7–0.5	f, l, m	NM	Brüning (1978), Igel (1984), Berggren et al. (1980)
Süssenborn, Germany	0.7–0.5	f, l, m	IQW	Kahlke (1969), Steinmüller (1972)

Table 9.1 (*cont.*)

Site	Approximate age (my BP)	Dating methods	Collections	References
Siniaya Balka, Russia	1.2–1.0	f, m	PIM	Dubrovo (1964) (personal communication)
Upper Valdarno, Italy	1.6–1.7	f	UF	Azzaroli et al. (1988), Masini (personal communication)
E. Runton & Sidestrand, England	~1.7	f, l, m, p	NHM	Gibbard et al. (1990), Lister (in press-a)
Senèze, France	1.6–2.0	f, m	UL	Bout (1970), Thouveny & Bonifay (1984), Azzaroli et al. (1988)
Khapry/Livensovka, Russia	2.0–2.5	f, m	GIN	Baigusheva (1971), Dubrovo (1989), E. Vangengeim (personal communication)

Note: Dating methods: a, amino acid epimerization; e, electron spin resonance; f, fauna; i, industry; l, lithostratigraphy; m, paleomagnetism; p, pollen; r, radiocarbon; t, thermoluminescence. Locations of collections: AIB, Anthropos Institute, Brno; BGS, British Geological Survey, Keyworth; GCE, G. Carpentier, Elbeuf; GIN, Geological Institute, Moscow; IQW, Institut für Quartärpaläontologie, Weimar; NCM, Castle Museum, Norwich; NHM, Natural History Museum, London; NM, Naturhistorisches Museum, Mainz; PIM, Paleontological Institute, Moscow; UF, University of Florence; UL, University of Lyon; UMZC, University Museum of Zoology, Cambridge; ZIN, Zoological Institute, Leningrad.

was distributed in Alaska, Canada, and the northeastern United States [see the map by Harington and Ashworth (1986)], so that this species and the more southern, less cold-adapted *M. columbi* (including *M. jeffersonii*) roughly divided the continent between them in the late Pleistocene.

Acknowledgments

I thank Jeffrey Saunders for helpful comments on the manuscript. The following curators kindly provided access to collections in their care (abbreviations as in Table 9.1): G. Baryshnikov and V. Garutt (ZIN), G. Carpentier (GCE), A. Currant (NHM), I. Dubrovo (PIM), W. Igel (NM), H. Ivimey Cook (BGS), H.-D. and R.-D. Kahlke (IQW), M. Oliva (AIB), M. Mazzini (UF), M. Prieur (UL), A. J. Stuart (NCM), and E. Vangengeim (GIN). Thanks are also due to A. Sher and W. von Koenigswald for pictures reproduced in Figures 9.1 and 9.2. This research was supported by the Natural Environment Research Council (grant no. GR3/6924) and the Science and Engineering Research Council (grant no. GR/H/21890).

References

Adam, K. D. (1961). Die Bedeutung der pleistozänen Säugetier-Faunen Mitteleuropas für die Geschichte des Eiszeitalters. *Stuttgarter Beiträge zur Naturkunde*, 78:1–34.
  (1988). Ueber pleistozäne Elefanten-Funde im Umland von Erzurum in Ostanatolien. Ein Beitrag zur Namengebung von *Elephas armeniacus* und *Elephas trogontherii*. *Stuttgarter Beiträge zur Naturkunde*, B146:1–89.
Agenbroad, L. D., A. M. Lister, D. Mol, and V. L. Roth (in press). *Mammuthus primigenius* remains from the mammoth site of Hot Springs, South Dakota. In L. D. Agenbroad & J. I. Mead (eds.), *The Hot Springs Mammoth Site: A Decade of Field and Laboratory Research on Paleontology, Geology, and Paleoecology*. Hot Springs, S.D.: Mammoth Site, Inc.
Aguirre, E. E. (1969). Revisión sistemática de los Elephantidae por su morfología y morfometría dentaria. Parts I-III. *Estudios Geologicos*, 24:109–67; 25:123–77, 317–67.
Aguirre, E. and J. Morales (1990). Villafranchian faunal record in Spain. *Quartärpaläontologie*, 8:7–11.
Azzaroli, A. (1952). L'alce di Senèze. *Palaeontographia Italica*, 47:133–41.
  (1953). The deer of the Weybourne Crag and Forest Bed of Norfolk. *Bulletin of the British Museum (Natural History), A: Geology*, 2:3–96.
  (1977). Evolutionary patterns of Villafranchian elephants in central Italy. Atti

della Accademia Nazionale dei Lincei. Memorie. *Classe di Scienze Fisiche, Matematiche e Naturali,* Sez. 2a 14 (Ser. 8):149–68.

(1981). On the Quaternary and recent cervid genera *Alces, Cervalces, Libralces. Bollettino della Società Paleontologica Italiana,* 20:147–54.

Azzaroli, A., C. De Guili, G. Ficcarelli, and D. Torre (1988). Late Pliocene to early mid-Pleistocene mammals in Eurasia: faunal succession and dispersal events. *Palaeogeography, Palaeoclimatology, Palaeoecology,* 66:77–100.

Baigusheva, V. S. (1971). [The fossil theriofauna of the Liventsov quarry, northeastern Azov area.] *Trudy Zoologicheskogo Instituta, Akademiya Nauk SSSR,* 49:35–42 (in Russian).

Berggren, W. A., L. H. Burckle, M. B. Cita, H. B. S., Cooke, B. M. Funnell, S. Gartner, J. D. Hays, J. P. Kennett, N. D. Opdyke, L. Pastouret, N. J. Shackleton, and Y. Takayanagi (1980). Towards a Quaternary time scale. *Quaternary Research,* 13:277–302.

Bout, P. (1970). Absolute ages of some volcanic formations in the Auvergne and Velay areas and chronology of the European Pleistocene. *Palaeogeography, Palaeoclimatology, Palaeoecology,* 8:95–106.

Bowen, D. Q., S. Hughes, G. A. Sykes, and G. H. Miller (1989). Land-sea correlations in the Pleistocene based on isoleucine epimerisation in nonmarine molluscs. *Nature,* 340:49–51.

Brandon, A., and M. G. Sumbler (1991). The Balderton Sand and Gravel: pre-Ipswichian cold stage fluvial deposits near Lincoln, England. *Journal of Quaternary Science,* 6:117–38.

Brüning, H. (1978). Zur Untergliederung der Mosbacher Terrassenabfolge und zum klimatischen Stellenwert der Mosbacher Tierwelt im Rahmen des Cromer-Komplexes. *Mainzer Naturwissenschaftliches Archiv,* 16:143–90.

Bubenik, A. B. (1990). Epigenetical, morphological, physiological, and behavioral aspects of evolution of horns, pronghorns, and antlers. In G. A. Bubenik and A. B. Bubenik (eds.), *Horns, Pronghorns, and Antlers* (pp. 3–113). New York: Springer.

Chaix, J., and L. Desse (1981). Contribution à la connaissance de l'élan (*Alces alces,* L.) postglaciaire du Jura et du Plateau suisse. Corpus de mesures. *Quartär,* 31–2:139–90.

Churcher, C. S. (1986). A mammoth measure of time: molar compression in *Mammuthus* from the Old Crow Basin, Yukon Territory, Canada. *Current Research in the Pleistocene,* 3:61–4.

(1991). The status of *Giraffa nebrascensis,* the synonymies of *Cervalces* and *Cervus,* and additional records of *Cervalces scotti. Journal of Vertebrate Paleontology,* 11:391–7.

Churcher, C. S., and J. D. Pinsof (1987). Variation in the antlers of North American *Cervalces* (Mammalia; Cervidae): review of new and previously recorded specimens. *Journal of Vertebrate Paleontology,* 7:373–97.

Coxon, P. (1982). The terraces of the River Waveney. In P. Allen (ed.), *Field Guide to the Gipping and Waveney Valleys, Suffolk* (pp. 80–94). Cambridge, U.K.: Quaternary Research Association.

Descombes, J. C, and G. Carpentier (1987). La faune de grands mammifères de Tourville-la-Rivière. *Bulletin du Centre de Géomorphologie du C.N.R.S.*, 32:19–23.

Dubrovo, I. A. (1964). [Elephants of the genus *Archidiskodon* on the territory of the USSR.] *Paleontologicheskii Zhurnal*, 3:82–94 (in Russian).

— (1977). A history of elephants of the *Archidiskodon-Mammuthus* phylogenetic line on the territory of the USSR. *Journal of the Palaeontological Society of India*, 20:33–40.

— (1989). [Systematic position of Khapry elephants.] *Paleontologicheskii Zhurnal*, 1:78–87 (in Russian).

Flerov, K. K. (1952). Musk deer and deer. In *Fauna of the USSR: mammals, vols. 1–2.* Jerusalem: Israel Program for Scientific Translation.

Garutt, W. E. (1977). [Dentition of elephants in ontogenesis and phylogenesis.] *Trudy Zoologicheskogo Instituta, Akademiya Nauk SSSR*, 73:3–36 (in Russian).

Geist, V. (1987). On the evolution and adaptations of *Alces. Swedish Wildlife Research Supplement*, 1:11–23.

Geraads, D. (1983). Artiodactyles (Mammalia) du Pléistocène moyen de Vergranne (Doubs). *Géologie*, 5:69–81.

Gibbard, P. L., R. G. West, W. H. Zagwijn, P. S. Balson, A. W. Burger, B. M. Funnell, D. H. Jeffery, J. de Jong, T. van Kolfschoten, A. M. Lister, T. Meijer, P. E. P. Norton, R. C. Preece, J. Rose, A. J. Stuart, C. A. Whiteman, and J. A. Zalasiewicz (1990). Early and early middle Pleistocene correlations in the southern North Sea basin. *Quaternary Science Reviews*, 10:23–52.

Gould, S. J. (1974). The origin and function of "bizarre" structures: antler size and skull size in the "Irish elk" *Megaloceros giganteus. Evolution*, 28:191–220.

Graham, R. (1986). Taxonomy of North American mammoths. In G. C. Frison and L. C. Todd (eds.), *The Colby Mammoth Site – Taphonomy and Archaeology of a Clovis Kill in Northern Wyoming* (pp. 165–9). Albuquerque: University of New Mexico Press.

Harington, C. R. (1977). Pleistocene mammals of the Yukon Territory. Ph.D. dissertation, University of Edmonton.

Harington, C. R., and A. C. Ashworth (1986). A mammoth (*Mammuthus primigenius*) tooth from late Wisconsin deposits near Embden, North Dakota, and comments on the distribution of woolly mammoths south of the Wisconsin ice sheets. *Canadian Journal of Earth Science*, 23:909–18.

Harland, W. B., R. L. Armstrong, A. V. Cox, L. E. Craig, A. G. Smith, and D. G. Smith (1990). *A Geologic Time Scale 1989.* Cambridge University Press.

Heintz, E., and F. Poplin (1981). *Alces carnutorum* (Laugel, 1862) du Pléistocène de Saint-Prest (France). Systématique et évolution des Alcinés (Cervidae, Mammalia). *Quartärpaläontologie*, 4:105–22.

Heptner, V. G., A. A. Nasimovich, and A. G. Bannikov (1988). *Mammals of the Soviet Union, Volume 1: Artiodactyla and Perissodactyla.* Washington D.C.: Smithsonian Institution.

Igel, W. (1984). Paläogeographie durch Sediment-Morphoskopie, Grundlagen, Methodik und Anwendung. Ph.D. dissertation, University of Mainz.

Kahlke, H.-D. (1956). *Die Cervidenreste aus den altpleistozänen Ilmkiesen von Süssenborn bei Weimar.* Berlin: Akademie-Verlag.

(1960). Die Cervidenreste aus den altpleistozänen Sanden von Mosbach (Biebrich-Wiesbaden). *Abhandlungen der Deutschen Akademie der Wissenschaften, Klasse für Chemie, Geologie und Biologie,* 7:1–75.

(ed.) (1969). Das Pleistozän von Süssenborn. *Paläontologische Abhandlungen,* A3(3/4).

(1975). Die Cerviden-Reste aus den Travertinen von Weimar-Ehringsdorf. *Abhandlungen des Zentralen Geologischen Instituts,* 23:201–49.

(1976). Die Cervidenreste aus den Travertinen von Taubach. *Quartärpaläontologie,* 3:113–22.

(1990). On the evolution, distribution and taxonomy of fossil elk/moose. *Quartärpaläontologie,* 8:83–106.

Kahlke, R.-D. (1990). Zur Festlegung des Lectotypus von *Mammuthus trogontherii* (Pohlig, 1885) (Mammalia, Proboscidea). *Quartärpaläontologie,* 8:119–24.

Klein, R. G. (1969). *Man and Culture in the Late Pleistocene.* San Francisco: Chandler.

Kubiak, H. (1982). Morphological characters of the mammoth: an adaptation to the arctic-steppe environment. In D. M. Hopkins and D. Moody (eds.), *Palaeoecology of Beringia* (pp. 281–9). New York: Academic Press.

Kurtén, B., and E. Anderson (1980). *Pleistocene Mammals of North America.* New York: Columbia University Press.

Lautridou, J. P., D. Lefebvre, F. Lecolle, G. Carpentier, J. C. Descombes, C. Gaquerel, M. F. and Huault (1984). Les terrasses de la Seine dans le méandre d'Elbeuf. Corrélations avec celles de Mantes. *Bulletin de l'Association Française pour l'Etude du Quaternaire,* 1–3:27–32.

Lister, A. M. (1981). Evolutionary studies on Pleistocene deer. Ph.D. dissertation, University of Cambridge.

(1990). Critical reappraisal of the middle Pleistocene deer species "*Cervus*" *elaphoides* Kahlke. *Quaternaire,* 1:175–92.

(in press-a). The stratigraphical distribution of deer species in the Cromer Forest-bed Formation. In P. L. Gibbard and C. Turner (eds.), *The Early Middle Pleistocene in Europe.* Rotterdam: Balkema.

(in press-b). History and evolution of mammoths in Eurasia. In J. Shoshani and P. Tassy (eds.), *The Proboscidea: Trends in evolution and palaeoecology.* Oxford: University Press.

Lister, A. M., and A. Brandon (1991). A pre-Ipswichian cold stage mammalian fauna from the Balderton sand and gravel, Lincolnshire, England. *Journal of Quaternary Science,* 6:139–57.

Lister, A. M., and K. A. Joysey (in press). Scaling effects in elephant dental evolution – the example of Eurasian *Mammuthus.* In P. Smith and E. Tchernov (eds.), *Proceedings of the 1989 Dental Morphology Symposium.* Jerusalem: Freund.

Lundelius, E. L., C. S. Churcher, T. Downs, C. R. Harington, E. H. Lindsay,

G. E. Schultz, H. A. Semken, S. D., Webb, and R. J. Zakrzewski, (1987). The North American Quaternary sequence. In M. O. Woodburne (ed.), *Cenozoic Mammals of North America; Geochronology and Biostratigraphy* (pp. 211–35). Berkeley: University of California Press.

Madden, C. T. (1981). Mammoths in North America. Ph.D. dissertation, University of Colorado, Boulder.

Maglio, V. J. (1973). Origin and evolution of the Elephantidae. *Transactions of the American Philosophical Society*, 62: 1–149.

Musil, R. (1968). Die Mammutmolaren von Předmostí. *Paläontologische Abhandlungen*, A3:1–192.

Repenning, C. A. (1987). Biochronology of the microtine rodents of the United States. In M. O. Woodburne (ed.), *Cenozoic Mammals of North America* (pp. 236–68). Berkeley: University of California Press.

Roth, V. L. (1982). Dwarf mammoths from the Santa Barbara, California Channel Islands: size, shape, development, and evolution. Ph.D. dissertation, Yale University.

Saunders, J. J. (1970). The distribution and taxonomy of *Mammuthus* in Arizona. M.Sc. dissertation, University of Arizona, Tucson.

(in press). Vertebrates of the San Pedro Valley 11,000 B.P. In C. V. Haynes (ed.), *The Clovis Hunters*. Tucson: University of Arizona Press.

Schultz, C. B., and L. D. Martin (1970). Quaternary mammalian sequence in the central Great Plains. In *Pleistocene and Recent Environments of the Central Great Plains* (pp. 341–53). Department of Geology, University of Kansas, Special Publication 3.

Schultz, C. B., L. G. Tanner, and L. D. Martin (1972). Phyletic trends in certain lineages of Quaternary mammals. *Bulletin of the University of Nebraska Museum*, 9:183–95.

Scott, W. B. (1885). *Cervalces americanus*, a fossil moose, or elk, from the Quaternary of New Jersey. *Proceedings of the Academy of Natural Sciences of Philadelphia*, 1885:181–202.

Sher, A. V. (1974). Pleistocene mammals and stratigraphy of the far Northeast USSR and North America. *International Geological Review*, 16:1–206.

(1987). History and evolution of moose in USSR. *Swedish Wildlife Research Supplement*, 1:71–97.

Steinmüller, A. (1972). Die Schichtenfolgen von Süssenborn und Voigtstedt und die Gliederung des Mittelpleistozäns. *Geologie*, 21:149–65.

Stuart, A. J. (1976). The history of the mammal fauna during the Ipswichian/last interglacial in England. *Philosophical Transactions of the Royal Society of London*, B276:221–50.

(1981). A comparison of the Middle Pleistocene mammal faunas of Voigtstedt (Thuringia, GDR) and West Runton (Norfolk, England). *Quartärpaläontologie*, 4:155–63.

(1982). *Pleistocene Vertebrates in the British Isles*. London: Longman.

Sutcliffe, A. J., and K. Kowalski (1976). Pleistocene rodents of the British Isles. *Bulletin of the British Museum (Natural History)*, A. Geology, 27:33–147.

Thouveny, N., and E. Bonifay (1984). New chronological data on European Plio-Pleistocene faunas and hominid occupation sites. *Nature*, 308:355–8.

Vangengeim, E. A., and C. C. Flerov (1965). [The broad-antlered elk (*Alces latifrons*) in Siberia.] *Byulleten' Komissii po Izuchenigu Chetvertichnogo Perioda, Akademiya Nauk SSSR*, 30:161–71 (in Russian).

Webb, S. D. (1993). Early mammoths in North America. In J. Shoshani and P. Tassy, (eds.) *The Proboscidea: Trends in Evolution and Paleoecology.* Oxford: Oxford University Press.

10

Evolution of hypsodonty and enamel structure in Plio-Pleistocene rodents

LARRY D. MARTIN

The structure of the plant community has profound effects on rates of tooth wear in herbivorous mammals. Tooth abrasion results from three factors. The first, and in many cases the most important, is tooth-to-tooth wear. Functionally, tooth-to-tooth wear seems to produce and maintain sharp edges. The other two factors involve the abrasiveness of the diet. As a natural by-product of certain metabolic processes, plants precipitate opaline silica within their tissues as phytoliths (plant stones). During the life of a plant, phytoliths become progressively larger and more abundant in certain leaves and stems, but phytoliths are not as important components in plant reproductive parts or in the early growth of foliage (Piperno, 1988). If an animal eats mainly new growth, the abrasive effects of phytoliths on dentition can be largely avoided, but phytoliths can be significant contributors to tooth wear in species that concentrate on grazing (Piperno, 1988). More tooth wear results from the consumption of plant parts that are covered with dust (Janis, 1988), which naturally results when there are exposed land surfaces and is especially common in arid regions. In fact, the increase in seasonal aridity during the late Cenozoic may have been a major contributor to increased hypsodonty in Tertiary mammals. The abundance of dust on the surfaces of plants also depends on how close the plant parts are to the ground surface. Grasses not only incorporate large quantities of biogenic opal but also grow close to the surface, and hence grazing is an especially abrasive activity. The winds that are common in open plant communities can effectively coat low vegetation with dust, and both grazers and browsers tend to be hypsodont in savannahs, steppes, and deserts (Janis, 1988).

205

Seasonal climates with periods of aridity and frequent fires may favor plants that store energy underground in roots, tubers, and bulbs (Walter, 1979), and many insects and mammals utilize these underground resources. Eating while burrowing necessarily results in chewing a considerable amount of dirt that rapidly abrades the teeth. Therefore, fossorial mammals, almost without exception, increase their tooth hypsodonty.

Because the decreasing global temperatures during the late Cenozoic and the subsequent reduction of forest cover favored the development of steppe and savannah, there was coevally an increase in the contribution of hypsodont forms to mammalian diversity (Figure 10.1). If the climatic changes had been absolutely continuous, we would expect to see one pattern of seemingly orthogenic changes toward more hypsodont species, but instead the climate fluctuated, resulting in multiple radiations of species achieving hypsodonty and then becoming extinct, only to be replaced by unrelated lineages that independently developed high-crowned teeth (Martin, 1984, 1985). For instance, modern pocket gophers are not derived from the highly hypsodont "Miocene gophers" (now totally extinct) belonging to the Entoptychinae, but are derived from late Miocene lower-crowned heteromyids. As a result of this repetitive evolutionary pattern, we have many examples of independent lineages acquiring hypsodonty and an excellent laboratory for studying evolutionary processes.

Crown heights can be categorized as brachydont (long roots and little elevation of the crown), hypsodont (increased crown height, but retention of the roots, as in *Ondatra*), and superhypsodont (loss of roots and extreme elevation of the crown, as in *Microtus*).

Function of hypsodonty

Although specific aspects of acquiring hypsodonty may be morphologically complex, the functions of most of the changes seem clear. The purpose of increasing the crown height is to ensure a useful grinding surface throughout the reproductive lifespan of the animal. Minor exceptions to this include the hypsodont canines of saber-toothed cats and various systems that involve sharpening surfaces between teeth (the superhypsodont canines of the marsupial saber-tooth *Thylacosmilus*, etc.). The fact that dental wear can be a limiting factor on an animal's potential success is beautifully illustrated by the artificial dentures that have been implanted in cattle to extend their reproductive lifespan for

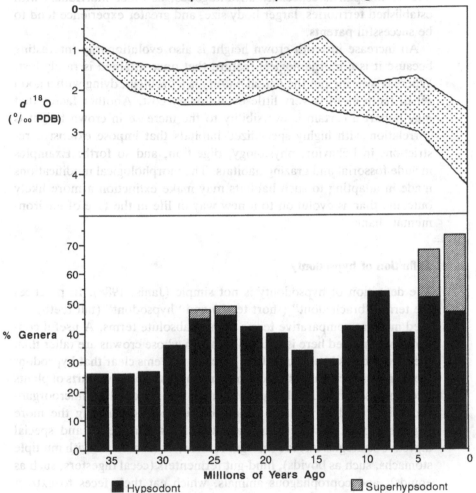

Figure 10.1. Percentages of rodents (bar graph) with hypsodont and superhypsodont teeth for each land-mammal age (Martin, 1985) compared with oxygen-isotope values (top graph) for benthic foraminifera plotted in 5-my intervals. Temperature values decrease as the delta values increase toward the Recent. Adapted from Prentice and Matthews (1988).

several years (Johnson, Hall, and Dorn, 1980). Extending the reproductive lifespan is especially advantageous, as older individuals with established territories, larger body size, and greater experience tend to be successful parents.

An increase in tooth crown height is also evolutionarily interesting because it is a morphological feature that once attained is rarely lost. Whereas wearing out the dental battery is a problem, dying with a good set of teeth extracts very little evolutionary cost. Another factor that may impart a certain irreversibility to the increase in crown height is correlation with highly specialized habitats that impose extensive restrictions in behavior, physiology, digestion, and so forth. Examples include fossorial and grazing habitats. The morphological modifications made in adapting to such habitats may make extinction a more likely outcome than is evolution to a new way of life in the face of environmental change.

Definition of hypsodonty

The definition of hypsodonty is not simple (Janis, 1988). In practice, the terms "brachydont" (short teeth) and "hypsodont" (tall teeth) are used more as comparative terms than as absolute terms. A useful convention employed here is to consider teeth whose crowns are taller than they are wide as hypsodont. In most cases it seems clear that hypsodont teeth are primarily suited for cutting the fibrous vegetative parts of plants into small sections. Further processing in the gut employs microorganisms that can deal with the cellulose skeleton surrounding the more digestible plant tissues. Mammals with hypsodont teeth and special digestive strategies include fore-gut fermenters (ruminants with multiple stomachs, such as bovids), hind-gut fermenters (cecal digesters, such as equids), and coprophagous animals, which eat their feces to extract nutrients missed the first time through (such as lagomorphs). In all three kinds of hypsodont animals, the increased crown height is correlated with an increase in the length of the shearing edges (lophs) that connect cusps; therefore hypsodont teeth usually are lophodont. The more complex crown pattern that results may allow processing of the same amount of vegetative material with fewer chewing strokes than would a less complex pattern, thereby reducing the rate of tooth wear and extending the lifespan of the dentition. Increased numbers of lophs might also decrease the size of the food particles entering the gut and increase the efficiency of digestion.

Most tooth crowns can be readily divided into two sections: first, the surficial part, which through moderate wear has evidence of lophs, cusps, and so forth; second, the base of the crown, composed essentially of a simple ring of enamel surrounding dentine. Hypsodonty may occur in one of these regions without greatly affecting the other. Hershkovitz (1967) noted, but did not fully understand, these differences. His tubercular hypsodonty would be a special case of the elongation of the surface detail (the cusps and lophs) or coronal features. Arvicolid rodents, mylagaulid rodents, proboscideans, and horses all show elongation of these coronal features. In this type of hypsodonty, high ridges are formed and then supported by the addition of crown cementum. In contrast, crown cementum usually does not characterize the second type of hypsodonty that progresses by elongation of the crown base. In these teeth the crown is complex in the unworn tooth, but as the tooth wears, complexity soon disappears, and the tooth becomes essentially an elongated cylinder. Living and fossil geomyoids are notable examples of this kind of hypsodonty, which Hershkovitz (1967) termed "cylindrification." Superhypsodonty can be defined as extreme elongation of the crown and loss of roots, as, for example, in *Geomys*.

Development of hypsodonty

The evolutionary development of even superhypsodont (evergrowing) teeth seems fairly simple. (Hypsodonty is mostly a mammalian feature, so only the development of mammalian hypsodonty is discussed here. An interesting special case of nonmammalian hypsodonty is found in the duck-billed dinosaurs, hadrosaurs, in which there was real tooth occlusion and rapid tooth wear. The whole battery of replacement teeth was mobilized to function as a single superhypsodont slicing and grinding tooth.) The mammalian jaw joint is a type II lever that works much like an ordinary nutcracker. The nearer the food object is to the fulcrum, the more force that can be exerted on it. The premolars usually are specialized for slicing and food manipulation. Crushing and grinding functions require more force and are associated with the molar teeth nearer to the fulcrum of the jaw. These functions result in more tooth wear, and hypsodonty is often most developed in the molar teeth. In many artiodactyls, increased hypsodonty is almost restricted to the last two molars (Janis, 1988). The deciduous dentitions (or "milk teeth") generally are not hypsodont because they are exposed to less abrasion (the animal is still suckling when it has them) and because they reside in the jaw for only a short time.

Mammalian teeth can generally be divided into two distinct regions: the real and potential grinding surface (covered with enamel), called the crown, and the portion below the gum line that usually anchors the tooth (lacks enamel), called the root. The enamel is usually a thin cap over a dentine core. In the living tooth the dentine surrounds and is produced by a bud of mesodermal tissue, the dental papilla. The enamel is initially laid down on the internal surface of a cup-shaped extension of ectodermal tissue that encloses the dental papilla. The whole crown is usually formed before the eruption of the tooth. The dental papilla continues to form tooth as the crown erupts, but this later dentine is not enamel-covered and hence constitutes the roots. Some portions of the enamel organ may persist longer than others, creating irregularities in the juncture between the tooth crown and the root. The most extreme example of this phenomenon occurs in rodent incisors, where ameloblasts persist on the anterior edge of the tooth, laying down an enamel strip only on the anterior surface. The periodontal sac lays down a compacted bony tissue on the roots called root cementum. In mammals the tooth is anchored to the tooth socket by periodontal ligaments running to the root cementum.

During ontogeny, the grinding surfaces of a tooth are covered with enamel before the walls of the crown are. Therefore, it is possible in a high-crowned tooth for the crown to begin to erupt before maturation of the lower collar of ameloblasts surrounding the sides of the tooth. The juncture between the enamel and the root indicates the maturation of the ameloblasts in that region. As the crown erupts, the roots form behind it. Initially they are thin-walled and contain large extensions of the dental papilla; eventually they thicken to enclose a narrow root canal. In evergrowing teeth this root canal is not seriously constricted, and the tooth is often referred to as open-rooted. In these superhypsodont teeth, normal roots fail to form, as at least a portion of the ameloblasts continue producing enamel throughout the life of the animal, and the enamel of the tooth extends to the apical foramen.

The balance between wear and eruption of the tooth crown must maintain an occlusal contact line between the upper and lower teeth at a nearly constant position. Changes in this contact will affect the size of the buccal cavity as well as the mechanics and occlusion of the jaws. Because superhypsodont teeth grow continuously, it might at first seem that they would have a fairly consistent relative tooth length from one taxon to another. However, we see variability related to the rate of tooth abrasion. Rodents that dig with their incisors, or have highly abrasive diets, have longer teeth than do forms experiencing less tooth wear (e.g., the living beaver *Castor* and the extinct fossorial beaver *Euhapsis*) (Martin, 1987). This may be due to a need to provide a longer surface area for tooth production in teeth that experience more rapid wear. One result is that the jaws are deepened and modified to accommodate the elongated crowns. Even in rooted teeth, usually in early wear only a small portion of the crown will be exposed above the gum line.

One problem facing high-crowned teeth is mechanical stability. Hypsodont teeth may erupt before the formation of significant roots, and because the periodontal ligaments usually attach to roots, this leaves the teeth loosely attached. At the same time, the high crowns increase the lever arm on the tooth. In many hypsodont forms this problem is solved by the production of dentine tracts. A dentine tract is formed when there is an interruption of the ameloblast girdle around the developing crown. As the crown continues to be formed, a strip of dentine is exposed and covered with root cementum. Periodontal ligaments are attached to these dentine tracks, which are, in many respects, inverse roots for teeth with small roots or none at all.

Enamel microstructure

Koenigswald (1980) described three types of arrangements for the enamel prisms in arvicolid rodents. The most basic of these types has the enamel prisms (Figure 10.2D) parallel to each other and crossed by the interprismatic substance (Figure 10.2D) at right angles. The interprismatic substance radiates away from the dentine-enamel junction perpendicularly, forming radial enamel (Figure 10.2D). Because they place the long axis of the enamel structures directly against both the horizontal and vertical impacts of chewing, they offer the greatest resistance to crushing of the enamel fabric and resulting tooth wear (Rensberger and Koenigswald, 1980). This is the first enamel organization to appear in most hypsodont lineages, but it does have a serious flaw. The parallel arrangement is vulnerable to the development of impact fractures between the prisms (Koenigswald and Martin, 1984a). In early forms, this vulnerability is compensated by thick enamel edges, but such edges are dull. Later forms solve the fracturing problem by twisting the prisms into other planes. This has the effect of "weaving" the enamel structures

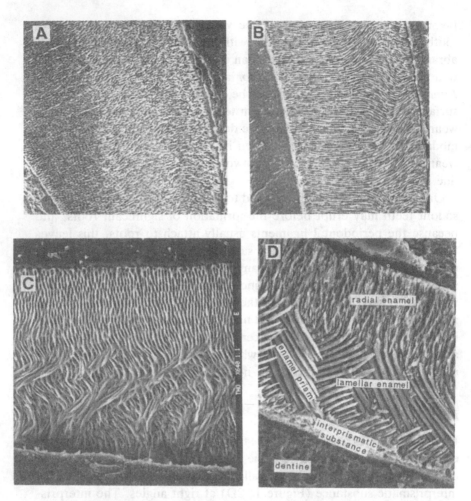

Figure 10.2. Scanning electron microscope (SEM) photography of crown sections of cheek teeth taken across the occlusal plane. A–C show pocket gophers. (A) *Pliogeomys* (5 my BP) P_4 showing thick radial enamel. (B) *Geomys* from the Borchers local fauna (2 my BP) from the same region of P_4 showing enamel thinning and the beginning of enamel structure. (C) Modern *Thomomys* molar showing advanced enamel structure. (D) *Microtus paroperarius* M_1 from the Cudahy local fauna (0.6 my BP) showing radial and lamellar enamel, as compared with the structure in *Thomomys*. C and D at twice the magnification of A and B.

together and thereby preventing cracks. Vulnerability to compressional force still exists, and radial enamel is always retained in combination with the other forms. The distribution of the various enamel types on a single tooth was called its *Schmelzmuster* by Koenigswald (1980, 1982).

The two common enamel types that are coupled with radial enamel are tangential enamel and lamellar enamel (Figure 10.2D). Tangential enamel has the prisms and interprismatic substance more or less parallel to the occlusal plane. In other words, the structure is rotated about 90° to that in radial enamel. Lamellar enamel (Figure 10.2D) is composed of single layers of prisms (lamellae) that are nearly parallel to the chewing surface. Within each lamella the prisms are parallel to each other, but form an angle of about 90° with the prisms of the adjacent lamella.

Because these structures are designed to resist the highly directional impacts of proplinal chewing, their arrangement reflects the distribution of impact points. In arvicolids, anterior edges of lower teeth impact posterior edges of uppers. Thus, the "leading" edges of lower teeth are anterior, and those of upper teeth are posterior (Figure 10.4D). The anterior ridges of uppers and posterior edges of lowers are called "trailing edges" (Koenigswald, 1980; Koenigswald and Martin, 1984b).

The radial enamel faces the direction of force, and in some forms it "crosses" the enamel band so that it is the outer layer on the leading edge and the inner layer on the trailing edge (Koenigswald, 1980). This close relationship between enamel structure and the direction of force is a remarkable proof of a functional basis for the enamel structures. We can further show that this arrangement is a result of the arrangement and shape of ameloblasts in the developing tooth and hence is the result of selection on a genetically controlled system.

It is apparently very easy to develop enamel prism modifications, and a large number of independent developments are known. Each development has the potential of further clarifying the relative roles of phylogenetic history and natural selection. In North America, all of the late Miocene arvicolines had thick, dominantly radial enamel on their molars. During the Pliocene in North America, thick, dominantly radial enamel was restricted to the muskrats *Pliopotamys–Ondatra* and the extinct arvicolid *Loupomys*. All North American Pleistocene arvicolines had advanced enamel. The development of these advanced *Schmelzmusters* permitted a general thinning of molar enamel in most lineages. No late Miocene arvicolids (although they are hypsodont) had evergrowing teeth, crown cementum, or high dentine tracts. During the Pliocene in North America, only the bog lemmings *Guildayomys* and

Pliolemmus had evergrowing teeth, and only *Ondatra* and *Loupomys* showed significant crown cementum. During the Pleistocene, rooted teeth were confined to *Clethrionomys, Phenacomys,* and *Ondatra,* and only *Phenacomys* lacked crown cementum. The genetic independence of these variables from each other is demonstrated by the full development of some without equivalent development of the others. For instance, *Pliolemmus* developed superhypsodont teeth without significant crown cementum, and *Loupomys* developed crown cementum without advanced enamel structure.

Hypsodonty in arvicolids and geomyines

In the modern fauna of North America the two rodent groups most noted for their hypsodonty are the arvicolids (voles and lemmings) and the geomyines (pocket gophers). Both groups are composed of small rodents that have achieved superhypsodont teeth during the past 5 my. Arvicolids and geomyines belong to separate superfamilies (Muroidea and Geomyoidea, respectively), and so phylogenetically they are not very close. They also differ greatly in habitat (surface grazing versus fossorial root consumption). Comparison of two such disparate examples can elucidate the roles of natural selection and environmental change in evolution.

Arvicolids may have originated in the Holarctic, and that region certainly is the center of their radiation (Martin, 1975, 1979). Several genera show nearly simultaneous first appearances in North America and Eurasia (*Prosomys, Predicrostonyx, Microtus*). This pattern of appearance requires a Holarctic distribution before their first occurrence farther south. Geomyoids are endemic to North America and may have originated in middle latitudes. At no time in their history do they seem to have dispersed far enough north to have used the Bering land bridge.

Arvicolids are small grazing rodents that seem to have originated in a water-marginal habitat. This is based on finding the primitive sister group of the Arvicolinae, the Microtoscoptinae, with the primitive giant beaver *Dipoides* in ponded water sediments (Martin, 1975). Many of the more primitive living members of the family (*Ondatra, Phenacomys, Clethrionomys*) favor moist habitats. Arvicolids develop a complex crown pattern of alternating triangles (Figure 10.4C) on the molars. This pattern persists for most of the life of the tooth. In this case, the crown of the tooth is elongated so that the surface complexity is extended as long thin ridges down the side of the tooth ("tubercular" hypsodonty).

In most species the valleys (reentrants) between these ridges are filled with crown cementum (Figure 10.4C and 10.4D).

The pocket gophers are fossorial analogues of the arvicolids in the evolution of superhypsodont teeth. Unlike the arvicolids, they have lengthened the base of the crown below the surface detail. The surface complexity is soon worn off, leaving a cylindrical tooth with a simple ovoid crown pattern ("cylindrification"). This is structurally a strong shape and is not buttressed with cementum.

Both arvicolids and geomyines originated and evolved during the climatic reorganization that culminated in the Pleistocene Ice Age. The earliest arvicoline, *Prosomys,* and the earliest geomyines, *Pliosaccomys, Parapliosaccomys,* and *Pliogeomys,* were all late Miocene in age. In both groups the major subdivisions appeared in the Pliocene, and superhypsodonty became established in the Plio-Pleistocene. Also in both groups, dentine tracts evolved to compensate for the reduction of the roots that anchored the teeth, and there was a tendency to reorganize the structure of the tooth enamel to resist wear and prolong the life of the tooth.

The three late Miocene genera *Pliogeomys, Pliosaccomys,* and *Parapliosaccomys* all had rooted teeth (Zakrzewski, 1969). *Pliogeomys* had two grooves on the anterior face of the upper incisors and is considered a likely progenitor to *Geomys.* It persisted into the Hagerman deposits of Idaho (Zakrzewski, 1969) at about 3.5 my BP, where it was contemporaneous with the earliest *Thomomys. Thomomys* at that level already had unrooted teeth. *Geomys* with unrooted teeth are also known from Blancan faunas, including the Rexroad, Broadwater, and Sand Draw local faunas. Rootless teeth were thus established at an earlier stage than was generally the case for arvicolines. After the development of superhypsodonty, additional evolutionary changes in these lineages were less profound and included a deepening of the temporal fossa on the lower jaw and slight increases in size in some lineages.

Geomyines were like arvicolines in showing progressive increases in dentine tract height and enamel prism structure. Late Miocene *Pliogeomys* (Figure 10.2A) had thick molar enamel with a radial structure. A slight amount of disorganized central enamel did occur on the tips of the crests. The *Geomys* in the Borchers local fauna from about 2 my BP shows development of poorly organized lamellar enamel (Figure 10.2B). In modern *Geomys* the pattern is more strongly developed, and the enamel is thinned. In modern *Thomomys* the enamel modification is more strongly developed (Figure 10.2C). The enamel changes are

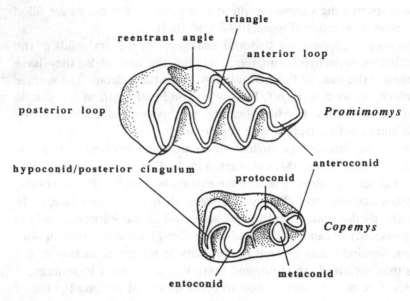

Figure 10.3. Primitive arvicolid (*Promimomys*) left M₁ compared with the same tooth in the cricetid *Copemys* showing homologous structures. *Promimomys* modified from Fejfar et al. (1990); *Copemys* modified from Lindsay (1972).

very similar to those in arvicolines (Figure 10.2D), except the enamel is thicker, and the bent prisms are more sinuous, with a tendency toward a three-layered structure. Radial enamel is retained on leading edges, and *Thomomys* develops tangential enamel on the enamel tip.

The primitive arvicolid lower M₁ can be derived directly from the typical cricetid pattern through increased hypsodonty (Figure 10.3). The anteroconid becomes the arvicolid anterior loop; the first three alternating triangles are the metaconid, protoconid, and entoconid, and the posterior loop is formed by the hypoconid plus the posterior cingulum. In primitive arvicolids the anterior loop is highly crenulated, and as the M₁ becomes more complex, the posteriormost crenulations are recruited to become new alternating triangles; thus the evolution of tooth complexity in the M₁ of the arvicolids mostly involves the anteroconid.

Tooth hypsodonty has increased in essentially all arvicoline lineages, and usually this is reflected in progressively higher dentine tracts through time. In most forms, cement is added to the crown and appears in geologically older forms first as small deposits at the base of the reentrant angles. Enamel prism reorganization shows a similar pattern, with thick radial enamel in earlier forms. The reorganized enamel usually appears

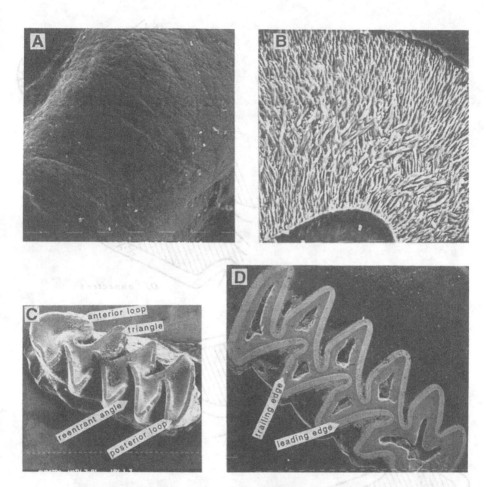

Figure 10.4. *Ondatra annectens* from the Wathena local fauna (?1 my BP). (A) Occlusal surface at the apex of a triangle showing wear striations and the ends of enamel prisms. (B) Crown section of a triangle apex showing a central layer of disorganized lamellar enamel. (C) Occlusal surface of right M^1. (D) Ground section of right M_1 showing pulp cavities and crown cementum.

initially at the apices of the triangle and gradually extends toward the apices of the reentrant angles (Figures 10.4 and 10.5). By the time the tooth becomes superhypsodont, the enamel reorganization is usually completed. Cementum is fully formed, and the dentine tracts reach the surface throughout the useful life of the tooth. Minor changes in tooth size and complexity may occur, but basically most of the easily observable evolution in tooth structure is over. The genus *Ondatra* (muskrats)

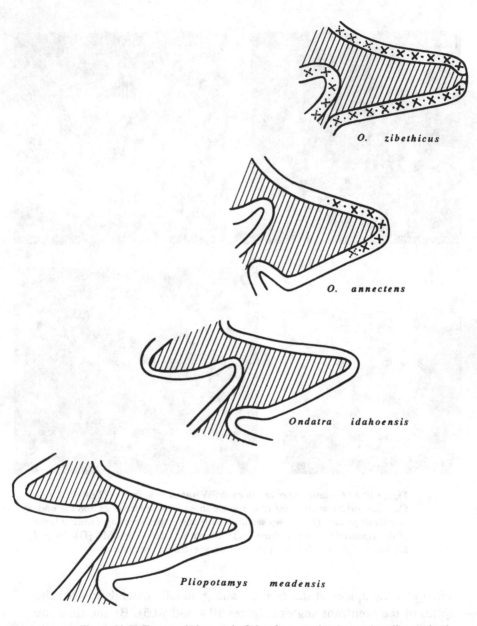

O.   zibethicus

O.   annectens

Ondatra   idahoensis

Pliopotamys   meadensis

Figure 10.5. Temporal changes in *Schmelzmuster* for the muskrat lineage leading from *Pliopotamys meadensis* (~3 my BP) to modern *Ondatra zibethicus* (~1 my BP). Unshaded enamel represent radial enamel, the pattern of crosses represents lamellar enamel.

demonstrates this transition over 3.5 my (Nelson and Semken, 1970; Martin, 1979). *Ondatra* is endemic to North America and shows little evidence of ever having consisted of more than one evolving clade (Martin, 1979). It is thus one of the best subjects for the study of a species lineage. Most workers agree that this lineage consists of *Pliopotamys minor* (3.5 my BP), *P. meadensis* (3 my BP), *Ondatra idahoensis* (2.5 my BP), *O. annectens* (1–0.6 my BP), *O. nebracensis* (0.4 my BP), and *O. zibethicus* (Recent species) (Nelson and Semken, 1970; Martin, 1979). The ages listed here are estimates (Martin, 1979) for the best-known samples. Progressive size increases in this lineage are demonstrable through comparison of individual M_1s from sequential faunas (Figure 10.6). When this sequence of size changes is graphed on an absolute time scale (Figure 10.7), an exponential curve develops, with the rate of change much faster in the last quarter of the total time interval. This pattern is even more evident when dentine tracts are used as a measure of hypsodonty. Crown complexity gives a more gradual pattern, with the most obvious changes in the first quarter of the lineage (Figure 10.7).

Enamel histology follows a similar pattern (Figure 10.5), with *Pliopotamys minor, P. meadensis,* and *O. idahoensis* showing only radial enamel in the alternating triangles. *Ondatra annectens* shows some lamellar enamel (Figure 10.4B and 10.5), and this is extended into the reentrant angles in *O. zibethicus* (Figure 10.5). Thus, enamel reorganization in *Ondatra* is confined to the same time interval as the greatest increase in hypsodonty.

Conclusions

The lifespan of mammalian dentitions can be extended by a wide range of evolutionary strategies, many of which may be used in combination by any single taxon. Behavioral strategies may precede functional ones and include food selection (young shoots and foliage well above the ground). Increased physiological efficiency may also reduce the amount of food that must be chewed to produce the same amount of energy. Janis (1988) shows that digestively more efficient fore-gut fermenters (ruminant artiodactyls) have a lower tooth volume than do hind-gut fermenters (grazing equids) of similar mass. Increases in tooth complexity can also favor digestion by producing smaller food particles, and in any case should reduce the food more effectively on each stroke. Rensberger (1988, p. 352) defines the functional efficiency of cheek teeth

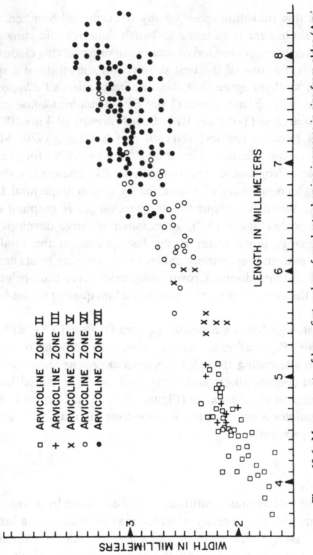

Figure 10.6. Measurements of lower first molars of muskrats showing increases in size from older (Arvicoline Zone II) to younger (Zone VII) arvicoline zones in North America. Note that the size increases are overlapping, except for that area where no sample was available (Zone IV). From Martin (1979, 1984).

ARVICOLINE ZONE II
ARVICOLINE ZONE III
ARVICOLINE ZONE V
ARVICOLINE ZONE VI
ARVICOLINE ZONE VII

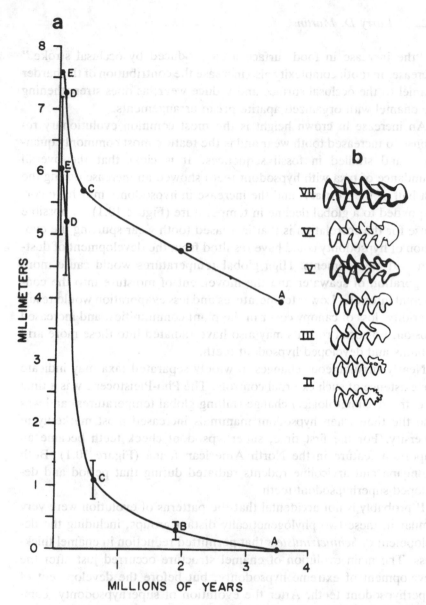

Figure 10.7. Graph (a) shows changes in dentine tract heights on M_1s (lower curve) and lengths of M_1s (upper curve) for muskrats [measurements adapted from Nelson and Semken (1970)]; A, Hagerman local fauna; B, Borchers, and Grandview local faunas; C, Cudahy local fauna; D, Wisconsinan faunas; and E, modern muskrats. Part (b) shows left M_1s of muskrats arranged in stratigraphic sequence, with oldest at the bottom, and North American arvicoline zone listed on the left. From bottom to top: *Pliopotamys minor*, *P. meadensis*, *Ondatra idahoensis*, *O.* cf. *annectens*, *O. nebracensis*, *O. zibethicus*. From Martin (1979, 1984).

as "the increase in food surface area produced by occlusal stroke." Increases in tooth complexity also increase the contribution of the harder enamel to the occlusal surface and reduce wear, as does strengthening the enamel with organized apatite prism arrangements.

An increase in crown height is the most common evolutionary response to increased tooth wear and is the feature most commonly quantified and studied in fossil sequences. It is clear that the overall abundance of taxa with hypsodont teeth showed an increase during the last half of the Tertiary, and the increase in hypsodonty may have corresponded to a global decline in temperature (Figure 10.1). A possible cause for the correlation is that increased tooth wear spurring the evolution of hypsodonty could have resulted from the development of dust-rich arid environments. High global temperatures would cause more evaporation of seawater and the movement of moisture into the continental interiors. Lower temperatures and less evaporation would result in aridity, loss of canopy cover in the plant communities, and increased erosion. Fossorial rodents may also have radiated into these more arid habitats and developed hypsodont teeth.

Nearly simultaneous changes in widely separated taxa may indicate the existence of such external controls. The Plio-Pleistocene was a time of extreme climatological change (falling global temperatures) and was also the time when hypsodont mammals increased most markedly in diversity. For the first time, superhypsodont cheek teeth became an important feature in the North American fauna (Figure 10.1). Both geomyine and arvicoline rodents radiated during that period and developed superhypsodont teeth.

It probably is not accidental that the patterns of evolution were very similar in these two phylogenetically distant groups, including the development of *Schmelzmusters* that permitted reduction in enamel thickness. The main evolution of enamel structure occurred just after the development of extreme hypsodonty, but before the development of superhypsodont teeth. After the evolution of superhypsodonty, compensation for tooth wear could be achieved through an increased rate of crown generation, and in some cases trailing-edge enamel was further thinned, with the loss of some of the complex enamel layers. In extreme cases (Koenigswald and Martin, 1984) such reduction may result in only a thin layer of radial enamel. Unlike the evolutionary pattern seen in the evolution of increased complexity of structural enamel, where new prism arrangements start at the apices of triangles and extend into the reentrants, thinning starts at the apices of the reentrants and extends outward. Thinning can even eventually result in the loss of enamel in entire regions of the tooth, as in some pocket gophers.

The rather exact correlation among increasing crown complexity, hypsodonty, and enamel histology suggests a common overall selective pressure or a linked genetic complex. Differences in rates may exist between functional units, but the common pattern remains, and it can be duplicated in great detail from one arvicoline taxon to another. This suggests external controls or genetic linkage of the different traits. External controls seem more likely in view of the similar evolutionary patterns in geomyines and the independent development of the traits in different taxa of arvicolines. The trends in both arvicolids and geomyines can be broadly explained as evolutionary responses to increasingly arid climates and open vegetation. Whereas the direction of evolutionary change in these rodents corresponds roughly with a long-term climatic trend, the rate and patterns of change are not as clearly linked to climate. However, the major shift to superhypsodont teeth in arvicolines did occur just after the inception of midlatitude glaciation (2.5 my BP) (Easterbrook and Boellstorff, 1984), and evolutionary rates do seem to have increased in that interval.

Acknowledgments

I have benefited greatly from discussions with W. Dort, E. Zeller, G. Dresshoff, C. B. Schultz, L. G. Tanner, T. M. Stout, and L. Smith. C. Burres and C. Cunningham critically read the manuscript. T. J. Meehan and A. Musser assisted with the figures. Part of the research was supported by University of Kansas general research grant 88-108.

References

Easterbrook, D. J., and J. Boellstorff (1984). Paleomagnetism and chronology of early Pleistocene tills in the central United States. In W. C. Mahaney (ed.), *Correlation of Quaternary Chronologies* (pp. 73–101). Norwich, U.K.: Geo Books.

Fejfar, O., P. Mein, and E. Moissenet (1990). Early arvicolids from the Ruscinian (early Pliocene) of the Teruel Basin, Spain. In O. Fejfar and W.-D. Heinrich (eds.), *International Symposium on the Evolution, Phylogeny and Biostratigraphy of Arvicolids* (pp. 133–64). Prague: Geological Survey.

Hershkovitz, P. (1967). *Dynamics of rodent molar evolution: a study based on New World Cricetinae, family Muridae. Journal of Dental Research*, 46(5, pt. 1):829–42.

Janis, C. M. (1988). An estimation of tooth volume and hypsodonty indices in ungulate mammals, and the correlation of these factors with dietary preference. In D. E. Russell, J. P. Santoro, and D. Sigogneau-Russell (eds.), *Teeth Revisited: Proceedings of the VIIth International Symposium on Dental Morphology, Paris, 1986* (pp. 361–87). Memoirs de Musée national d'Histoire naturelle (C), no. 53.

Johnson, J. H., B. L. Hall, and A. S. Dorn (1980). The mouth. In N. V. Anderson (ed.), *Veterinary Gastroenterology* (pp. 337–71). Philadelphia: Lea & Febiger.

Koenigswald, W. von (1980). Schmelzstruktur und Morphologie in den Molaren der Arvicolidae (Rodentia). *Abhandlungen der Senckenbergischen Naturforschenden Gesellschaft,* 539:1–129.

(1982). Enamel structure in the molars of Arvicolidae (Rodentia, Mammalia): a key to functional morphology and phylogeny. In B. Kurtén (ed.), *Teeth: Form, Function and Evolution* (pp. 109–22). New York: Columbia University Press.

Koenigswald, W. von, and L. D. Martin (1984a). Revision of the fossil and Recent Lemminae. In M. Dawson and R. M. Mengel (eds.), *Festschrift for R. W. Wilson* (pp. 122–37). Carnegie Museum of Natural History, Special Publications.

(1984b). The status of the genus *Mimomys* (Arvicolidae, Rodentia, Mammalia) in North America. *Neues Jahrbuch für Geologie Paläontologie,* 168(1):108–24.

Lindsay, E. H. (1972). Small mammal fossils from the Barstow Formation, California. *University of California Publications in Geological Sciences,* 93:1–104.

Martin, L. D. (1975). Microtine rodents from the Ogallala Pliocene of Nebraska and the early evolution of the Microtinae in North America. *University of Michigan Papers on Paleontology. (Hibbard Memorial),* 12:(3):101–10.

(1979). The biostratigraphy of arvicoline rodents in North America. *Transactions of the Nebraska Academy of Sciences,* 7:91–100.

(1984). Phyletic trends and evolutionary rates. In M. Dawson and H. Genoways (eds.), *Festshrift for J. Guilday* (pp. 526–38). Carnegie Museum of Natural History, Special Publications.

(1985). Tertiary extinction cycles and the Pliocene-Pleistocene boundary. *Institute for Tertiary-Quaternary Studies, TERQUA Symposium Series,* 1: 33–40.

(1987). Beavers from the Harrison Formation (early Miocene) with a revision of *Euhapsis.* In J. E. Martin and G. E. Ostrander (eds.), *Papers in Vertebrate Paleontology in Honor of Morton Green* (pp. 75–91). *Dakoterra 3,* South Dakota School of Mines and Technology.

Nelson, G., and H. Semken (1970). Paleoecological and stratigraphic significance of the muskrat in Pleistocene deposits. *Bulletin of the Geological Society of America,* 81:3733–8.

Piperno, D. R. (1988). *Phytolith Analysis: An Archaeological and Geological Perspective.* San Diego: Academic Press.

Prentice, M. L., and R. K. Matthews (1988). Cenozoic ice-volume history: development of a composite oxygen isotope record. *Geology,* 16:963–6.

Rensberger, J. M. (1988). The transition from insectivory to herbivory in mammalian teeth. In D. E. Russell, J. P. Santoro, and D. Sigogneau-Russell (eds.), *Teeth Revisited: Proceedings of the VIIth International Symposium on Dental Morphology, Paris, 1986* (pp. 351–61). Memoirs de Musée national d'Histoire naturelle (C), no. 53.

Rensberger, J. M., and W. V. Koenigswald (1980). Functional and phylogenetic interpretation of enamel microstructure in rhinoceroses. *Paleobiology,* 6(4):477–95.

Walter, H. (1979). *Vegetation of the Earth and Geological Systems of the Geobiosphere.* 2nd ed. New York: Springer-Verlag.

Zakrzewski, R. J. (1969). The rodents of the Hagerman local fauna, upper Pliocene of Idaho. *Contributions of the Museum of Paleontology, University of Michigan,* 23:1–36.

11

Patterns of variation and speciation in Quaternary rodents

ROBERT A. MARTIN

The tenth-anniversary issue of *Paleobiology* (Vol. 11, 1985) summarized much of what paleontology has contributed to evolutionary theory and also raised many provocative issues. Gould (1985), for example, proposed that manifestly different evolutionary processes operate at three temporal levels, which he called "tiers." He defined these tiers as (1) events measured in ecological moments, (2) evolutionary trends over millions of years, and (3) periodic mass extinctions. The theory of punctuated equilibrium was also a dominant theme of Gould's paper, and it remains an important and influential model today, despite Levinton's (1988) articulate criticisms. According to Gould and other proponents of punctuated equilibrium (e.g., Stanley, 1975, 1979, 1985; Vrba, 1983), microevolution and macroevolution are not necessarily reflections of a single underlying process variously expressed as a function of temporal scale; rather, the two have fundamentally different causative agents. Natural selection, for example, may have a higher-order analogue at the species level (Stanley, 1975; Vrba, 1983). In contrast, Gingerich (1983, 1985) and Levinton (1988) generally support the position that natural selection among individuals and adaptation are the primary driving forces at all hierarchical and temporal levels – the standard Darwinian paradigm. If the process of punctuated equilibrium dominates in the history of life, and if most significant morphological change is concentrated during speciation events, then phyletic change should be, as Stanley (1979) suggests, only the fine-tuning of an organism to its environment.

Paleontologists are in a unique position to aid in the resolution of these questions. Although Fortey (1985) proposed that only gradualism is falsifiable (when stasis is rejected), a compendium of case histories at all temporal levels in which stasis has been difficult to demonstrate and significant morphological change has not been coordinated with speciation events would certainly suggest that punctuated equilibrium, in the classic sense, was highly unlikely. The quality of the fossil record will, of course, impact on this controversy, and it seems likely that greater control and more examples can be generated from the Quaternary epoch than from any other period of time (e.g., Barnosky, 1987). Although certainly it will prove consequential to investigate lineages and clades of large animals, the statistical prevalence of small mammals in Quaternary sediments makes them better test subjects for most evolutionary studies.

Since the introduction of sediment screen-washing by C. W. Hibbard in 1928 (Hibbard, 1948, 1975) it has become clear that large samples of small mammals can be obtained from Quaternary deposits – deposits that in the past usually were rapidly excavated only for the larger, more visually impressive taxa. These "micromammals" predominantly include bats, insectivores, and rodents. The fossilization potential of bats is highly dependent on depositional context, and the sectorial nature of bat and insectivore dentitions makes accurate measurement difficult when compared with the flat and prismatic molar patterns of, for example, arvicolid rodents. Rodents are often the most common mammalian taxa represented in fossil accumulations and offer fewer problems in obtaining reliable measurement data. There is a large body of information on extant rodents available for consultation, and there have also been a few published studies of variation in late Pleistocene rodent sequences that can be used as controls for investigations of change in extinct taxa examined through longer periods (e.g., Barnosky, 1990; Martin and Prince, 1990; Nadachowski, 1990). In view of the ready availability of primary data, it is hardly surprising that many students interested in the tempo and mode of Quaternary mammalian evolution have investigated rodent lineages.

Following a brief section in which taxonomic philosophy is considered, I review some of the literature dealing with dental variation in modern and Quaternary species, examine long- and short-term changes in dental size and morphology, integrate these data into the context of modern evolutionary theory, and conclude with a brief description of a comprehensive model of phenotypic evolution at the species level.

Evolutionary rates in darwins (Haldane, 1949) are now known to be inversely related to the intervals over which they are measured (Gin-

Table 11.1. *Evolutionary rates for Quaternary rodents*

Species	I	d	d_t
Clethrionomys glareolus	0.000015	>50,000	0.94
Microtus pennsylvanicus	0.015	4.47	0.07
Microtus nivalis	0.020^a	3.35	0.07
Microtus nivalis	0.010	5.00	0.06
Ondatra zibethicus	3.75	0.20	0.73
Ondatra zibethicus	0.58^a	0.68	0.40
Ondatra zibethicus	0.32^a	0.19	0.06
Ondatra zibethicus	0.016	12.0	0.21
Arvicola terrestris	0.35^a	0.91	0.33
Onychomys, all lineages	3.4	0.21	0.70
Onychomys leucogaster	0.02^a	4.43	0.10
Sigmodon, all lineages	3.75	0.06	0.23

Note: d = rate in darwins; d_t = rate corrected to a common interval length of 1 my; I = interval length in millions of years.
^{a}Approximate.

gerich, 1983). This observation and its implications will be examined at some length later in this chapter, but a descriptive comment is necessary at this point. Evolutionary rates were calculated for a number of rodent lineages for this study and then standardized to an interval length of 1 million years (my) (Table 11.1). Throughout the text, rates scaled to the common 1-my interval length are identified as "d_t." Rates calculated directly from Haldane's equation are reported simply as "d."

Taxonomic philosophy

As with most fossil mammalian taxa, rodent dentition has been the skeletal component most intensively studied, and this is likely to continue. Rodent molars fossilize in abundance and very often demonstrate species-specific patterns. However, just as a closely focused view of the subatomic world reveals a dynamic and quirky system acting according to principles of quantum dynamics, so close scrutiny of dental variation among populations of extinct and extant rodents reveals not discrete dental entities, but morphotypic swarms. The investigator intent on studying fossil vole taxonomy or using fossil voles to test evolutionary hypotheses must expect to spend considerable time examining variation in modern species before the fossil ones will become comprehensible. For example, Paulson (1961), Zakrzewski (1985), and Graham and Semken (1987) considered the separation of *Microtus pinetorum* from *M. ochrogaster* to have been virtually impossible. In a recent paper (Martin, 1991) I presented a suite of dental features for separating these species. Certainly there is some overlap in morphology between *M. ochrogaster* and *M. pinetorum*, as there is between many species of *Microtus*. In spite of this, the variations expressed for most valid species that I have examined lie within bounds that are recognizable for each species.

The foregoing point has considerable implications for the study of evolutionary pattern and process. Most neontologists and paleontologists agree that species exist and have boundaries in time and space [for the extreme case, see Ghiselin (1974)]. Some paleontologists eschew the use of modern taxa to test taxonomic and phylogenetic hypotheses, but ignoring modern taxa is essentially the same as conducting an experiment without a control (Martin, 1987; Martin and Prince, 1990). Although there are limitations to the comparative method, particularly in its application to phyletic sequences, we need to determine the range of variations in modern organisms in order to make at least elementary sense of the fossil record.

Once character variations for modern taxa have been established (this is, of course, an ongoing process), there can be disagreement on the application of this information to fossil materials. Gingerich (1985) argues, for example, that the limits of variation in modern taxa be used to determine the positions of arbitrary species boundaries in a phyletic sequence. He makes the point as follows:

> Inclusion in the same species of organisms differing in size or form by factors greater than those characterizing living species reduces the functional comparability of species. This approach would be reasonable only if species in the faunal and ecological sense could be shown to be static morphological entities. . . . [1985, p. 38]

Rose and Bown (1986, p. 121) further testify as follows:

> To regard . . . successional forms as a single species simply because intermediates linking them are now known would greatly amplify the morphological limits of paleontological species compared to biospecies and would, in effect, reject widely-held criteria and procedures for establishing species in the fossil record.

Rose and Bown (1986) also cite a similar opinion by Simpson (1961). I have no argument with Gingerich's or Rose and Bown's proposition that considerable and significant change can come about through phyletic evolution; the muskrat fossil record that will be examined later provides another good example of this phenomenon. These authors are also consistent in the application of currently recognized methods for species discrimination in the fossil record. However, there is a confusion of ontological units in this treatment, and, I would contend, it results in a biased estimate of biological diversity and confounds the use of the fossil record to examine evolutionary pattern and process.

The standard and classical approach to fossil species, which dates back at least to Cuvier's development of the comparative method, has been to compare variation in a fossil sample with variation in modern relatives. If the expression of variation in the fossil sample falls outside that of geographic variation in the modern species, then the fossil sample is generally given a new name at the appropriate taxonomic level. Most paleontologists have recognized that a fossil sample may not be wholly analogous to a modern museum sample because of a variety of influences, mostly dealing with the original sampling and depositional environments. There has been little recognition that a museum sample usually also represents a mixture of generations. Nevertheless, to the extent that the coefficients of variation of fossil and modern comparative materials are not outrageously different (e.g., 5.0 versus 45.0), it is not unreasonable to make this comparison. However, through the years there has been a confusion of ontological units. Instead of confining the "principle of functional comparability" to like units, namely, modern samples compared to individual or multiple samples from a single temporal zone, there has been a tacit acceptance that the same static museum samples could be compared to an entire time-transgressive phyletic sequence. The rationalization has been, I suppose, that a phyletic sequence can be viewed as a time-static or "horizontal" ring species phenomenon turned vertical (time-transgressive). But this comparison cannot legitimately be made. A set of modern museum samples cannot be analogous *both* to a set of fossil samples from a single past interval and to the entire set of samples from a full phyletic lineage (Figure 11.1).

The analogous units for comparison with temporal variation can only be (1) samples of modern species examined through historical time and (2) samples of modern species examined through brief times during the Quaternary or some other time period. Where these investigations have been done, the results indicate that dental morphology is somewhat fluid. Corbet (1975) has shown that considerable variation can occur within a few decades. Differences expressed in the dentition of *Clethrionomys glareolus* from Loch Tay, Perthshire, between the years 1955

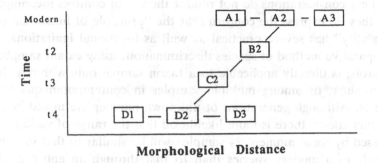

Figure 11.1. Comparison of samples in a phyletic sequence. A1–A3 represent three modern museum samples, each composed of *n* generations; B2, C2, and D1–D3 represent fossil samples from three temporal intervals (t2-t4), each also composed of *n* generations. Where more than one sample is represented at a given interval, this is understood to represent geographic variation during that period. Geographic samples from the modern interval and those from, for example, t4, if their coefficients of variation are not considerably different, are both analogous and roughly homologous ontological units. Samples from different geographic regions in modern time are neither analogous nor homologous to a set of samples taken through the entire phyletic sequence (that is, samples A1-A2-A3 cannot be homologous to *both* B2-C2-D2 and D1-D2-D3) and therefore cannot legitimately be used, by direct comparison, to determine the taxonomic status of sequence segments. As noted in the text, it is the author's considered opinion that such sequences represent a continuum within a single species.

and 1972 are at the level often used to recognize species in the fossil record. Martin and Prince (1990) and Barnosky (1990) also demonstrated directional change, sometimes of a mosaic nature, during a few thousand years of the Quaternary.

Furthermore, why should the expression of variation at one moment in time (the Recent period) be taken as the benchmark of variation for the entire species throughout its geological history? The unstated implication is that at no time in the past would the modern species have displayed a greater range of variation, but we know this is not true. All one has to do to prove this is to examine the observed range of measurements in any well-studied clade of organisms; see, for example, the ranges of variation in M_1 length in extinct and extant *Sigmodon* (Martin, 1986). For modern species that have experienced a late Pleistocene or early Holocene bottleneck resulting in genetic and morphological uniformity, we might predict variability to be considerably expanded in earlier samples, the reverse of the expectation under the principle of functional comparability.

These considerations do not render the use of controls meaningless, but they do express the concern that the "principle of functional comparability" has severe practical as well as functional limitations. The comparative method of species discrimination, using extant samples as controls, is directly applicable to a taxon sample only within a single paleofauna, or among multiple samples in contemporaneous paleofaunas. Although generations of unknown number are mixed in fossil accumulations, there is more likelihood that the range of variation expressed by these momentary samples will be similar to that in a given sample of a modern species than to that through an entire phyletic sequence. The arbitrary recognition of species in a phyletic series based on static geographic variation in modern, related species is inappropriate.

In spite of this, and with full knowledge that the comparison is among unlike quantities, modern samples must be used for rough comparison with samples in a phyletic series in order to give functional value to the word "significant" as used to express morphological change in such a sequence. Gould and Eldredge (1977) and Stanley (1979) have argued that variation exhibited in the phyletic series of Eocene condylarths and primates reported by Gingerich (1985) is not "significant." I have elsewhere (Martin and Prince, 1990) noted the additional, perhaps unresolvable, problems in dealing with entirely extinct taxa, but comparison of geographic variations in the modern muskrat allows me to conclude, unequivocally, as discussed later, that morphological change through the entire muskrat phyletic sequence is "significant." In other words, the term "significant," as applied to problems of interpretation in organismal lineages, must be defined as follows in order to have any utility and meaning: *Morphological change in a phyletic sequence can be considered significant if variation in the sequence is considerably greater than the expression of intraspecific morphological variation and/or at least equal to interspecific variation as demonstrated by the closest living relatives of the taxon under investigation.* In some cases, the appropriate comparison might be with a late Pleistocene sample, rather than a modern sample. There is some difficulty in choosing the proper units for comparison, because at least conceptually, it is unclear that either unit is wholly adequate. When dealing with the species concept, it might appear logical to consider only differences between related modern species, and discount intraspecific geographic variation. However, some modern species are more conservative in character divergence than others, or may have recently experienced a genetic bottleneck; one need only compare the minimal dental variation in *Microtus umbrosus* to the expansive variability in *M. pinetorum* and *Chionomys nivalis* to see this adequately. Thus, at least temporarily, it appears more reasonable to include both intraspecific and interspecific categories of morphological variation as benchmarks of significance. When dealing with an extinct group without close phylogenetic or functional ties with modern taxa, there are no "calibration" taxa, and significance can be determined only through comparison with general mammalian patterns of variation.

Species are not, as Gingerich (1985) correctly notes, static entities. They are serious players in a game for energy capture (Van Valen, 1973), and they function as members of one or more complex animal communities. As the environment changes, and through the actions of natural selection and stochastic origination, population morphologies are determined and manifested. Populations of modern species may have changed to measurable degrees between the time this article was written and its date of publication. Controls are both necessary and useful, but it is important to remember that evolution is a dynamic process, and that yesterday's controls are today's fossils.

The recognition of multiple species in a phyletic sequence also artificially bloats the record of biological diversity and confuses any objective attempt to determine evolutionary patterns in diversity. In the final analysis, a phyletic sequence is the story of a single taxon undergoing an "incomplete metamorphosis," to borrow a phrase from entomology. There has been no net increase of taxa. Once phyletic sequences are collapsed, new evolutionary principles may be recognized. I have elsewhere suggested, for example, that mammalian species of large size may have longer species lifespans than those of small size (Martin, 1992). This hypothesis would not have been obvious without a new conceptual approach based upon treating phyletic sequences as single-species lineages.

One can also argue that different taxonomic treatments are required for phyletic change and cladogenesis because two quite different evolutionary processes are at work. If, as has been suggested, cladogenesis is primarily a probability function of population fragmentation, perhaps driven by body size (Gould and Eldredge, 1977; Martin, 1992), then any unique and discontinuous morphological characters that develop through lineage splitting can and should be recognized as distinct from those that arise in phyletic sequences through intraspecific mosaic character evolution, as discussed later.

I recommend collapsing all known or highly suspected phyletic sequences into single-species lineages, and applying the "informal lineage segment" trinomial, as introduced by Krishtalka and Stucky (1985) and modified here. They provided this system to recognize chronocline segments in Eocene *Diacodexis*. In their scheme, an informal third name is given to a segment, separated from the species name by a hyphen. No formal type material is designated, but a representative sample from a single locality is identified as "typical" of the segment. I suggest using a slash (/) rather than a hyphen to isolate the third name; otherwise there can be considerable confusion when species names are listed in succession. The term *chronomorph* is here substituted for "informal lineage segment" because, as noted later, there is reason to expect, especially to the extent that intraspecific mosaic evolution dominates in a phyletic sequence, temporal overlap of populations with derived versus underived features. Just as it is apparent that ancestral species can survive in contemporaneity with their descendants, so morphological features should not be expected to spread instantaneously through all populations in a species. A progressive increase from a five- to a six-triangle M_1 morphotype in Great Plains *Microtus pennsylvanicus* during the past 250,000 to 300,000 years was documented by Davis (1987). At some future time the six-triangle M_1 form may dominate all populations, but for now there is a geographic pattern representing a range of variation between the two morphotypes.

After the full terminology is introduced, the chronomorph may be identified by its trinomial form (e.g.; /minor for *O. zibethicus* /minor; see the later section dealing with phyletic change in *Ondatra zibethicus*). In order for the reader to better compare the more standard taxonomic treatment of a lineage with that proposed here, the chronomorph approach is applied only to mammalian groups I have personally studied in detail.

Dental variation in modern and late Quaternary rodents

Despite the long reliance of paleontologists on dental structures to infer relationships and phylogenetic patterns among extinct mammals, neo-mammalogists have rarely examined dental variation. For example, no systematic treatment of dental variation in extant arvicolids has been published since Hinton's (1926) monograph. Some information can be gleaned from isolated papers, such as the following: *Geomys,* White and Downs (1961), Wilkins (1984); *Onychomys,* Carleton and Eshelman (1979); *Reithrodontomys,* Hooper (1952); *Peromyscus,* Hooper (1957);

Sigmodon, Martin (1979); the Lemminae, Agadzhanyan (1973), von Koenigswald and Martin (1984); the Arvicolinae, Goin (1943), Ursin (1949), Guilday and Bender (1960), Guilday (1982), Corbet (1964, 1975), Guthrie (1965), Oppenheimer (1965), Semken (1966), R. L. Martin (1973), Smartt (1977), von Koenigswald (1980), Nadachowski (1982, 1990), Kratochvil (1983), Davis (1987), Martin (1987, 1989a), Harris (1988), Nelson and Semken (1970); sciurids, Bryant (1945); zapodids, Martin (1989b); murids, Musser (1981), Freudenthal and Suarez (1990). This list is not meant to be exhaustive, and I apologize to authors whose work I may have overlooked. I do not doubt that many investigators currently working with fossil rodents are very knowledgeable of variation in modern taxa, but more of this information needs to be published.

Nomenclature

Rodent dental nomenclature has become a nightmare for the uninitiated, but much of the terminology is justified. Only with precise names applied to dental structures can rodent specialists study and communicate effectively about the evolution of those structures. Dental topology for some rodent groups is illustrated in Figures 11.2 and 11.3. Martin and Prince (1989) simplified the elaborate and cumbersome terminology of Hershkovitz (1962) for *Sigmodon* first lower molars, preferring instead to apply the terminology of arvicolid M_1s. Dental nomenclature for arvicolid molars follows van der Meulen (1973, 1978), Rabeder (1981), Martin (1987), and Heinrich (1990).

Various measurements and ratios introduced by van der Meulen (1973) have come to be important in the expression of evolutionary pattern in arvicolid rodent molars, and these are illustrated in Figure 11.2.

Features of the enamel banding of arvicolid molars have considerable utility, both in classification and in evolutionary analyses, and the reader is referred to von Koenigswald (1980), von Koenigswald and Martin (1984), and Heinrich (1990) for descriptions and illustrations of the characters studied from the histological and macroscopic perspectives.

In this chapter, the terms "monophyly," "paraphyly," and "polyphyly" follow the definitions given by Schoch (1986, p. 181).

Intraspecific variation in modern rodents

We probably know more about dental variation in the North American meadow vole, *Microtus pennsylvanicus,* than in any other rodent. Davis

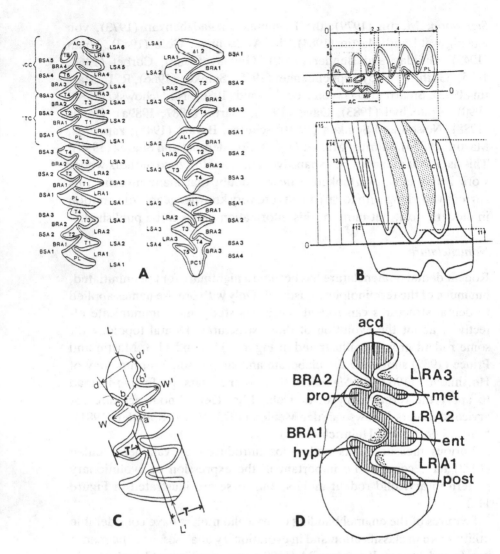

Figure 11.2. Examples of rodent dental nomenclature and measurements. (A) Topography of lower and upper *Ondatra zibethicus* molars. Adapted from van der Meulen (1973). (B) Measurements from a *Mimomys* molar. From Chaline and Laurin (1986). (C) Standard measurements used to construct ratios in arvicolid dental evolutionary studies. Adapted from van der Meulen (1978) and Martin (1987). (D) Arvicolid terminology applied to a cricetine M_1. From Martin and Prince (1989). See references for details and explanations.

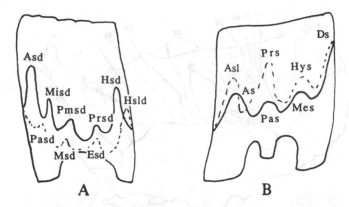

Figure 11.3. Terminology associated with the linea sinuosa of the lower (A) and upper (B) first molars in arvicolids: Asd, anterosinuid; Msd, mimosinuid (sinuid associated with the "*Mimomys*-kante"); Pmsd, primosinuid; Prsd, protosinuid; Hsd, hyposinuid; Pasd, parasinuid; Msd, metasinuid; Esd, entosinuid; Hsld, hyposinulid; As, anterosinus; Pas, parasinus; Mes, metasinus; Asl, anterosinulus; Prs, protosinus; Hys, hyposinus; Ds, distosinus. For further illustrations and discussion, see Rabeder (1981).

(1987) illustrated geographic variations in the complexity of the first lower molar, showing that populations on the Great Plains have the highest proportions of six or more triangles. The average numbers decrease to the north and east, as the populations frequent more heavily forested regions (Figure 11.4). Although it is tempting to speculate that the most complicated M_1s are also the largest, Martin and Prince (1990) found no statistical correlation between M_1 length and the five- or six-triangle morphotype, and Davis (1987, Figures 11–12) also showed that the largest molars were found in southwestern Nebraska and in the southern Appalachians, where six- and five-triangle morphotypes dominate, respectively. As the difference between the five- and six-triangle conditions is determined on closure of the dentine isthmus connecting these triangles, rather than on the addition of structures, it is possible that a correlation could be identified between complexity and M_1 length if one examined length as a function of odd- or even-numbered triangles (e.g., 5–7 in *M. pennsylvanicus;* 3–5–7 in *M. pliocaenicus* through one of its descendants with more complicated molars). In support of this idea, Morlan (1989, p. 153) noted that for remains of the singing vole *Microtus miurus* from the late Pleistocene Bluefish Cave localities, "the larger the molar, the less it resembles typical modern singing vole teeth and the more it tends to resemble modern meadow voles."

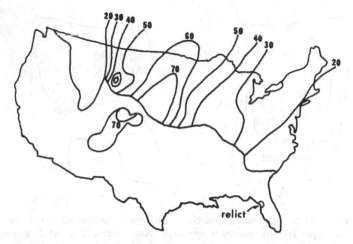

Figure 11.4. Contour map for distribution of *Microtus pennsylvanicus* first lower molars with six closed triangles. Adapted from Davis (1987).

Barnosky (1990) documented a north-south increase in M^3 length in modern meadow voles. Additionally, he showed that M^3s to the west are relatively narrower than those to the east. The perimeter of the posterior loop exhibits a more complicated geographic distribution – smaller from north to south and from west to east. LRA4 and BRA4 of M^3 were present and filled with cement in approximately 4–5% and 18% of the specimens examined, respectively (Guilday, 1982).

Guilday (1982) found that T5 (the "*pennsylvanicus* fold") on M^2 in *M. pennsylvanicus* was present at near 100% in all samples examined.

Geographic variation was summarized for a few European vole species by Kowalski (1970). The M^3 in *Clethrionomys glareolus* and *Microtus arvalis* varies between a "simplex" and "normal" type in a clinal fashion.

Nadachowski (1990) reported intraspecific variation in 23 subspecies of modern *Chionomys nivalis*. He examined the standard series of morphometric measures of van der Meulen (1973, 1978) and also developed a numerical formula for each sample of first lower molars based on nine observed tendencies. The snow vole demonstrates considerable variation in size and morphology, and clinal changes are only locally developed. There is, for example, an east-west cline in the European Alps, with the smallest specimens found in southwestern regions. However, averages from other samples appear to demonstrate a mosaic pattern, as has been documented for *M. pennsylvanicus* by Barnosky (1990, and

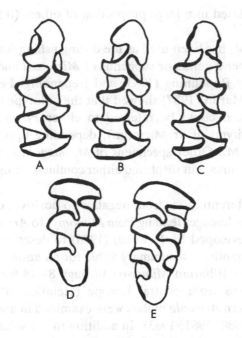

Figure 11.5. The extremes of geographic variation in the first lower molar (A–C) and third upper molar (D, E) of the snow vole, *Chionomys nivalis*. From Nadachowski (1990).

Chapter 3, this volume). Nadachowski speculates that disjunct distributions and competition influence size variation in *C. nivalis*. Populations from western Europe retain the most underived dental pattern, where a small proportion of each sample show the classic ratticepoid morphotype, with only four closed triangles and an asymmetrical anterior cap. The most advanced populations are found today in the Tatra and Caucasus mountains, where some M_1s may have six well-developed triangles, and T5-6 may be confluent and closed from the anterior cap (Figure 11.5).

Martin (1987, 1991) examined geographic variation in *Microtus pinetorum*. He found little difference in numbers of lingual and labial reentrant valleys in M_1 and M^3 or closure of T3-4 on M_2. T5 was occasionally developed on M^2 in a few specimens of *M. p. nemoralis,* and in this subspecies AC2 was closed from T4-5 in approximately 50% of the specimens (versus none in the other subspecies). As in *M. ochrogaster* (Harris, 1988; Martin, 1991), AC2 may be widely confluent with T4-5 on M_1 in certain subspecies (0.90 open in *M. p. parvulus, M.*

p. pinetorum) or tightly closed in a large proportion of others (0.50 in *M. p. nemoralis*).

Confluence of T4-5 on M_1 has been used as the distinguishing feature of *Pitymys* (treated as either a genus or subgenus of *Microtus*) and the tribe Pitymyini as used by Repenning (1983) and Repenning, Fejfar, and Heinrich (1990), but Martin (1987) showed that the holotype specimen of *Microtus oaxacensis* could be referred to either *Pitymys* or *Microtus* (or the tribes Pitymyini or Microtini), depending upon the choice of the left or right M_1. Other specimens of *M. oaxacensis* were more consistent in their features, but displayed either confluent or tightly closed T4-5 on M_1.

A change in enamel differentiation, from negative to positive, characterizes dental evolution in lineages leading from *Mimomys* to *Arvicola*. Enamel quotients were developed by Heinrich (1978) to describe this modification. The SDQp quotient, a summary value for an entire sample, ranges from 138.59 for Biharian *Mimomys* through 84.48 for late Toringian *Arvicola terrestris* from central Europe (Heinrich, 1990). Enamel quotients for modern *Arvicola* molars were examined in a series of papers by Kratochvil (1980, 1981, 1983). In addition to considerable variation, he found that the SDQp quotient increased with increasing altitude. For example, a population of *Arvicola terrestris exitus* from 1,160 m in Switzerland exhibited a quotient (110) that had been recorded in the fossil record only in *A. cantiana,* an extinct species preceding *A. terrestris.*

Developmental anomalies are also known. Kowalski (1970) illustrated an upper dentition of *Lagurus lagurus* from Siberia in which the left side had three molars, and the right four.

The genetic fields governing variation in arvicolid dentitions appear to be similar in that molars other than the M_1 and M^3 remain conservative and have only minimal taxonomic utility.

Rodent molars are notorious for changing occlusal patterns with wear (Figure 11.6). Stephens (1960) illustrated this process in a series of known-age *Ondatra* specimens, and further discussion and illustration may be found in Rabeder (1981), for a variety of arvicolids. Kratochvil (1980, 1981, 1983) reported developmental changes in the thickness of enamel in the dentition of modern European water voles, *Arvicola terrestris.* He noted that enamel quotients decreased with increasing age of the animals sampled.

The roots of rodent molars have some taxonomic utility, as one sees both fusion and addition through time (R. A. Martin, 1979; Rabeder,

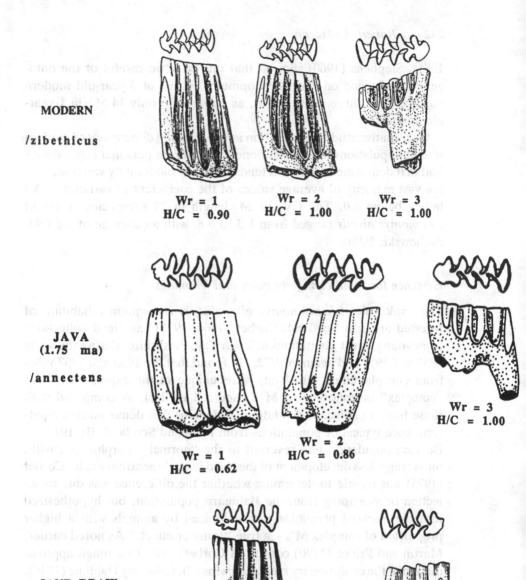

MODERN

/zibethicus

Wr = 1
H/C = 0.90

Wr = 2
H/C = 1.00

Wr = 3
H/C = 1.00

JAVA
(1.75 ma)

/annectens

Wr = 1
H/C = 0.62

Wr = 2
H/C = 0.86

Wr = 3
H/C = 1.00

SAND DRAW
(3.0 ma)

/meadensis

Wr = 1
H/C = 0.29

Wr = 2
H/C = 0.28

Figure 11.6. Variation of the first lower molar in part of the phyletic sequence leading to the modern muskrat, *Ondatra zibethicus*. Wr, wear stage; H/C, ratio of hyposinuid height to crown height; H/C = 1.0 when the hyposinuid breaks through at the occlusal surface. Sand Draw illustrations from Hibbard (1972); modern *O. zibethicus* modified from Stephens (1960).

1981). Stephens (1960) showed that one must be careful of the onto-genetic influence on root development; 56.2% of 3-year-old modern muskrats had three roots on M^1, as opposed to only 14.5% in 1-year-old animals.

Quantitative studies of variation in linear dental dimensions of modern rodent populations are almost nonexistent. From personal experience I find that dental measurements tend to be only moderately variable, with the vast majority of average values of the coefficient of variation (CV) falling below 6.0. The CV for M_1 length in 23 subspecies of extant *Chionomys nivalis* ranged from 1.3 to 6.8, with an average of 4.8 (Nadachowski, 1990).

Evidence for rapid change in voles and lemmings

The bank vole, *Clethrionomys glareolus*, is a common inhabitant of forested areas in Scotland. Corbet (1964, 1975) examined collections from young forest plantations at Loch Tay, Perthshire, during the years 1955 and 1957 and again in 1972. He found that the 1955 and 1957 voles from one plantation (Borland) were unique in the expression of the "complex" morphotype of M^3 (discussed earlier), as compared with those from an adjacent plantation (Balnearn) that demonstrated a pattern more typical of populations from mainland Scotland. By 1972 the Borland population had reverted to the "normal" morphotype, with, on average, less development of the third lingual reentrant angle. Corbet (1975) was unable to determine whether the difference was due to selection or swamping from the Balnearn population, but hypothesized that the Borland plantation was colonized by animals with a higher proportion of complex M^3s – a true "founder effect." As noted earlier, Martin and Prince (1990) converted Corbet's data to a rough approximation of the evolutionary rate in darwins (d), following Haldane (1949), finding that the rate change was in excess of 50,000 d, a level encountered only in laboratory selection experiments (Gingerich, 1983). Even when scaled to a common interval length of 1 my, the evolutionary rate for *C. glareolus* is rapid (0.94 d_t).

Morlan (1989) recorded intraspecific size reductions in the first lower molar in four species of arvicolines from Bluefish Cave 1, Yukon Territory in northwest Canada. The late Pleistocene component of the Bluefish Cave 1 sequence appears to represent about 12,000 years, from about 24,000 years BP to 12,000 years BP, based on unpublished AMS radiocarbon dates (Morlan, 1989). The taxa represented were *Dicrostonyx torquatus, Lemmus sibiricus, Microtus xanthognathus,* and *M. miurus.* The data for the collared lemming were the most impressive, registering a mean change in length from about 3.6 mm (Pleistocene levels 5, 6, and 7) to 3.0 mm in modern specimens (means are estimated from Morlan's Figure 3). Only one specimen of *D. torquatus* was recorded from the highest Pleistocene zone (level 4), and the length of this tooth is within the range for the smallest modern collared lemmings, and outside the range of the M_1s from the lower Pleistocene units. If this molar was not reworked from upper Holocene sediments, it suggests that the dwarfing event was relatively instantaneous, occurring during late Pleistocene time. This is in contrast to the findings for other species, in which changes were less dramatic and may, in fact, be indicative of stasis when the appropriate statistical tests are performed. Because radiometric dates were not published for specific sampled zones, I made no attempt to calculate rates of change.

Martin and Prince (1990) provided data for intraspecific size change in *Microtus pennsylvanicus* molars from successive levels in an Alabama cave spanning approximately 15,000 years of the latest Wisconsinan glacial period. The rate for mean length reduction of the first lower molar was 4.47 *d,* which, when scaled (0.07 d_t), is not particularly fast.

Nadachowski (1984) reported arvicolid rodents from a series of superposed strata in Bacho Kiro Cave, Bulgaria. Two radiocarbon dates for level BI average 30,925 years BP. The top layer, C, was deposited at an unknown time in the Holocene. Using a conservative figure of 10,000 years for level BI yields 0.02 Ma (million years ago) for the denominator in Haldane's equation. Although the number of specimens recovered from level C was not large for any of the species, none of the mean values of measurements are unsynchronized with trends developed through the preceding layers, and they are therefore considered reasonable. Mean values and numbers of specimens (in parentheses) for M_1 lengths of *Microtus nivalis* from levels BI and C are, respectively, 2.94 mm (45) and 2.75 mm (6). The rate of change for this size decrease is 3.35 *d.* According to Nadachowski (1982), the most pronounced change in M_1 length actually comes between levels BI and BII ($\bar{x}$ M_1 length = 2.78 mm, N = 6), where BII is a layer attributed to full-glacial conditions lying stratigraphically between levels BI and C. Unfortunately, there were no ^{14}C dates reported for level BII, and I have to this point ignored the data from that zone. If Nadachowski's evaluation

Table 11.2. *Average body mass (W) and average M_1 lengths for selected samples of modern North American arvicolid rodents*

Species	N^a	W (g)	L M_1 (mm)	Source
Microtus pinetorum	7	17.2	2.64	Original data
M. pennsylvanicus	100	38	3.1	Martin and Prince (1990)
M. richardsoni	13	106	3.6	Martin (1974); Ludwig (1984)
M. ochrogaster	98	51	3.0	Original data; McNab (1988)
Clethrionomys gapperi	63	23	2.1	Guilday et al. (1964); McNab (1988)
Synaptomys cooperi	25	36	2.8	Guilday et al. (1977); Linsey (1983)
Neofiber alleni	2	271	5.1	Original data; Birkenholz (1972)
Ondatra zibethicus	68	840	7.1	Martin and Tedesco (1976); McNab (1988)

$^a N$, number of specimens on which M_1 length was measured.

of level BII is correct, and assuming a full-glacial date of about 20,000 years BP, the rate change between levels BI and BII would then be 5.00 d. Scaled rates are presented in Table 11.1.

Long-term patterns of character and size change

Phyletic evolution in the late Pliocene and Pleistocene muskrats of North America

Until recently the muskrat fossil record was considered to represent a phyletic series including at least six species and two genera. Based on the taxonomic philosophy explained earlier in this chapter, I have collapsed the named forms into a single species, *Ondatra zibethicus* L., recognizing the following chronomorphs: *O. z. /minor, O. z. /meadensis, O. z. /idahoensis, O. z. /annectens,* and *O. z. /zibethicus* (hereafter, each chronomorph will be abbreviated by its trinomial form, e.g., */annectens*). I have also derived an equation that accurately estimates body mass in modern arvicolids based on M_1 length, as follows:

$$W = 0.71 \, L^{3.59}, \tag{11.1}$$

where W is mass in grams, and L is M_1 length in millimeters. The data used to generate this equation are provided in Table 11.2.

Muskrats increased in average mass from about 103 g in the latest Pliocene to over 1,600 g at the height of the Wisconsinan glaciation [this large form was named by Lawrence (1942) as an extinct subspecies of the modern muskrat, *Ondatra zibethicus floridanus,* based on material

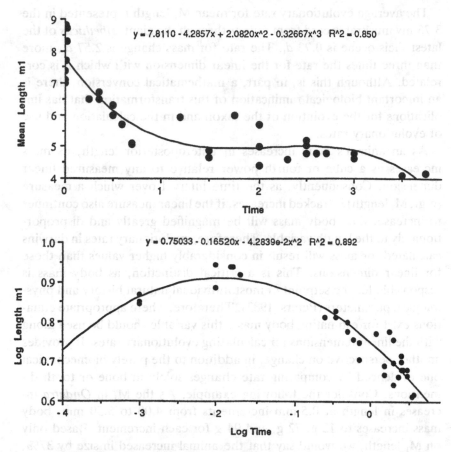

Figure 11.7. Polynomial fits of untransformed (A) and $\log_{10}$-transformed (B) mean first lower molar lengths for select North American samples of the *Ondatra zibethicus* phyletic sequence.

from the Ichetucknee River of Florida, but is here subsumed under the chronomorph *O. z. /zibethicus*]. They then decreased in average size to approximately 840 g sometime between the latest Pleistocene and Recent time (it is unclear whether the last dwarfing event was mediated by human culling or natural selection). This size change was somewhat episodic and is best characterized by a second- or third-order polynomial (Figure 11.7). Without an accepted underlying regularity (e.g., exponential decay of a radioisotope), there is no reason to prefer one mathematical construct over another. A third-order polynomial of the log-transformed data provides an expanded graphical expression of the late Pleistocene dwarfing event.

The average evolutionary rate for mean M_1 length represented in the 3.75-my muskrat record from /*minor* through the giant /*zibethicus* of the latest Pleistocene is 0.73 d_t. The rate for mass change is 2.57 d_t, more than three times the rate for the linear dimension with which it is correlated. Although this is, in part, a mathematical conversion, there is an important biological ramification of this transformation that has implications for the evolution of the taxon and in the calculation and use of evolutionary rates.

As an animal's body increases in anteroposterior length, its mass increases as a cube or fourth power relative to any measured linear dimension. Consequently, as the time interval over which a measure (e.g., M_1 length) is tracked increases, if the linear measure also continues to increase, then body mass will be magnified greatly and disproportionately to the length variable. Therefore, evolutionary rates in darwins calculated for mass will result in considerably higher values than those for linear dimensions. This is a critical distinction, as body mass is responsible for the setpoint of most individual natural history and physiological parameters (Peters, 1983). Therefore, where appropriate equations exist for estimating body mass, this variable should be used along with the linear dimensions for calculating evolutionary rates. It provides another perspective on change, in addition to the purely biomechanical one evidenced by comparing rate changes solely in bone or tooth dimensions. Consider the following example. As the M_1 in *Ondatra* increases in length at 0.5-mm increments from 4.00 to 5.50 mm, body mass increases to 12 g, 72 g, and 94 g for each increment. Based only on M_1 length, we would say that the animal increased in size by 37%, whereas its mass, the biologically important unit and most relevant indicator of body size, increased by 123%.

As noted first by L. D. Martin (1979) and as indicated in Figure 11.7, size change was not continuous throughout muskrat history; most of it has been concentrated in the past 600,000 years. The evolutionary rate for average M_1 change from Cudahy time (0.60 Ma) to the time of the latest Pleistocene giant muskrats is 0.40 d_t. This translates to 1.5 d_t for the correlated change in mass. Although the rapid dwarfing of muskrats during the past 16,000 years appears to have been a highly dramatic event (12 d), when rescaled to a common-interval length (0.21 d_t), the rate of change appears to have been only moderately pronounced.

The following character modifications in muskrat history have been documented by a number of investigators:

1. Number of triangles on M_1: five to seven
2. Height of crown and dentine tracts on M_1: low to high

3. Enamel atoll on M_1: present to absent
4. Cementum in reentrant angles: absent to present
5. Number of roots on M^1: three dominant to two dominant

According to Semken (1966) and L. D. Martin (1979), the number of triangles and the dentine tract heights on M_1 increase progressively as muskrat molars increase in size and become more hypsodont, but there are not enough data now to document the changes with the same precision as body mass. The changes in these characters do not appear to be geologically instantaneous, as the samples studied often show intermediate ranges of dental morphotypes (Figure 11.6).

The enamel atoll on M_1, a plesiomorphic feature for arvicolids, does not appear after the /*meadensis* chronomorph, although the pattern of its loss has not been quantified.

Thin, poorly developed interstitial cement is recorded in two M_1s of /*meadensis* from the Dixon local fauna (l.f.) (Eshelman, 1975), tentatively dated on biostratigraphic information by Lundelius et al. (1987) to 2.6 Ma, and complete cementum is reported on many molars from faunas dated by Lundelius et al. (1987) in the range from 1.9 to 2.5 Ma. As compared with other dental features, cementum appears to have been added relatively rapidly.

The number of roots on M^1 has not been examined in enough detail to allow even measured speculation regarding trends.

These patterns may be compared to changes over a period of approximately 315,000 years in the Hagerman fauna from the Glenns Ferry Formation of Idaho. Zakrzewski (1969) examined character change in /*minor* through 83.8 m of sediments. The mean length of the M_1, and thus body size, increased modestly, from an average of about 4.0 mm ($= 103$ g) to 4.25 mm ($= 128$ g). The evolutionary rate for change in mean M_1 length in /*minor* from the Glenns Ferry Formation is 0.06 d_t. This slight increase in size was accompanied by a tendency toward complexity of the M_1, with the anterior loop more tightly closed from T4-5 in molars from the quarry highest in the section. However, stasis is indicated for width of the M_1 and features normally associated with increased hypsodonty, such as dentine tract height and crown height.

Phyletic change in European Arvicola *and* Mimomys

A number of authors have examined character changes in *Arvicola*, from the British Isles through the western USSR (Fejfar and Heinrich, 1990). Only a few of these studies will be reported here.

Stuart (1982) postulated a phyletic relationship between *Mimomys*

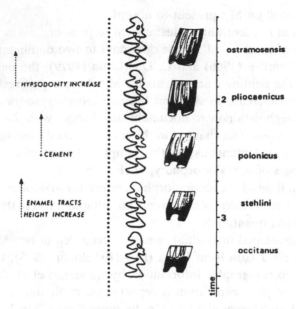

Figure 11.8. Dental evolution in the *Mimomys occitanus–Mimomys ostramosensis* phyletic sequence from Europe. Adapted from Chaline and Laurin (1986).

savini from the Cromerian Interglacial through *Arvicola cantiana* to the extant *A. terrestris* in the British Isles. This relationship supposedly spans about 350,000 years. The dental changes included increase in size, loss of roots, changes in enamel differentiation (negatively differentiated to undifferentiated) (Martin, 1987), and loss of the "*Mimomys* fold." According to Stuart, most of these characters changed progressively, but not simultaneously.

From equation (11.1), body mass estimates for this lineage are as follows: Cromerian *Mimomys savini* ($\bar{x}$ length M_1 = 3.2 mm; 21.6 g), late Devensian *Arvicola cantiana* ($\bar{x}$ length M_1 = 3.95 mm; 98.4 g), modern *A. terrestris* ($\bar{x}$ length M_1 = 4.35 mm; 139.1 g). Assuming for the moment that the 350,000-year date for *M. savini* is correct, the evolutionary rate for change in mean length of the M_1 from *M. savini* to *A. terrestris* is 0.33 d_t. As expected, the evolutionary rate for mass change is much higher: 1.04 d_t.

Phyletic change has been documented by Chaline and Laurin (1986) for the *Mimomys occitanus – M. ostramosensis* lineage in Europe during the latest Pliocene and early Pleistocene (Figure 11.8). Their report indicates a differential response between characters, as was noted earlier for muskrats and water voles. There was a continuous increase in hypsodonty through the lineage, and as in other arvicolids it was accompanied by an increase in lateral dentine tract height. As shown by Chaline and Laurin (1986), these modifications did not progress at a constant rate, but were somewhat episodic and may have been correlated with climatic changes. On the other hand, cement was added to the reentrant folds all at once, during the "*polonicus*" substage (= chronomorph). Occlusal patterns changed little, but according to Chaline and Laurin (1986), intermediate forms can be recognized. Prolonged stasis is not indicated, but shorter bouts of stasis would be obscured by the lack of temporal refinement.

Dental evolution in North American grasshopper mice

Grasshopper mice of the genus *Onychomys* are the only predominantly carnivorous rodents in North America. They are extremely aggressive toward other mammalian species and often include such species in their diet (McCarty, 1975, 1978). They have a rich repertoire of vocalizations and are reputed to "bay" in pursuit of their prey. Carleton and Eshelman (1979) reviewed the fossil record of *Onychomys* and determined the evolutionary rates of character modifications and a measure of variability: the coefficient of variation.

Onychomys leucogaster, the northern grasshopper mouse, is derived from the late Blancan *O. gidleyi*, appearing first in the Fox Canyon l.f. of Kansas, whereas *O. torridus* was descended from *O. bensoni* of the Benson l.f. of Arizona. The Benson l.f. has been dated to 3.1 Ma (Lundelius et al., 1987). The Fox Canyon l.f is considered to have been somewhat earlier in age, at about 3.6 Ma (Lundelius et al., 1987). Carleton and Eshelman (1979) estimated Haldane's (1949) evolutionary rates for raw mean character scores, mean proportions of some measurements (e.g., length of M_3/length of M_1), principal-component scores, and (transformed) coefficients of variation. The values for all dimensions and coefficients of the fossil samples average 0.70 d_t, except for the change from large late Wisconsinan *O. leucogaster* to modern *O. leucogaster*, which averages 0.10 d_t.

Carleton and Eshelman (1979; p. 40) also demonstrated a relationship between the coefficient of variation and the degree of evolution: "Specifically, those dimensions that have undergone substantial reduction in relative size . . . consistently exhibit both higher coefficients of variation and rates of evolution." Interestingly, Van Valkenburgh (1990) found the greatest variation in linear measurements among small rather than large carnivores.

Phyletic evolution and character change in the ochrogaster species group of Microtus (Pitymys)

The prairie vole, *Microtus ochrogaster,* is widely distributed on the central Great Plains of the United States and southern Canada (Hall, 1981). It displays a reverse Bergmann's response, with the smallest subspecies, *M. o. minor,* found at the northern limits of its geographic range (Martin, 1991). Although some investigators have had considerable difficulty in distinguishing the dentitions of *M. ochrogaster* from those of *M. pinetorum,* a substantial character matrix is now available to facilitate this process (Martin, 1987, 1991; Pfaff, 1990). The overlap in measurements and characters between these species serves to demonstrate a close phylogenetic relationship, and for this reason I (Martin, 1974) synonomized *Pedomys* with *Pitymys,* and now treat *Pitymys* as a subgenus of *Microtus,* as did Miller (1896). There are other taxonomic scenarios in the literature, but they will not be discussed here (Repenning, 1983; Chaline and Graf, 1988; Pfaff, 1990). This section summarizes studies of dental evolution in the *ochrogaster* species group of *Pitymys* (Martin, 1991, and unpublished information).

Klippel and Parmalee (1982) reported a small *Microtus* species from late Pleistocene deposits in Cheekbend Cave, Tennessee, that they were unable to identify to species. They noted its similarity to *M. ochrogaster,* and I have since identified it from 10 localities of late Pleistocene age from the eastern United States. Initially, I assumed that the taxon represented populations of *M. o. minor* pushed south in front of the advancing Laurentide ice sheet, because size measurements showed the fossil taxon to be indistinguishable from *M. o. minor.* However, the occlusal pattern was, on the average, much simpler, virtually identical with that of an Irvingtonian taxon, *M. llanensis.* I informally named the diminutive late Pleistocene form "Parmalee's steppe vole" (Martin, 1991).

Figure 11.9 shows my current views of the early evolution of dental morphotypes in the *ochrogaster* species group of the subgenus *Pitymys.* I agree with van der Meulen (1978) that *M. ochrogaster* was descended, through phyletic evolution, from North American populations of *M. pliocaenicus. Microtus guildayi* from the Cumberland Cave fauna of Maryland displays elaboration of dental features in line with this trend,

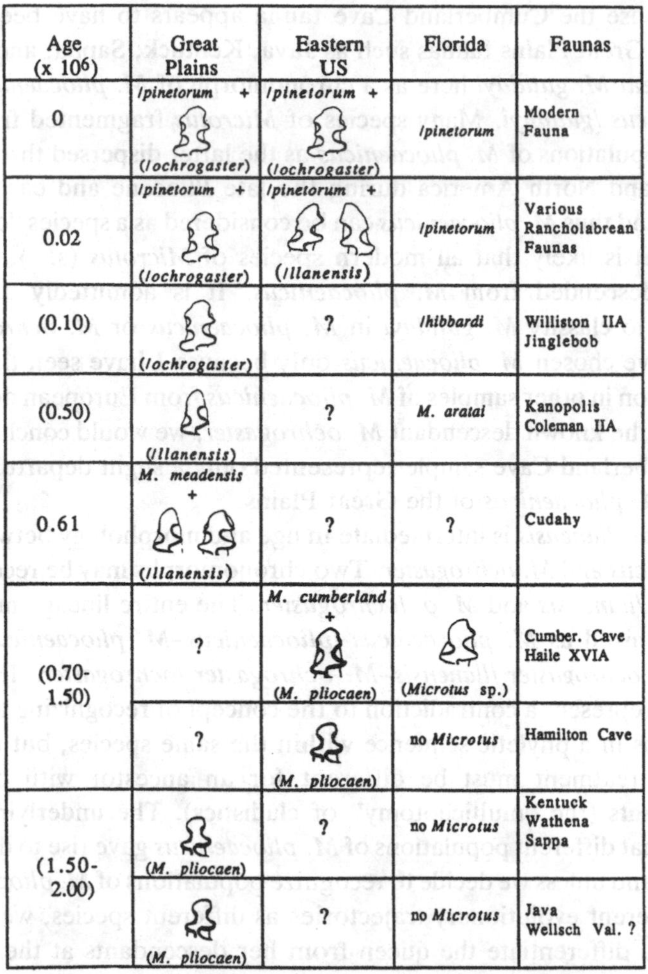

Age (x 10⁶)	Great Plains	Eastern US	Florida	Faunas
0	/pinetorum + (/ochrogaster)	/pinetorum + (/ochrogaster)	/pinetorum	Modern Fauna
0.02	/pinetorum + (/ochrogaster)	/pinetorum + (/llanensis)	/pinetorum	Various Rancholabrean Faunas
(0.10)	(/ochrogaster)	?	/hibbardi	Williston IIA Jinglebob
(0.50)	(/llanensis)	?	M. aratai	Kanopolis Coleman IIA
0.61	M. meadensis + (/llanensis)	?	?	Cudahy
(0.70–1.50)	?	M. cumberland + (M. pliocaen)	(Microtus sp.)	Cumber. Cave Haile XVIA
	?	(M. pliocaen)	no Microtus	Hamilton Cave
(1.50–2.00)	(M. pliocaen)	?	no Microtus	Kentuck Wathena Sappa
	(M. pliocaen)	?	no Microtus	Java Wellsch Val. ?

$Age (x 10^6)$

Figure 11.9. Replacement chronology for first lower molar morphotypes in the *Pitymys–Pedomys* complex from the Great Plains and eastern North America. Time periods in parentheses are approximations, unsupported by radioisotope dates; /pinetorum = *M. pinetorum* /pinetorum; /ochrogaster = *M. o.* /ochrogaster; /llanensis = *M. o.* /llanensis; /hibbardi = *M. pinetorum* /hibbardi; pliocaen = pliocaenicus; Cumber. = Cumberland. *Cumberland = Cumberlandensis*. The llanensis M_1 morphotype characterizes Parmalee's steppe vole from Wisconsinan faunas of the Appalachian chain.

and because the Cumberland Cave fauna appears to have been older than the Great Plains faunas such as Java, Kentuck, Sappa, and Wathena, I treat *M. guildayi* here as a chronomorph of *M. pliocaenicus, M. pliocaenicus /guildayi*. Many species of *Microtus* fragmented from different populations of *M. pliocaenicus* as the latter dispersed throughout Eurasia and North America during the late Pliocene and early Pleistocene, and thus *M. pliocaenicus* can be considered as a species "queen." Indeed, it is likely that all modern species of *Microtus* (s.s.) are ultimately descended from *M. pliocaenicus*. It is admittedly arbitrary whether to classify *M. guildayi* in *M. pliocaenicus* or *M. ochrogaster,* and I have chosen *M. pliocaenicus* only because I have seen this level of variation in other samples of *M. pliocaenicus* from European deposits. Without the known descendant *M. ochrogaster,* we would conclude that the Cumberland Cave sample represented only a slight departure from earlier *M. pliocaenicus* of the Great Plains.

Microtus llanensis is intermediate in age and morphology between *M. pliocaenicus* and *M. ochrogaster*. Two chronomorphs may be recognized as *M. o. /llanensis* and *M. o. /ochrogaster*. The entire lineage may thus be recognized as *M. pliocaenicus /pliocaenicus–M. pliocaenicus /guildayi–M. ochrogaster /llanensis–M. ochrogaster /ochrogaster*. This may seem to represent a contradiction to the concept of recognizing all chronomorphs in a phyletic sequence within the same species, but the taxonomic treatment must be different for an ancestor with multiple descendants (the "multichotomy" of cladistics). The underlying principle is that different populations of *M. pliocaenicus* gave rise to different species, and unless we decide to recognize populations of *M. pliocaenicus* with different evolutionary trajectories as different species, we will be forced to differentiate the queen from her descendants at the species level. Even if the majority of Great Plains populations of *M. pliocaenicus* were transformed into *M. ochrogaster,* and one or two other species evolved from *M. pliocaenicus* by population fragmentation from small peripheral isolates, it is essential that the queen maintain her identity. I suspect that were we to follow muskrat evolution back far enough, that sequence also would reduce to a Pliocene queen, albeit of another lineage.

Prolonged stasis is not observed in the prairie vole sequence, but short intervals may have existed and may be masked by the currently coarse temporal framework.

Radiocarbon dates place Parmalee's steppe vole in full-Wisconsinan glacial faunas of the Appalachian region of the eastern United States roughly at the same time (ca 10,000–20,000 years BP) that more complex M_1s referable to *M. ochrogaster /ochrogaster* were deposited on the central Great Plains (Zakrzewski, 1985). It is highly unlikely that the advanced dental morphotype of *M. o. /ochrogaster* went through a geologically instantaneous simplification back to the */llanensis* morphotype. Martin (1991) showed that dental complexity in this group was unrelated to molar size. Rather, Parmalee's steppe vole appears to have evolved as a dwarf relict of *M. o. /llanensis,* confined to the Appalachian Mountains of eastern North America. It may be a distinct species or a relict chronomorph that intergraded with *M. o. /ochrogaster* to the west, in which case it can also be considered to have been a typical geographic subspecies during late Pleistocene time. The steppe vole became extinct during the early Holocene.

Character change in Quaternary pocket gophers

Samples of fossil pocket gophers are well represented in Quaternary sites on the Florida peninsula and have been studied and reported by Wilkins (1984). As there are no radiocarbon dates associated with any of the materials or localities considered, I have not calculated evolutionary rates from his results. My studies (e.g., Martin, 1974, 1979) confirm Wilkins's (1984) Irvingtonian sequence, from Inglis IA–Haile XVIA–Coleman IIA, but their exact temporal placement remains to be determined [see Lundelius et al. (1987) for an alternative chronology]. Nevertheless, examination of qualitative and quantitative information presented by Wilkins is instructive in determining character change and relative linkage of traits in the phyletic sequence represented by the Inglis IA *Geomys propinetis* through modern *G. pinetis.*

Both the upper fourth premolar and upper third molar demonstrate reductions in enamel banding patterns and overall shape changes (Figure 11.10). The posterior border of P^4 evidences various morphologies of an interrupted enamel pattern in the Inglis IA and Haile XVIA samples, and this enamel was completely absent by Coleman IIA time. An enamel band that wraps around the posterior border of M^3 displays both complete and interrupted patterns in the Inglis IA sample, whereas it is interrupted by the Haile XVIA interval and is perpetuated in that form through Coleman IIA and modern time. Thus, these trends are not exactly chronologically linked.

Wilkins (1984) also examined size changes in eight dentary measurements. Only two, width of the retromolar fossa and width of the posterior

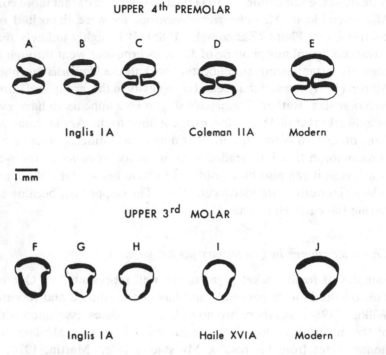

Figure 11.10. Temporal variations in *Geomys propinetis* (Inglis IA, Haile XVIA) and modern *G. pinetis* dental features. From Wilkins (1984).

border of P⁴, demonstrate continual size increase. The length of the anterior border of P⁴ effectively shows stasis, and the average length of P⁴ is actually slightly less in modern *G. pinetis* than in the Inglis IA *G. pinetis*. The remaining four features increased in size from Inglis IA to Coleman IIA time, but showed a variety of responses from that time on. The overall expression of these meristic characters is a classic mosaic pattern.

Evolutionary patterns in Quaternary cotton rats

Cotton rats, genus *Sigmodon,* have a long and reasonably well documented 4-my fossil history in North America (Martin, 1979, 1986, 1990; Czaplewski, 1987). Modern representatives of the genus are small, aggressive, runway-making rodents in riparian oldfield communities of the southern United States. Cotton rats are the ecological analogues of their northern arvicolid cousins, members of the genus *Microtus*. They generally replace *Microtus* wherever the two are sympatric. With the exception of a late Blancan sequence in Kansas documenting a phyletic decrease in body size from *S. minor /medius* through *S. minor /minor*, only rough macroevolutionary patterns are comprehensible.

The diminutive late Blancan *Sigmodon minor* first appeared in North American faunas at approximately 4.0 my BP (Vallecito-Fish Creek beds of southern California and Verde Formation, Arizona). It dispersed throughout North America and has been reported, for example, from generally contemporaneous late Blancan faunas in Kansas, Arizona, and Florida. In the late Blancan and early Irvingtonian, three new, large species replaced *S. minor* throughout its range: *S. curtisi, S. hudspethensis,* and *S. lindsayi* (Martin and Prince, 1989). The dentition of these species is more hypsodont and better supported by the addition of accessory roots, as compared with that of *S. minor.* For a brief time in the early Irvingtonian Curtis Ranch l.f. of Arizona, *S. minor* and *S. curtisi* were sympatric. This is the only fossil record of sympatry among cotton rat species, although Martin and Prince (1989) speculated that *S. lindsayi* and *S. minor* were sympatric for a limited time during the early Irvingtonian in southern California. Two additional species, *S. libitinus* and *S. bakeri,* are known from the early and late Irvingtonian of Florida, respectively (Figure 11.11). The former appears to have been descended from *S. curtisi* of earlier Florida faunas, such as Inglis IA, but it is currently unclear whether this was a phyletic or speciational relationship. All Rancholabrean cotton rat records have been reported as *S. hispidus,* but this is certainly a trashbasket taxon for modern sibling species such as *S. arizonae, S. mascotensis,* and *S. ochrognathus* (Stangl and Dalquest, 1990).

Seven modern species are recognized in North America. Six of these, plus the extinct *S. bakeri,* are included in Martin's (1979) *hispidus* species group, characterized by four well-developed roots on M_1. The extinct *S. hudspethensis, S. libitinus,* and *S. lindsayi* demonstrate a combination of three- and four-rooted M_1s and, with *S. curtisi* and the extant *S. leucotis* that have only three roots on M_1, are included by Martin (1979) in an aggregation of less derived species, the *leucotis* species group. It is not yet clear whether the *hispidus* species group is descended from a single member or multiple members of the *leucotis* group.

Although Barnosky (1987) suggested that the trends in cotton rat evolution support a gradual, phyletic mode of character modification through time, the only secure phyletic change in cotton rat fossil history

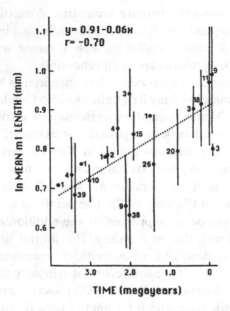

Figure 11.11. Changes in mean length of M₁ through time in cotton rats. Many species are represented; see Martin (1986) for details. The slope of this line, 0.06, represents the overall rate change in darwins. See Table 11.1 for scaled values.

is the dwarfing in Kansas *S. minor* at the end of the Blancan, and this is trivial. From the Rexroad location 3 l.f. at about 3.5 my BP to the Borchers l.f. at 1.9 my BP, *S. minor* decreased in mass from an average of about 39 g to 33 g, which translates to an evolutionary rate of 0.18 d_t (= ca. 0.05 d_t for the linear dimensions on which the mass estimates are based). The overall change in average M₁ length since Rexroad location 3 time, including many samples of most extinct species and the full range of sizes for extant ones, is 0.23 d_t (Martin, 1986) (Figure 11.12). We do not yet have the refined fossil record for cotton rats that we have for some arvicolids, and it is impossible now to determine if phyletic evolution was important in the appearance of larger size for any of the *leucotis* or *hispidus* group species. Nevertheless, the conservative nature of cotton rat dental and skeletal changes over their 4.0-my history (notwithstanding the origin of several new species) expresses something profound about the relationship between cladogenesis and morphological change, and this relationship will be discussed later.

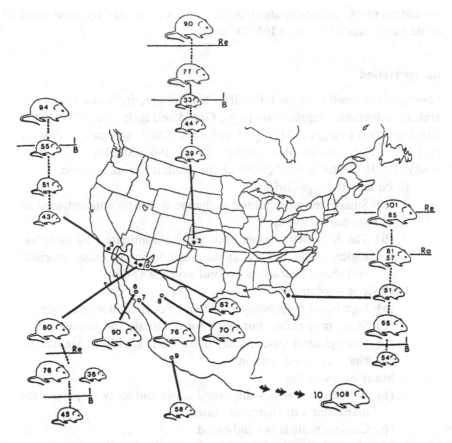

Figure 11.12. Average body mass estimates for fossil cotton rats from four depositional basis (1–4) compared to mean body mass in modern samples from the same and other locations. See Martin (1986) for details.

Morphological stability in Cosomys primus

The extinct *Cosomys primus* was a small vole, confined to Blancan (late Pliocene) deposits of western North America. It is particularly common in lower sediments of the Glenns Ferry Formation of Idaho. Mammals recovered from this site are part of the Hagerman l.f., which also includes the earliest muskrat, *Ondatra zibethicus /minor*. Lich (1990) reported stasis in three metric and five qualitative features of 10 samples of first lower molars of *C. primus* from the 3,295–3,000-foot (1,004–914-m) levels. On the basis of radiometric data, and depending on sedimentation rates, Lich estimates that the time represented by these units could range from 45,000 to 164,000 years.

Interpretation

Four general models of evolutionary change currently exist in the literature: punctuated equilibrium (e.g., Gould and Eldredge, 1977), phyletic evolution (anagenesis) (e.g., Gingerich, 1985), staircase evolution (Stanley, 1985), and mosaic evolution (Mayr, 1963; Stanley, 1979; Barnosky, 1990). Testable predictions of the models are as follows:

1. Punctuated equilibrium
   (a) Significant morphological change is mostly concentrated at speciation events.
   (b) The histories of most lineages are dominated by morphological stasis, defined as the meandering of linear meristic and shape characters around average values.
2. Phyletic evolution
   (a) Significant morphological change occurs in a single lineage.
   (b) Stasis may occur, but does not dominate the sequence.
   (c) Cladogenesis (speciation) often occurs without significant character modification.
3. Staircase evolution
   (a) A lineage that has displayed stasis suddenly displays dramatic shifts in character states.
   (b) Cladogenesis is not indicated.
4. Mosaic evolution (redefined at the species level)
   (a) Different characters in a given population evolve at different rates.
   (b) The same characters in different populations of a single species may evolve at different rates. This can result in phenotypic regionalism and at least a temporary overlap of relatively underived and derived character states in a given species (i.e., overlap of chronomorphs, depending on characters examined).

Phyletic evolution in its most extreme and metaphysical form as orthogenesis probably is not worth examining further. As Levinton (1988) and others note, there are very few, if any, evolutionary biologists who accept this simplistic view of evolution or character modification. The concept of phyletic evolution is useful, however, as a model embracing the notion that significant and directional character changes (or changes in paleontologically unmeasurable quantities, such as behavior, for that matter) usually occur without species multiplication or a rapid reorganization of the genotype. Significant change must be operationally defined as modification beyond the limits seen in closely related living species treated as controls, as noted earlier. Phyletic evolution may be episodic and may feature temporary reversals (in size, but generally not in structure). It is not inconsistent to speak of periods of stasis within a phyletic sequence. This is to be expected if the scale of resolution is fine enough.

Speciation (= cladogenesis = lineage splitting in this chapter) owes allegiance neither to phyletic evolution nor to punctuated equilibrium, but I am aware of no evidence from the fossil record for "gradual" speciation. Lineage splitting appears to be in all cases a geologically instantaneous process. Perhaps the most impressive examples of rapid speciation are the cichlid species "flocks" in the lakes of the African Great Rift valley. According to Meyer et al. (1990), more than 200 species in Lake Victoria and associated impoundments are monophyletic, genetically uniform, and they may have speciated in less than 200,000 years. It is important to realize that phyletic evolution does not deny cladogenesis; after all, the diversity of life cannot be explained without it. Alternatively, the punctuated equilibrium model does not require that significant change be associated with all speciation events.

An interesting play on both punctuated equilibrium and phyletic evolution was proposed by Stanley (1985) as "staircase" evolution, in which a species that has demonstrated stasis for a considerable length of time then quickly undergoes considerable morphological change, perhaps because of a genetic bottleneck. This pattern has also been labeled "punctuated gradualism" by Malmgren and Berggren (1984). Stanley (1985) was careful not to label this as a speciation process, and although he described it as a subset of punctuated equilibrium because of the stasis requirement, in the absence of lineage splitting it could be considered an extreme form of phyletic evolution. Staircase evolution could, of course, represent an extreme case of punctuated equilibrium in which most of the parent populations quickly become extinct (E. S. Vrba, personal communication). The extinction event might not be expressed in the fossil record. Additionally, we should expect a staircase *pattern* to appear with some statistical regularity in phyletic sequences over brief periods of time, and it is therefore critical to distinguish between the pattern and the process, the latter of which I use in a restricted sense to indicate a significant shift in morphology. Flynn (1986) suggested that Miocene rodents of the genus *Kanisamys* demonstrated staircase evolution, and Barnosky (Chapter 3, this volume) reports an example of intraspecific staircase variation in meadow vole dentitions over a brief period of the late Quaternary.

Mosaic evolution is actually a pluralistic model that can incorporate almost any tempo and requires a multivariate analysis of samples from different geographic localities. Its primary assumption is that characters may evolve at different rates in both a single population and separate populations. That is, molar length may continue to increase at a faster rate (or in a different manner altogether, e.g., gradual vs. episodic) than does width in all samples, but tooth shapes may differ from locality to locality. Thus, if tooth shape at locality 1 evolves at a different rate than shape at locality 2, then at any point in time underived and derived features may occur simultaneously (e.g., the five- and six-triangle M_1 morphotypes of modern *M. pennsylvanicus*) (Davis, 1987). A predictable consequence of this model would be the overlap of what I have defined as chronomorphs, depending on the characters and samples chosen for comparison. A continuous, directional trend would not occur for all characters for all localities.

In the sections immediately following, I shall (1) examine and reject the obligatory linkage of cladogenesis and significant character change in favor of a pluralistic model, (2) summarize the data on evolutionary tempo, particularly in light of the recent discovery by Gingerich (1983) that evolutionary rates measured in darwins are not independent of the interval over which they are measured, (3) consider higher-order properties of evolving systems, such as trends within clades, and (4) briefly discuss coordinated character changes in geographically widespread species.

Speciation rates and character divergence

One need go no further than the modern mammalian fauna to see the lack of linkage between character modification and cladogenesis. As I have shown for the 4 my of *Sigmodon* history in North America (Martin, 1986), and as also seems abundantly clear for the genera *Microtus, Reithrodontomys, Peromyscus,* and other highly speciose generic clades composed of small mammals, the component species appear to be minor plays on the same theme, the result of second-order speciation events (Martin, 1986) producing no radical morphological differences. If speciation is mediated to a great extent by the potential of population fragmentation as it relates to body size, expressed as the inverse relationship between speciation rate and body mass (Martin, 1992), then there is no inherent biological rationale for a coupling of cladogenesis and morphological change (Wright, 1982). The primary driving force behind cladogenesis may be the probability of isolation.

As Stanley (1985) observed, size is probably the most labile feature in mammalian evolution, but it is also one of the most important. Muskrats changed from animals about the size of the modern European water vole, *Arvicola terrestris,* to a giant form of more than 1.6 kg in the latest Pleistocene. Following a rapid dwarfing event during the latest Pleistocene, muskrats stabilized at about 0.84 kg. The dental modifications discussed earlier probably relate to this size increase in one way or another. It is admittedly debatable whether or not the loss of an enamel atoll on M_1, the addition of two triangles on M_1, and the addition of dental cement represent biologically meaningful changes, but they certainly are changes that fall outside the realm of intraspecific variation in modern populations. Regardless, I would propose that the change in size was so profound, indicating as it does major differences in lifestyle between the earliest and later forms, that the specific modifications of dental form (other than size) become trivial by comparison.

Many papers and books attest to the high correlations between body mass and most of the important physiological and natural history parameters that characterize mammalian species (e.g., Peters, 1983; Martin, 1986; Damuth and MacFadden, 1990). Muskrat populations at 103 and 1,600 g are only remotely the same animals (it is unknown if the earlier forms were aquatic). A size difference of this magnitude equals about a sevenfold difference in total basal metabolic rates and considerable differences in home range size, lifespan, and almost every other measurable biological variable (Table 11.3).

The earliest arvicolids, although demonstrating a diversity of dental patterns, as they do today, probably were not that different in overall appearance (= shape) from one another or from their descendants. We would classify the species ancestral to all arvicolids as a cricetid [or murid, according to Carleton and Musser (1984)], but structurally the dentition would have some tendencies toward the planed and prismatic molars of arvicolids. Such species are known (Repenning, 1968, 1987). There are no examples of speciation in rodent evolution of which I am aware in which there was rapid and distinctive change in form to the extent that we would recognize the descendant species as something drastically different from its ancestor. Width and length of the M_1 did not continue to increase at the same rate in muskrat evolution (Martin

Table 11.3. *Comparative biological data for* Ondatra zibethicus/minor (3.75 Ma) *and two samples of O. z. /zibethicus (W, late Wisconsinan; R, Recent)*

Parameter[a]	/minor	/zibethicus (W)	/zibethicus (R)
W (kg)	0.12	1.60	0.84
M_b (kcal/day)	14.3	99.6	61.4
H (ha)	4.4	22.5	14.5
D (no./km²)	780	161	238
L (years)	3.9	6.0	5.4
O	4.7	5.9	5.6

[a]W, mass; M_b, basal metabolic rate; H, home range; D, population density; L, lifespan; O, number of offspring/litter. The equations used for calculations are from Martin (1990), this study, and Maiorana (1990).

and Tedesco, 1976), resulting in length/width ratios that differed after a certain overall size had been attained. Barnosky (1990) documented intraspecific differences in meadow vole dental features that varied through time in a mosaic fashion. I suspect these are good examples of the manner by which mammalian features change in shape through time, and the changes are not correlated with speciation events.

Evolutionary rates in rodent evolution

It has become well known that evolutionary rates are inversely related to the temporal interval over which they are measured (Gingerich, 1983). This relationship has been referred to as "Gingerich's law" by Webb and Barnosky (1989). The original presentation of this pattern was severely criticized by Gould (1984), and later by Stanley (1985) and Stanley and Yang (1987). Gould (1984) concluded that the correlation was an artifact, being little more than a plot of time against its reciprocal. He also explained the value of 1.2, the average ratio between mean final and beginning character states reported by Gingerich, as a property of human perception; not a characteristic of the real world. Scientists, according to Gould (1984, p. 994), "rarely notice smaller degrees of change, while larger amounts are accompanied by so much uncertainty about actual ancestors and descendants that we do not identify lineages with confidence and do not make the evolutionary links."

However, Levinton (1988) supported Gingerich's analysis, pointing out, as had Gingerich (1984), that the numerator in Haldane's ratio is free to wander over a considerable range. Levinton also challenged Gould's conclusion that slow or negligible rates were ignored in the short-term studies, proposing that studies at all temporal scales were biased, if at all, toward uncovering maximum rates. The inverse relationship was reported also by Kurtén (1960) and Martin and Prince (1990) for a variety of Quaternary taxa. Levinton (1988) additionally reported unpublished data of Michael Bell for Miocene sticklebacks, *Gasterosteus doryssus,* showing that the inverse relationship held for a series of characters measured over a period of 100,000 years as determined by lake varves.

Intuitively we know that the denominator in Haldane's equation is free to change almost infinitely, but the numerator is not. When we make comparisons within or between related species lineages, we constrain the numerator considerably. The average value of 1.2 for x_2/x_1 does represent the kinds of comparisons that evolutionary biologists make, but not because they are biased. Comparisons beyond certain limits are either biologically inappropriate or absurd. As Gingerich (1983, 1984) aptly noted, the upper limit of proportional change allowed by his data set is approximately 1,600, which just happens to be about the proportional difference between any linear dimension of a large blue whale and the least shrew.

The $\log_e$-transformed rates reported here for Quaternary rodents are plotted against $\log_e$ interval lengths in Figure 11.13. They clearly follow the inverse function noted by Gingerich, and the slope is mathematically indistinguishable from -1.0. In spite of published criticisms, Gingerich's (1983) main point appears secure: In order to be directly compared, evolutionary rates must be calculated over the same time interval or rescaled to a common interval length.

Rescaling may be accomplished by generating new regressions from data sets and then extrapolating beyond the set (e.g., examining an average point for invertebrates for an interval length of 1 my even though the data cover only 10–20-my intervals). Rescaling individual values within a single data set is done by adding the residuals for each point to the value predicted by the regression at an interval of 1 my. The assumption in this latter operation is that each point represents a theoretical population of points that would produce a regression slope statistically indistinguishable from that of the entire data set. This may not be true, and until this proposition is tested, the values presented here must be considered only tentative.

There are some surprises in the rescaled evolutionary rates in Table

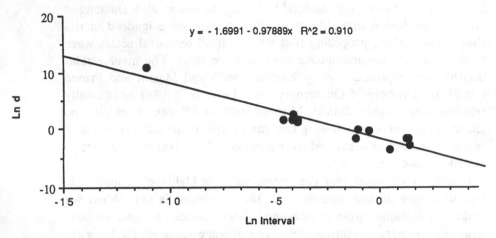

Figure 11.13. Natural-log plot of evolutionary rate in darwins against interval length for the Quaternary rodent data from Table 11.1.

11.1. The rate change for the bank vole, *Clethrionomys glareolus,* remains by far the fastest, although it is not as dramatic as the original 50,000-*d* figure indicated (although we do not have enough values to tell at this point; perhaps among a population of rescaled points, 0.94 *is* extraordinarily fast). High rates for other voles during the late Pleistocene are not particularly unusual when rescaled, but the overall change in muskrat size, as indicated by the increase in the first lower molar, is dramatic, as are dental modifications in all lineages of *Onychomys*. The apparently rapid dwarfing of modern muskrats from their late Pleistocene giants is not as consequential an event when seen in the context of change in the entire lineage through 1 my.

However, the muskrat data presented here warn that the assumption of linearity or a simple exponential decay function in morphological change may not be warranted, even in a phyletic sequence. The response in muskrats was episodic, and evolutionary rates vary considerably depending on the chosen interval, even when scaled to a common interval length (Table 11.1). That is, at least in muskrats, there is no single, set time interval that can adequately represent character change for the entire sequence. Evolutionary rates change both as a function of time interval and as a function of time itself. This information does not negate the use of Haldane's equation or Gingerich's conclusions, but simply provides another facet of evolutionary rate analysis that must be considered and suggests that a multivariate, perhaps three-dimensional analysis including evolutionary rate, interval length, and the specific time interval might yield the comparative data we seek.

Morphological trends in rodent evolution

Bookstein (1987), like Raup (1977) before him, warns that apparent evolutionary trends can be mimicked effectively by symmetrical random walks. Bookstein further argues that evolutionary rates do not exist for random walks and that the random walk model must first be rejected prior to any further consideration of evolutionary rate. Unfortunately, many time series of paleontological data can be mirrored by a random model of some sort, particularly those with small samples distributed over brief periods of time [but see Stanley (1975) and Vrba (1987) for rejection of random models with fossil materials]. The discrimination, then, will often be made on a conceptual rather than statistical basis. That is, if we assume that the observed directionality has as its basis the process of natural selection, then it is legitimate to consider the results in terms of evolutionary rates. On the other hand, even if we choose not to reject the random explanatory pattern, it is, as Bookstein (1987) agrees, reasonable to examine rate of change within the sequence. Bookstein suggests that we refer to these velocities as "Brownian speeds" rather than evolutionary rates, but this point seems debatable.

I have proposed that the overall trend toward large size in *Sigmodon*, the only one of any significance I can identify in the 4.0-my history of cotton rats in North America, was the result of interspecific competition favoring species with larger, more aggressive individuals (Martin, 1986). At least one evolutionary reversal was noted, suggesting further that speciation events were random with respect to the trend. Cotton rat populations of varying sizes originated in a stochastic pattern, and selection at the individual level, but essentially between species, determined the success of those populations. Competition in the classic sense. The trend developed through extinction of small species. A higher-order speciation process, such as species selection (Stanley, 1975, 1979), in which selection acts on characters at the focal level of the species (Vrba, 1984a), is not required to explain the pattern [Mayr (1988) included interspecific competition as a mode of species selection, but this was not the original intent for the concept (Stanley, 1975; Gould and Eldredge, 1977; Vrba, 1984b), as discussed later].

Macroevolution in some rodent groups may also be powered by anagenesis. In this scenario, significant change occurs through intraspecific natural selection. The selective agent may be competition, predation, or some other factor that operates on individuals within species. A given species may split, but daughter species often do not differ in any appreciable manner from their ancestor. Sorting among species in a clade (e.g., by competition or differential speciation rates) is not necessary to explain the morphological shifts. This process may dominate morphological change and species diversity in some arvicolids, within which species "fecundity" can truly be labeled explosive.

Various names have been applied to processes that create trends resulting from selection at the organismal level; "speciational trends" (Grant, 1963, 1989) and the "effect hypothesis" (Vrba, 1980, 1984b) are two of these. As explained by Grant (1989), speciational trends are the results of a time series of speciations in which each daughter species diverges from its ancestor in the same general direction. Grant (1989, p. 604) explicitly states that orientation of the trend is adaptive: "The trend runs from an ancestral position on the adaptive landscape to a new and different adaptive peak." If each species in a successional set is better adapted than its predecessor, then this is a valid theoretical construct, because a trend can surely result in this manner. In the fossil record, two criteria would need to be satisfied in order for this model to be at least tentatively accepted: (1) Cladogenesis would have to be observed, represented by contemporaneity of each parent and daughter pair. (2) Adaptive structural modifications among a series of parent-daughter speciations would have to occur in the same direction. I am unaware of any examples in the fossil record that appear to correspond with this scenario.

"Effect macroevolution" was applied by Vrba (1980, 1983, 1984b, 1989) to trends among species that develop by upward causation (Campbell, 1974) from selection acting at the organismal level. Organismal properties, according to this model, determine speciation, speciation rate, and extinction rate, the latter two of which are regarded as adaptively neutral. Perhaps the most important contribution of this model is the idea that macroevolutionary properties such as speciation, speciation rate, and extinction rate are not adaptations, but *effects*. Let us examine its limits in cotton rat evolution.

If there is a high, inverse correlation between speciation rate and body mass in mammals (Martin, 1992), then the diversity of cotton rat species may be heavily influenced by the probability of cladogenesis as a function of their size. That is, body mass would predict that a certain number of speciation events would occur at random, and the expectation would be that some species would be larger than others. Interspecific competition would then act to select against small species, creating the trend toward large size. The trend conforms to "effect macroevolution" to the extent that speciation and extinction rates arise from size relationships and competition among individuals.

Opposed to "effect macroevolution" is the rival hypothesis of "species selection," the label coined by Stanley (1975, 1979) for the concept of sorting among species, as described by Gould and Eldredge (1977). As later formulated by Vrba (1984a,b), Vrba and Eldredge (1984), and Vrba and Gould (1986), in order for species selection to hold, there must be emergent, heritable character variation at the focal level of the species that is responsible for sorting among species in a clade. That is, there must be species aptations as opposed to organismic aptations. If the trend toward large size in the cotton rat clade was mediated not by interspecific competition among individuals but by a heritable species-level property such as the tendency toward greater patchiness in distribution, then species selection could be said to be operating. The distinction between effect macroevolution and species selection will be difficult to determine with fossil materials, and may have a fuzzy boundary in any case. The overall small size and relative stenotopy of cotton rats dictate that population fragmentation can be expected (e.g., Selander, 1970); thus, every cotton rat species has, to a greater or lesser extent, the species-level property of relatively high fragmentation potential. Yet interspecific competition among individuals is a logical candidate for the sorting agent among species and therefore dictates the trend toward larger size.

How do we explain character modification throughout the geographic distribution of a widespread species?

Chaline and Laurin (1986) proposed phyletic evolution as the dominant pattern of change in the *Mimomys occitanus–ostramosensis* lineage of Europe. However, they were rightfully concerned about simultaneous character changes over the wide geographic area represented by their samples. How, they asked, could coordinated morphological changes occur throughout the entire species' range over such a long period of time? After all, rodent distributions, particularly those of species with small body size, are characteristically patchy. One would expect more of a mosaic character response, with some populations at a given time expressing advanced morphotypes, and some less derived morphotypes.

This phenomenon is also recorded to some extent for the North American muskrat, but I think it results from three influences: (1) an aquatic lifestyle, (2) the assumption of gradual and simultaneous changes among all characters, and (3) the choice of correlated characters. As I have noted elsewhere (Martin, 1992), aquatic mammals, regardless of body size, appear to speciate rarely. They also tend to be distributed over extremely wide geographic ranges, undoubtedly promoted by their watery runways. Whereas individual tributaries may harbor distinct species of fishes, aquatic mammals show no such isolationist tendencies. An aquatic existence may therefore foster rapid exchange of genes between populations. Although *Mimomys* also includes the ancestor for *Microtus* and probably other genera of fully terrestrial voles, Chaline (1987) recognizes *M. ostramosensis* as part of the ancestral lineage leading to the aquatic *Arvicola sapidus*.

On the other hand, there are limited radioisotopic dates associated with samples in the *M. ostramosensis-A. sapidus* sequence; almost all of the samples are derived from isolated cavern or fissure fills sequenced by biostratigraphic information. I suspect that as more deposits are dated, we shall see more overlap of chronomorphs. If the change in proportion from five- to six-triangle M_1 morphotypes in Great Plains *Microtus pennsylvanicus* is any indication, we should expect to see overlap of chronomorphs on a scale of roughly 250,000 years or less (Davis, 1987) (Figure 11.14).

Additionally, those dental features classically examined in *Mimomys* and *Ondatra* evolution are generally highly correlated with each other and closely associated with larger body size. To the extent that the fossil lineage expresses a trend toward large size it is expected that the associated character matrix will follow in a predictable fashion. Rapid exchange of genetic information, in other words, may not be the dominant control variable. Nevertheless, we still must explain the size trend throughout the full species range, and I think, as noted earlier, that the finer the temporal resolution, the more overlap will be discernible in size, as well as in other features.

Conclusions: toward a comprehensive theory of phenotypic evolution at the species level

The Quaternary rodent record supports the contention that dentition in rodent species may exist for millions of years basically unchanged (*Sigmodon minor*), or it may change continually, albeit at varying rates

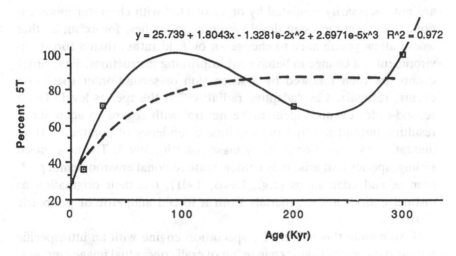

$$y = 25.739 + 1.8043x - 1.3281e\text{-}2x^2 + 2.6971e\text{-}5x^3 \quad R^2 = 0.972$$

Figure 11.14. Changes in proportions of five-triangle (5T) and six triangle M_1 morphotypes in the meadow vole, *Microtus pennsylvanicus*, during approximately the past 300,000 years on the Great Plains of North America. Points represented are for the following faunas: Sandahl (ca. 300 ky), average of Mt. Scott/Duck Creek (ca. 200 ky), Jones (29 ky), Robert (11 ky), and modern Great Plains (0.001 ky). Data from Davis (1987). Although the data have been fitted well with a complex polynomial (solid line and formula), other curvilinear and linear models are possible, perhaps even more likely. One (dashed line) is informally drawn as an example.

(*Ondatra zibethicus, Mimomys savini, M. occitanus*), over the same length of time. Characters in phyletic sequences that change continually over long periods of time may demonstrate stasis over periods of time measured in thousands of years (*O. zibethicus, M. pennsylvanicus*). Where multiple characters have been examined, a mosaic pattern is evidenced (*Microtus pennsylvanicus, Geomys*); some features change directionally, and some show stasis or a staircase pattern, and often there is only moderate linkage between the modifications. Punctuated equilibrium is not the most common pattern displayed by dental characters in the Quaternary rodent fossil record. Neither is monotonic phyletic gradualism. The combination of studies reported in this review points to a complex mosaic pattern of episodic phyletic evolution as the prevalent theme of dental character change.

Elsewhere (Martin, 1986) I proposed that two classes of speciation events should be recognized in mammalian evolution: first-order events that transferred a descendant species into a new adaptive zone, and second-order events within the same adaptive zone. First-order events are not necessarily mediated by or associated with character modification. They may represent developmental anomalies, for example, that would allow young mice to choose an oldfield rather than a forest environment – a change in behavioral imprinting instructions. First-order events are then followed by a succession of second-order speciation events, recognized as "adaptive radiation" at the species level. These second-order events appear to be neutral with regard to adaptation, resulting instead mostly from the innate tendency of population fragmentation as a function of body mass (Martin, 1992). The new, often sibling, species that arise may demonstrate regional environmental preferences and adaptations (e.g., Nevo, 1991), but their origination as distinct entities arises generally from a forced allopatric or peripatric distribution.

If we couple this model of a speciation engine with an intraspecific mosaic pattern of phyletic change, an overall conceptual image emerges. The combination of a genetic/epigenetic program and natural selection maintains intraspecific phenotypic modifications within certain tolerable boundaries. Because there are wide ranges of variation for most characters, this is expressed as the intraspecific geographic variation in modern and late Pleistocene mammal dentitions documented by a number of authors, especially Avery (1982), Davis (1987), Barnosky (1990), and Nadachowski (1990). Thus, at any given moment in geological time, a species may have embedded within it a series of populations with independent morphological and evolutionary trajectories. These trajectories may or may not have adaptive consequences, and we observe them oscillating through time as the mosaic pattern illustrated by Barnosky (1990). In a very real sense there is little going on here in the short run other than the "fine tuning" to which most punctuationists refer. But this appears also to be the stuff of large scale change.

Just as the data for *Microtus pennsylvanicus* provide us with an expectation of morphological change in the range of a few thousand years, information from other arvicolid sequences, such as *Ondatra zibethicus*, illustrate possibilities over longer time intervals. When expanded to a period of about 315,000 years (*O. zibethicus*), we see a level of change comparable to that of the meadow vole over a shorter period. When further extended to 3.75 my, however, we see that the muskrat has increased in size by more than a factor of 10 at one point, and dental modifications have become pronounced. This was not a speciation process, but a transformation of a population following the unique phenotypic trajectories inherited from its ancestor.

The extent to which new features appear during speciation is unclear from the fossil record and is unresolved generally in the theoretical literature (Wright, 1982; Levinton, 1988; Allmon, 1990; Ross and Allmon, 1990). Even among the haplochromine cichlids of the African Great Rift valley lakes, where 200 species may have diverged since the late Pleistocene, tremendous intraspecific phenotypic plasticity is recognized in modern species (Meyer, 1987), and it is conceivable that each currently distinct species may have inherited its unique attributes as at least tendencies from the ancestral complex. Seven hundred species of Drosophilidae have evolved on the Hawaiian archipelago during the past 400,000 years, but the majority of species differ little in phenotypic architecture (Carson and Kaneshiro, 1976; Levinton, 1988). Nevo's (1991) research program with the *Spalax ehrenbergi* "superspecies," a set of four mole rat sibling species with narrow hybrid zones in Israel, reveals no qualitative differences in phenotypic characters among the taxa.

The model presented here requires no unique contribution to phenotypic variation by speciation and in that sense fully supports the standard Darwinian paradigm that large-scale morphological change results directly from population-level phenomena. Evolutionary novelties appear at the population level as minor tendencies in an ancestral species and later become magnified through phyletic evolution in isolated population fragments (daughter species). Thus, with regard to morphological change at the species level, macroevolution and microevolution are not decoupled.

This model is supported by dental changes in the Quaternary rodent fossil record, but can it be extrapolated to account for the major biological shifts in Quaternary rodent morphology? That is, can it account for major *shape* changes, such as the proportions of the face, diminution of eyes and ears in fossorial species, pelage color, number of mammae, short versus long tails, webbed feet, and so forth? As noted in a previous section, shape changes, at least in the dentition, seem to be guided by shifts in allometric relations and developmental limitations that have been recorded in phyletic sequences. After a certain point in *Ondatra* dental evolution, dental width becomes less "flexible" than length, resulting in a longer, narrower first lower molar. On a minor scale, this shift reflects a process that can account for shape changes in other skeletal areas and systems and is consistent with the manner by which epigenetic systems are presumed to operate (Løvtrup, 1984; Saunders, 1984).

To the extent that we choose to recognize organismal effects on speciation and extinction rates as phenomena distinct from intraspecific processes and as components of macroevolution rather than microevolution, then a macroevolutionary influence is necessary to explain long-term trends and complete the comprehensive model. This is an active and controversial area of evolutionary investigation, as reviewed in an earlier section. Although we lack the detailed record for cotton rats that we have for some arvicolids, the available evidence does suggest that the higher extinction rate in small species, perhaps mediated by interspecific competition, is necessary to explain the overall trend toward large size within the clade. Other forces besides competition may affect speciation and extinction rates, and one of the current, healthy trends in the field of paleobiology is an examination of the potential influence of environmental factors (climate, resource partitioning, etc.) at different temporal scales (e.g., Vrba, 1985, 1987; Barnosky, Chapter 3, this volume).

Acknowledgments

My sincerest appreciation is extended to Anthony Barnosky, Philip Gingerich, and Elisabeth Vrba, whose competent reviews greatly increased the scientific merit of this contribution.

References

Agadzhanyan, A. K. (1973). The history of collared lemmings in the Pleistocene. In V. L. Kontrimavichus (ed.), *Beringia in the Cenozoic Era* (pp. 379–88). New Delhi: Oxonian Press.

Allmon, W. D. (1990). [Review of] D. Otte and J. A. Endler (eds.), *Speciation and Its Consequences*. Sunderland, Mass: Sinauer Associates. *Historical Biology*, 4:143–9.

Avery, D. M. (1982). Micromammals as palaeoenvironmental indicators and an interpretation of the late Quaternary in the southern Cape Province, South Africa. *Annals of the South African Museum, Cape Town*, 85:183–374.

Barnosky, A. D. (1987). Punctuated equilibrium and phyletic gradualism: some facts in the Quaternary mammalian record. *Current Mammalogy*, 1: 107–47.

    (1990). Evolution of dental traits since latest Pleistocene in meadow voles (*Microtus pennsylvanicus*) from Virginia. *Paleobiology*, 16:370–83.

Birkenholz, D. E. (1972). *Neofiber alleni. Mammalian Species*, 15:1–4.

Bookstein, F. L. (1987). Random walk and the existence of evolutionary rates. *Paleobiology*, 13:446–64.

Bryant, M. D. (1945). Phylogeny of Nearctic Sciuridae. *American Midland Naturalist*, 33:257–390.

Campbell, T. (1974). "Downward causation" in hierarchically organized biological systems. In F. J. Ayala and T. Dobzhansky (eds.), *Studies in the Philosophy of Biology* (pp. 179–86). San Francisco: University of California Press.

Carleton, M. D., and R. E. Eshelman (1979). *A Synopsis of Fossil Grasshopper Mice, Genus* Onychomys, *and Their Relationships to Recent Species*. C. W. Hibbard Memorial Volume 7, Museum of Paleontology, University of Michigan.

Carleton, M. D., and G. G. Musser (1984). Muroid rodents. In S. Anderson and J. K. Jones, Jr. (eds.), *Genera and Families of Recent Mammals of the World* (pp. 289–379). New York: Wiley.

Carson, H. L., and K. Y. Kaneshiro (1976). *Drosophila* of Hawaii: systematics and ecological genetics. *Annual Review of Ecology and Systematics*, 7: 311–46.

Chaline, J. (1987). Arvicolid data (Arvicolidae, Rodentia) and evolutionary concepts. *Evolutionary Biology*, 21:237–310.

Chaline, J., and J.-D. Graf (1988). Phylogeny of the Arvicolidae (Rodentia): biochemical and paleontological evidence. *Journal of Mammalogy*, 69: 22–33.

Chaline, J., and B. Laurin (1986). Phyletic gradualism in a European Plio-Pleistocene *Mimomys* lineage (Arvicolidae, Rodentia). *Paleobiology*, 12:203–16.

Corbet, G. B. (1964). Regional variation in the bank-vole *Clethrionomys glareolus* in the British Isles. *Proceedings of the Zoological Society, London*, 143:191–217.

  (1975). Examples of short- and long-term changes of dental pattern in Scottish voles (Rodentia; Microtinae). *Mammalian Review*, 5:17–21.

Czaplewski, N. (1987). Sigmodont rodents (Mammalia; Muroidea; Sigmodontinae) from the Pliocene (early Blancan) Verde Formation, Arizona. *Journal of Vertebrate Paleontology*, 7:183–99.

Damuth, J., and B. J. MacFadden (1990). *Body Size in Mammalian Paleobiology*. Cambridge University Press.

Davis, L. C. (1987). Late Pleistocene/Holocene environmental changes in the Central Plains of the United States: the mammalian record. In R. W. Graham, H. A. Semken, Jr., and M. A. Graham (eds.) *Late Quaternary Mammalian Biogeography and Environments of the Great Plains and Prairies* (pp. 88–143). Volume 22, Scientific Papers, Illinois State Museum, Springfield.

Eshelman, R. E. (1975). *Geology and Paleontology of the Early Pleistocene (Late Blancan) White Rock Fauna from North-Central Kansas*. C. W. Hibbard Memorial Volume 4, Museum of Paleontology, University of Michigan.

Fejfar, O., and W.-D. Heinrich (eds.) (1990). *International Symposium on the Evolution, Phylogeny and Biostratigraphy of Arvicolids (Rodentia, Mammalia)*. Prague: Geological Survey.

Flynn, L. J. (1986). Species longevity, stasis, and stairsteps in rhizomyid rodents. *Contributions to Geology, University of Wyoming, Special Papers,* 3: 273–85.

Fortey, R. A. (1985). Gradualism and punctuated equilibria as competing and complementary theories. In J. C. W. Cope, and P. W. Skelton (eds.), *Evolutionary Case Histories from the Fossil Record* (pp. 17–28). Special Papers in Paleontology 33, The Palaeontological Association, London.

Freudenthal, M., and E. M. Suarez (1990). Size variation in samples of fossil and recent murid teeth. *Scripta Geologica,* 93:1–40.

Ghiselin, M. T. (1974). A radical solution to the species problem. *Systematic Zoology* 23:536–44.

Gingerich, P. D. (1983). Rates of evolution: effects of time and temporal scaling. *Science,* 222:159–61.

   (1984). Response to S. J. Gould. *Science,* 226:995.

   (1985). Species in the fossil record: concepts, trends, and transitions. *Paleobiology,* 11:27–41.

Goin, O. B. (1943). A study of individual variation in *Microtus pennsylvanicus pennsylvanicus. Journal of Mammalogy,* 24:212–23.

Gould, S. J. (1984). Smooth curve of evolutionary rate: a psychological and mathematical artifact. *Science,* 226:994–6.

   (1985). The paradox of the first tier: an agenda for paleobiology. *Paleobiology,* 11:2–12.

Gould, S. J., and N. Eldredge (1977). Punctuated equilibria: the tempo and mode of evolution reconsidered. *Paleobiology,* 3:115–51.

Graham, R. W., and H. A. Semken, Jr. (1987). Philosophy and procedures for paleoenvironmental studies of Quaternary mammalian faunas. In R. W. Graham, H. A. Semken, Jr., and M. A. Graham (eds.), *Late Quaternary Mammalian Biogeography and Environments of the Great Plains and Prairies* (pp. 1–7). Volume 22, Scientific Papers, Illinois State Museum, Springfield.

Grant, V. (1963). *The Origin of Adaptations.* New York: Columbia University Press.

   (1989). The theory of speciational trends. *American Naturalist,* 133:604–12.

Guilday, J. E. (1982). Dental variation in *Microtus xanthognathus, M. chrotorrhinus,* and *M. pennsylvanicus* (Rodentia, Mammalia). *Annals, Carnegie Museum of Natural History,* 51:211–30.

Guilday, J. E., and M. S. Bender (1960). Late Pleistocene records of the yellow-cheeked vole, *Microtus xanthognathus* (Leach). *Annals, Carnegie Museum of Natural History,* 35:315–30.

Guilday, J. E., H. W. Hamilton, and A. D. McCrady (1969). The Pleistocene vertebrate fauna of Robinson Cave, Overton County, Tennessee. *Palaeovertebrata,* 2:25–75.

Guilday, J. E., P. S. Martin, and A. D. McCrady (1964). New Paris No. 4: a Pleistocene cave deposit in Bedford County, Pennsylvania. *Bulletin of the National Speleological Society,* 26:121–94.

Guilday, J. E., P. W. Parmalee, and H. W. Hamilton (1977). The Clark's Cave bone deposit and the late Pleistocene paleoecology of the central Appalachian Mountains of Virginia. *Bulletin Carnegie Museum of Natural History*, 2:1–87.

Guthrie, R. D. (1965). Variability in characters undergoing rapid evolution: an analysis of *Microtus* molars. *Evolution*, 19:214–33.

Haldane, J. B. S. (1949). Suggestions as to quantitative measurement of rate of evolution. *Evolution*, 3:51–6.

Hall, E. R. (1981). *The Mammals of North America*. New York: Wiley.

Harris, A. H. (1988). Late Pleistocene and Holocene *Microtus* (*Pitymys*) (Rodentia: Cricetidae) in New Mexico. *Journal of Vertebrate Paleontology*, 8:307–13.

Heinrich, W.-D. (1978). Zur biometrischen Erfassung eines Evolutionstrends bei *Arvicola* (Rodentia, Mammalia) aus dem Pleistozan Thuringens. *Saugetierkdl. Informationen*, 2:3–21.

(1990). Some aspects of the evolution and biostratigraphy of *Arvicola* (Mammalia, Rodentia) in the central European Pleistocene. In O. Fejfar and W. -D. Heinrich (eds.), *International Symposium on the Evolution, Phylogeny and Biostratigraphy of Arvicolids (Rodentia, Mammalia)* (pp. 165–82). Prague: Geological Survey.

Hershkovitz, P. (1962). Evolution of neotropical cricetine rodents (Muridae) with special reference to the phyllotine group. *Fieldiana, Zoology*, 46:1–524.

Hibbard, C. W. (1948). Techniques for collecting microvertebrate fossils. *Bulletin of the Geological Society of America*, 59:1330.

(1972). Class Mammalia. In M. F. Skinner and C. W. Hibbard (eds.), *Early Pleistocene Pre-glacial and Glacial Rocks and Faunas of North-Central Nebraska* (pp. 77–116). *Bulletin of the American Museum Natural History*.

(1975). Letter to W. G. Kuhne. In G. R. Smith and N. E. Friedland (eds.), *Studies on Cenozoic Paleontology and Stratigraphy* (pp. 135–8). C. W. Hibbard Memorial Volume 3, Museum of Paleontology, University of Michigan.

Hinton, M. A. C. (1926). Monograph of the voles and lemmings (Microtinae) living and extinct. *British Museum of Natural History Publications*, 1:1–488.

Hooper, E. T. (1952). A systematic review of the harvest mice (genus *Reithrodontomys*) of Latin America. *Miscellaneous Publications, Museum of Zoology, University of Michigan*, 77:1–255.

(1957). Dental patterns in mice of the genus *Peromyscus*. *Miscellaneous Publications, Museum of Zoology, University of Michigan*, 99:1–59.

Klippel, W. E., and P. W. Parmalee (1982). *The Paleontology of Cheek Bend Cave: Phase II Report*. Tennessee Valley Authority.

Koenigswald, W. von (1980). Schmelzstruktur und Morphologie in dem Molaren der Arvicolidae (Rodentia). *Abh. Senckenb. Naturforsch. Ges.*, 529:1–129.

Koenigswald, W. von, and L. D. Martin (1984). Revision of the fossil and Recent Lemminae (Rodentia, Mammalia). In H. H. Genoways and M. R. Dawson (eds.), *Contributions in Quaternary Vertebrate Paleontology: A Volume in Memorial to John E. Guilday* (pp. 122–37). Carnegie Museum of Natural History, Special Publication no. 8.

Kowalski, K. (1970). Variation and speciation in fossil voles. *Symposium, Zoological Society, London,* 26:149–61.

Kratochvil, J. (1980). Zur Phylogenie und Ontogenie bei *Arvicola terrestris* (Rodentia, *Arvicola). Folia Zoologica,* 29:109–24.

—— (1981). *Arvicola cantiana* vit-elle encore? *Folia Zoologica,* 30:289–300.

—— (1983). Variability of some criteria in *Arvicola terrestris* (Arvicolidae, Rodentia). *Acta Sc. Nat. Brno, N.S.,* 17:1–40.

Krishtalka, L., and R. K. Stucky (1985). Revision of the Wind River faunas, early Eocene of central Wyoming. Part 7: Revision of *Diacodexis* (Mammalia, Artiodactyla). *Annals of the Carnegie Museum of Natural History,* 54:413–86.

Kurtén, B. (1960). Rates of evolution in fossil mammals. *Cold Spring Harbor Symposium in Quantitative Biology,* 24:205–15.

Lawrence, B. (1942). The muskrat in Florida. *Proceedings of the New England Zoological Club,* 19:17–20.

Levinton, J. (1988). *Genetics, Paleontology and Macroevolution.* Cambridge University Press.

Lich, D. K. (1990). *Cosomys primus,* a case for stasis. *Paleobiology,* 16:384–95.

Linsey, A. V. (1983). *Synaptomys cooperi. Mammalian Species,* 210:1–5.

Løvtrup, S. (1984). Ontogeny and phylogeny. In M.-W. Ho and P. T. Saunders (eds.), *Beyond Neo-Darwinism* (pp. 159–92). London: Academic Press.

Ludwig, D. R. (1984). *Microtus richardsoni. Mammalian Species,* 288:1–6.

Lundelius, E. L., Jr., C. S. Churcher, T. Downs, C. R. Harington, E. L. Lindsay, G. E. Schultz, H. A. Semken, S. D. Webb, and R. J. Zakrzewski (1987). The North American Quaternary sequence. In M. O. Woodburne (ed.), *Cenozoic Mammals of North America: Geology and Biostratigraphy* (pp. 211–35). Berkeley: University of California Press.

McCarty, R. (1975). *Onychomys torridus. Mammalian Species,* 59:1–5.

—— (1978). *Onychomys leucogaster. Mammalian Species,* 87:1–6.

McNab, B. K. (1988). Complications inherent in scaling the basal rate of metabolism in mammals. *Quarterly Review of Biology,* 63:25–54.

Maiorana, V. C. (1990). Evolutionary strategies and body size in a guild of mammals. In J. Damuth and B. J. MacFadden (eds.), *Body Size in Mammalian Paleobiology* (pp. 69–102). Cambridge University Press.

Malmgren, B. A., and W. A. Berggren (1984). Species formation through punctuated gradualism in planktonic foraminifera. *Science,* 225:317–19.

Martin, L. D. (1979). The biostratigraphy of arvicoline rodents in North America. *Transactions of the Nebraska Academy of Sciences,* 7:91–100.

Martin, R. A. (1974). Fossil mammals from the Coleman IIA Fauna, Sumter County. In S. D. Webb (ed.), *Pleistocene Mammals of Florida.* Gainesville: University of Florida Press.

—— (1979). Fossil history of the rodent genus *Sigmodon. Evolutionary Monographs,* 2:1–36.

—— (1986). Energy, ecology and cotton rat evolution. *Paleobiology,* 12:370–82.

—— (1987). Notes on the classification and evolution of some North American fossil *Microtus* (Mammalia: Rodentia). *Journal of Vertebrate Paleontology*, 7:270–83.

(1989a). Arvicolid rodents of the early Pleistocene Java local fauna from north-central South Dakota. *Journal of Vertebrate Paleontology*, 9:438–50.

(1989b). Early Pleistocene zapodid rodents from the Java local fauna of north-central South Dakota. *Journal of Vertebrate Paleontology*, 9:101–9.

(1990). Estimating body size and correlated variables in extinct mammals: travels in the fourth dimension. In J. Damuth and B. J. MacFadden (eds.), *Body Size in Mammalian Paleobiology* (pp. 49–68). Cambridge University Press.

(1991). Evolutionary relationships and biogeography of late Pleistocene prairie voles from the eastern United States. *Scientific Publications, Illinois State Museum*, 23:251–60.

(1992). Generic species richness and body mass in North American mammals: support for the inverse relationship of body size and speciation rate. *Historical Biology*, 6:73–90.

Martin, R. A., and R. H. Prince (1989). A new species of early Pleistocene cotton rat from the Anza-Borrego Desert of southern California. *Bulletin, Southern California Academy of Sciences*, 88:80–7.

(1990). Variation and evolutionary trends in the dentition of late Pleistocene *Microtus pennsylvanicus* from three levels in Bell Cave, Alabama. *Historical Biology*, 4:117–29.

Martin, R. A., and R. Tedesco (1976). *Ondatra annectens* (Mammalia, Rodentia) from the Java local fauna of South Dakota. *Journal of Paleontology*, 50:846–50.

Martin, R. L. (1973). The dentition of *Microtus chrotorrhinus* and related forms. *Occasional Papers, Biological Science Series, University of Connecticut*, 2:183–201.

Mayr, E. (1963). *Animal Species and Their Evolution.* Cambridge: Harvard University Press.

(1988). *Toward a New Philosophy of Biology.* Cambridge: Harvard University Press.

Meulen, A. J. van der (1973). Middle Pleistocene smaller mammals from the Monte Peglia (Orvieto, Italy) with special reference to the phylogeny of *Microtus* (Arvicolidae, Rodentia). *Quaternaria*, 17:1–144.

(1978). *Microtus* and *Pitymys* from Cumberland Cave, Maryland, with a comparison of some New and Old World species. *Annals of the Carnegie Museum of Natural History*, 47:101–45.

Meyer, A. (1987). Phenotypic plasticity and heterochrony in *Cichlosoma managuense* (Pisces: Cichlidae) and their implications for speciation in cichlid fishes. *Evolution*, 41:1357–69.

Meyer, A., T. D. Kocher, P. Basasibwaki, A. C. Wilson (1990). Monophyletic origin of cichlid fishes suggested by mitochondrial DNA sequences. *Nature*, 347:550–3.

Miller, G. (1896). The genera and subgenera of voles and lemmings. *North American Fauna*, 12:1–17.

Morlan, R. E. (1989). Paleoecological implications of late Pleistocene and Holocene microtine rodents from the Bluefish Caves, North Yukon Territory. *Canadian Journal of Earth Science*, 26:149–56.

Musser, G. G. (1981). The giant rat of Flores and its relatives east of Borneo and Bali. *Bulletin of the American Museum of Natural History*, 169:69–175.

Nadachowski, A. (1982). Late Quaternary rodents of Poland with a special reference to morphotype dentition analysis of voles. *Polska Academia Nauk, Panstwo we Wydawnictwo Naukowe, Krakow:* pp. 1–108.

(1984). Morphometric variability of dentition of the late Pleistocene voles (Arvicolidae, Rodentia) from Bacho Kiro Cave (Bulgaria). *Acta Zoologica Cracoviensia*, 27:149–76.

(1990). Comments on variation, evolution and phylogeny of *Chionomys* (Arvicolidae). In O. Fejfar and W.-D. Heinrich (eds.), *International Symposium on the Evolution, Phylogeny and Biostratigraphy of Arvicolids (Rodentia, Mammalia)* (pp. 353–68). Prague: Geological Survey.

Nelson, R. S., and H. A. Semken (1970). Paleoecological and stratigraphic significance of the muskrat in Pleistocene deposits. *Bulletin of the Geological Society of America*, 81:3733–8.

Nevo, E. (1991). Evolutionary theory and processes of active speciation and adaptive radiation in subterranean mole rats, *Spalax ehrenbergi* superspecies, in Israel. *Evolutionary Biology*, 24:1–124.

Oppenheimer, J. R. (1965). Molar cusp pattern variations and their interrelationships in the meadow vole, *Microtus p. pennsylvanicus* (Ord). *American Midland Naturalist*, 74:39–49.

Paulson, G. R. (1961). The mammals of the Cudahy fauna. *Papers of the Michigan Academy of Science, Arts and Letters*, 46:127–53.

Peters, R. H. (1983). *The Ecological Implications of Body Size*. Cambridge University Press.

Pfaff, K. S. (1990). Irvingtonian *Microtus, Pedomys*, and *Pitymys* (Mammalia, Rodentia, Cricetidae) from Trout Cave No. 2, West Virginia. *Annals of the Carnegie Museum of Natural History*, 59:105–34.

Rabeder, G. (1981). Die Arvicoliden (Rodentia, Mammalia) aus dem Pliozan und dem alteren Pleistozan von Niederösterreich. *Beitrage Paläontologie und Geologie Oesterreich*, 8:1–373.

Raup, D. M. (1977). Stochastic models in evolutionary paleontology. In A. Hallam (ed.), *Patterns of Evolution* (pp. 59–68). Amsterdam: Elsevier.

Repenning, C. A. (1968). Mandibular musculature and the origin of the subfamily Arvicolinae (Rodentia). *Acta Zoologica Cracoviensia*, 13:29–72.

(1983). *Pitymys meadensis* from the Valley of Mexico and the classification of North American species of *Pitymys* (Rodentia: Cricetidae). *Journal of Vertebrate Paleontology*, 2:471–82.

(1987). Biochronology of the microtine rodents of the United States. In M. O. Woodburne (ed.), *Cenozoic Mammals of North America: Geology and Biostratigraphy* (pp. 236–68). Berkeley: University of California Press.

Repenning, C. A., O. Fejfar, and W.-D. Heinrich (1990). Arvicolid rodent biostratigraphy of the northern hemisphere; pp. 385–417. In O. Fejfar and W.-D. Heinrich (eds)., *International Symposium on the Evolution, Phy-*

logeny and biostatigraphy of Arvicolids (Rodentia, Mammalia). Prague: Geological Survey.

Rose, K. D., and T. M. Bown (1986). Gradual evolution and species discrimination in the fossil record. *Contributions to Geology, University of Wyoming, Special Papers* 3:119–30.

Ross, R. M., and W. D. Allmon (eds.) (1990). *Causes of Evolution: A Paleontological Perspective*. Chicago: University of Chicago Press.

Saunders, P. T. (1984). Development and evolution. In M.-W. Ho and P. T. Saunders (eds.), *Beyond Neo-Darwinism* (pp. 243–66). London: Academic Press.

Schoch, R. M. (1986). *Phylogeny Reconstruction in Paleontology*. New York: Van Nostrand Reinhold.

Selander, R. K. (1970). Behavior and genetic variation in natural populations. *American Zoologist*, 10:53–66.

Semken, H. A., Jr. (1966). Stratigraphy and paleontology of the McPherson *Equus* beds (Sandahl local fauna), McPherson County, Kansas. *Contributions, Museum of Paleontology, University of Michigan*, 20:121–78.

Simpson, G. G. (1961). *Principles of Animal Taxonomy*. New York: Columbia University Press.

Smartt, R. A. (1977). The ecology of late Pleistocene and recent *Microtus* from south-central and southwestern New Mexico. *Southwestern Naturalist*, 22:1–19.

Stangl, F. B., and W. W. Dalquest (1990). Discrimination of two species of cotton rats (*Sigmodon*) using characters of the upper first molar. *Texas Journal of Science*, 42:333–8.

Stanley, S. M. (1975). A theory of evolution above the species level. *Proceedings of the National Academy of Sciences, USA*, 72:646–50.

(1979). *Macroevolution: Pattern and Process*. San Francisco: Freeman.

(1985). Rates of evolution. *Paleobiology*, 11:13–26.

Stanley, S. M., and X. Yang (1987). Approximate evolutionary stasis for bivalve morphology over millions of years: a multivariate, multilineage study. *Paleobiology*, 13:113–39.

Stephens, J. J. (1960). Stratigraphy and paleontology of a late Pleistocene basin, Harper County, Oklahoma. *Bulletin of the Geological Society of America*, 71:1675–702.

Stuart, A. J. (1982). *Pleistocene Vertebrates in the British Isles*. London: Longman.

Ursin, E. (1949). Variation in number of enamel loops in the two anterior upper cheekteeth in Danish *Microtus agrestis* (L.). *Videnskabelige Meddelelser fra Dansk Naturhistorisk Forening*, 111:257–61.

Van Valen, L. (1973). A new evolutionary law. *Evolutionary Theory*, 1:1–30.

Van Valkenburgh, B. (1990). Skeletal and dental predictors of body mass in carnivores. In J. Damuth and B. J. MacFadden (eds.), *Body Size in Mammalian Paleobiology: Estimation and Biological Implications* (pp. 181–206). Cambridge University Press.

Vrba, E. S. (1980). Evolution, species and fossils: How does life evolve? *South African Journal of Science*, 76:61–84.

(1983). Macroevolutionary trends: new perspectives on the roles of adaptation and incidental effect. *Science*, 221:387–9.

(1984a). What is species selection? *Systematic Zoology*, 33:318–28.

(1984b). Patterns in the fossil record and evolutionary processes. In M.-W. Ho and P. T. Saunders, (eds.), *Beyond Neo-Darwinism* (pp. 115–42). London: Academic Press.

(1985). Environment and evolution: alternative causes of the temporal distribution of evolutionary events. *South African Journal of Science*, 81: 229–36.

(1987). Ecology in relation to speciation rates: some case histories of Miocene-Recent mammal clades. *Evolutionary Ecology*, 1:283–300.

(1989). Levels of selection and sorting with special reference to the species level. In P. H. Harvey and L. Partridge (eds.), *Oxford Surveys in Evolutionary Biology*, Vol. 6 (pp. 111–68). Oxford: Oxford University Press.

Vrba, E. S., and N. Eldredge (1984). Individuals, hierarchies and processes: towards a more complete evolutionary theory. *Paleobiology*, 10:146–71.

Vrba, E. S., and S. J. Gould (1986). The hierarchical expansion of sorting and selection cannot be equated. *Paleobiology*, 12:217–28.

Webb, S. D., and A. D. Barnosky (1989). Faunal dynamics of Pleistocene mammals. *Annual Review of Earth and Planetary Sciences*, 17:413–38.

White, J. A., and T. Downs (1961). A new *Geomys* from the Vallecito Creek Pleistocene of California. *Contributions in Science, Los Angeles County Museum*, 42:1–34.

Wilkins, K. E. (1984). Evolutionary trends in Florida Pleistocene pocket gophers (genus *Geomys*), with description of a new species. *Journal of Vertebrate Paleontology*, 3:166–81.

Wright, S. (1982). The shifting balance theory and macroevolution. *Annual Reviews of Genetics*, 16:1–19.

Zakrzewski, R. J. (1969). The rodents from the Hagerman local fauna, upper Pliocene of Idaho. *Contributions, Museum of Paleontology, University of Michigan*, 23:1–36.

(1985). The fossil record. In R. H. Tamarin, (ed.), *Biology of New World Microtus* (pp. 1–51). American Society of Mammalogists, Special Publication no. 8.

12

Decrease in body size
of white-tailed deer (*Odocoileus virginianus*)
during the late Holocene
in South Carolina and Georgia

JAMES R. PURDUE AND ELIZABETH J. REITZ

Body size is an important determinant for many biological functions and characters closely associated with fitness (e.g., Peters, 1983; Calder, 1984). Many of the genes that determine traits like body size probably are invariant in a population (Fisher, 1930) and have low heritability due to past selection. Indeed, body size in mammals has been shown to have low heritability (Falconer, 1989). Body size also tends to be quite responsive to changes in certain environmental factors that in turn serve as the ultimate sources of selection (Falconer, 1989). Studies of animals that have changed body size rapidly in the fossil record could be useful to an understanding of evolution in response to variations in paleoecology.

White-tailed deer (*Odocoileus virginianus*) have varied greatly in body size through space (Rees, 1969; Koch, 1986) and time (Purdue, 1986, 1989, 1991). In the American Midwest, the shifting size through time was associated with changes in climate and vegetation (Purdue, 1989, 1991). Small deer prevailed during the warm and dry middle Holocene period that produced prairie expansion (~8.5–4 ky BP). In the late Holocene, climatic conditions became more mesic, and deer became large.

Late Holocene environments on the coastal plain in the Southeast contrasted sharply with those in the Midwest (Delcourt and Delcourt, 1985, 1987; Webb, 1988). Southern pine forests began to replace oaks and hickories after 8 ky BP, with the trend continuing through the late Holocene (Delcourt and Delcourt, 1985; Webb, 1988). Other environmental changes were also evident in the late Holocene, in particular the development from 4 ky BP to the present of the active floodplain of the

281

Savannah River in the upper coastal plain (Brooks and Sassaman, 1990). The higher river terraces also continued to aggrade, but at a slower rate than in the early to middle Holocene (Brooks et al., 1986). Few workers question that fire was heavily involved in causing or prolonging the environmental changes in the late Holocene, but there is less agreement about the role of climate (Delcourt and Delcourt, 1985; Webb et. al, 1987). An analysis of the size of white-tailed deer may help to resolve the issue.

The premise that best explains body size in deer focuses on the amount and duration of high-quality summer forage (Guthrie, 1984; Klein, 1985; Geist, 1987; Purdue, 1989). When high-protein, low-fiber food is abundantly available for a prolonged period into the summer, somatic growth in deer is maximized (Geist, 1987). In essence, deer body size (excluding fat) is sensitive to summer forage conditions and insensitive to ecological parameters during the rest of the year (Purdue, 1989).

Computer models of paleoclimate indicate a reduction in summer monsoon rains in the Southeast during the late Holocene. (Kutzbach, 1987; Webb et al., 1987). If that is true, the drier summers would have provoked plants to curtail the production of new growth, thus reducing a critical resource for deer. Over time, smaller body size in deer would be expected.

Delcourt and Delcourt (1985) offer an alternative hypothesis to explain the continuing shift to pine in the late Holocene. They suggest that the shift was the result of human-set fires in the sandy uplands of the coastal plain. [Delcourt and Delcourt (1985) also offer a climate-based model, but they do not state which hypothesis they prefer.] Previously, Küchler (1964) and Wahlenberg (1946) also suggested that anthropogenic fires were important for the persistence of pine. Fire releases nutrients for vigorous new plant growth and creates more forest-edge vegetation (Crawford, 1984). In the absence of climate change, forage conditions for deer would be enriched, and body size would be, at worst, unchanged or, more likely, increased. One objective of the current study is to determine deer size and evaluate the competing models that seek to explain important vegetational patterns in the Southeast.

The vegetational shift was not the only significant environmental change that occurred in the Southeast during the Holocene. On the coast, barrier islands came into being at 4 ky BP when sea level ceased its rise from Pleistocene levels (DePratter, 1979a; Dolan, Hayden, and Lins, 1980). Through time and because of wave action, the detached landforms became increasingly isolated from the mainland, eventually developing the resource-poor environments typical of islands. No doubt the selective regime for deer living on the islands was also altered. A second objective of this study is to document through time the body size variations of deer subjected to developing island environments.

Methods

White-tailed deer were selected from 22 analytical units. An analytical unit represents one or more samples of archaeological or modern specimens from similar places and times. Eight of these units are from archaeological sites representing a temporal sequence from central Illinois ranging from 5.3 ky BP to the present (Purdue, 1989). Archaeological sites and modern collections provided two sequences of four analytical units each from the Southeast (Table 12.1). One sequence, 3,600 years in duration, was from the barrier islands of St. Catherines and Ossabaw off the coast of Georgia (Figure 12.1). A mainland sequence that spanned 4,400 years was centered on and around the Savannah River Site located in Aiken and Barnwell counties, South Carolina, near the Savannah River on the upper coastal plain (Figure 12.1).

The remaining seven analytical units in the study were samples of modern white-tailed deer from widely separated localities in the eastern United States [for a map, see Figure 3-1 of Purdue (1986)]. These units were used to determine the sex of archaeological specimens, as discussed later, and three of the seven units were used as modern points in the temporal sequences.

The astragalus, a bone in the hindfoot, is a good indicator of adult body size (Purdue, 1987) and is often preserved whole in archaeological sites. To discriminate subtle differences in the sizes and shapes of these elements, we made six measurements on each astragalus (Figure 12.2). Only whole astragali were measured. The occasional missing value was estimated from the reference specimens using a regression based on a highly correlated (typically, $r > 0.7$) measurement. Ontogenetic variation was ignored because the astragalus is near adult size in 6-month-old fawns (Purdue, 1986, 1987). In contrast, the bones differ according to sex because the species is dimorphic. Purdue (1983, 1986) reported alterations of standard statistical procedures that estimated sex for each specimen and then controlled the effect of sex on measurements used for studying clinal variation. Briefly, measurements were transformed into z-scores separately for each of the analytical units. As a result,

Table 12.1. *Analytical units from Georgia and South Carolina*

Analytical unit	Abbreviation	Date[a] (years BP)	Source
Savannah River Site, Aiken and Barnwell counties, South Carolina	SRS	0	Purdue (1986, 1989)
Lewis Site (38AK228), Aiken County, South Carolina	LS	1,750	Sassaman et al. (1989)
Stallings Island Site (9CB1), Columbia County, Georgia	ST	3,700	Sassaman (1991)
Fennel Hill Site (38AL2), Allendale County, South Carolina	FH	4,400	Sassaman (1991)
Ossabaw Island, Chatham County, Georgia	OSS	0	Purdue (1986, 1989)
Santa Catalina de Guale, St. Catherines Island, Liberty County, Georgia	STC	325	Reitz (1989)
Irene, Savannah, St. Catherines, and Wilmington period sites, St. Catherines Island, Liberty County, Georgia	SCI	950	Reitz (1989)
St. Simons Period sites, St. Catherines Island, Liberty County, Georgia, and Cane Patch Site (9CH35), Ossabaw Island, Chatham County, Georgia	ILA	3,600	DePratter (1979a,b)

[a]A representative date in middle of range.
Source: See Purdue (1986, 1989) for detailed information about analytical units used for sex determinations and for archaeological samples from Illinois.

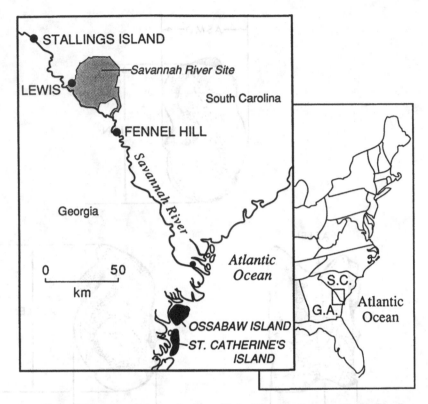

Figure 12.1. Map of the southeastern United States and parts of South Carolina and Georgia.

females from all units were negative, and males were positive (with some overlap), even though the mean size of deer varied considerably between analytical units. These and subsequent statistical procedures were insensitive to skewed sex ratios in the samples (Purdue, 1983). Standardized scores were then entered into a single discriminant function analysis, using modern deer as reference groups for sex, and archaeological specimens as unknowns. Each astragalus was assigned (for deer of unknown sex) or reclassified (known sex) to a sex (Purdue, 1983). In all modern units, and presumably all prehistoric samples, ~20% of the deer were misclassified. To correct the influence of these misclassifications on the descriptive statistics for each unit, regressions were calculated for the reference analytical units that compared the known values of $\overline{X}$, SD, and N with those based on the same specimens sexed by discriminant function analysis. The regressions then estimated a set

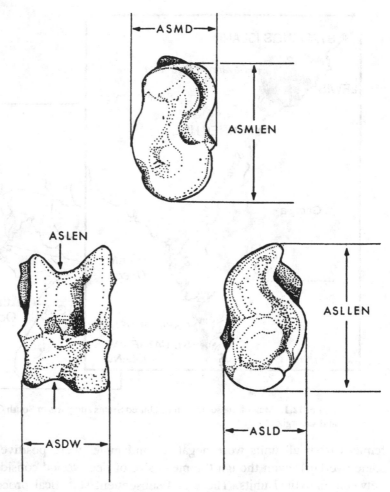

Figure 12.2. Measurements of the astragali of *Odocoileus virginianus*. Abbreviations: ASMD, medial depth; ASMLEN, medial length; ASLEN, length; ASDW, distal width; ASLLEN, lateral length; ASLD, lateral depth.

of "corrected" descriptive statistics for the archaeological units. For modern test samples, corrected means were within 0.5% of the actual means (Purdue, 1986).

With sexes treated independently, the southeastern temporal sequences were tested by analyses of variance (ANOVA). Student-Newman-Keuls (SNK) procedures were employed to detect homogeneous groups of analytical units. For ease of reference and for graphic display, male and female body weights for each analytical unit were

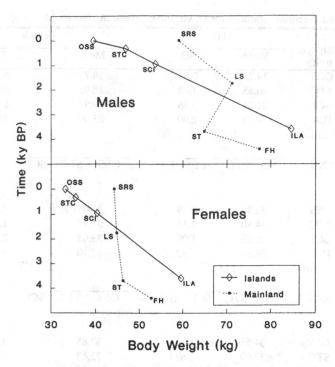

Figure 12.3. Temporal changes in estimated total body weights of *Odocoileus virginianus* in the southeastern United States. See Table 12.1 for abbreviations of analytical units.

calculated from regressions that utilized the means of the medial depth, medial length, and distal width as independent variables (Purdue, 1987). The variability of the weight estimate could not be calculated because of unknown interactions between the several variances in the complex algorithm.

Results

Both of the southeastern temporal sequences showed significant decreases in the size of white-tailed deer (Figure 12.3). The reduction was more dramatic for the island deer, which at 3.6 ky BP were slightly larger than conspecifics inhabiting the upper coastal plain (ILA and ST, analytical units at approximately the same time, were significantly different, at $p < 0.05$ or less, for three measurements out of six on males and all six on females; Tables 12.2 and 12.3). The island deer steeply

Table 12.2. *Corrected descriptive statistics for* O. virginianus *from the barrier islands of Georgia*

Measurement	Analytical unit	Male		Female	
		Mean	SD	Mean	SD
Medial depth	OSS	19.75	0.51	18.75	0.79
	STC	20.55	1.42	18.92	0.67
	SCI	21.24	0.66	19.97	1.03
	ILA	24.73	2.00	23.60	1.93
	F ratio	10.00***		31.68***	
	SNK	OSS STC SCI ILA		OSS STC SCI ILA	
Medial length	OSS	32.44	0.99	31.05	1.17
	STC	34.50	1.33	31.97	1.21
	SCI	35.63	1.09	33.63	2.69
	ILA	38.80	2.52	37.10	1.52
	F ratio	13.66***		21.24***	
	SNK	OSS STC SCI ILA		OSS STC SCI ILA	
Lateral length	OSS	34.55	1.13	32.85	1.15
	STC	37.09	1.68	33.43	1.53
	SCI	37.62	1.40	35.26	2.52
	ILA	43.20	3.10	40.45	1.04
	F ratio	14.14***		35.73***	
	SNK	OSS STC SCI ILA		OSS STC SCI ILA	
Distal width	OSS	22.02	0.96	20.57	0.84
	STC	22.78	1.17	21.32	1.16
	SCI	23.76	1.03	22.24	1.39
	ILA	26.82	1.12	26.41	0.84
	F ratio	10.71***		43.83***	
	SNK	OSS STC SCI ILA		OSS STC SCI ILA	

and steadily continued to decline in size to the present time (Table 12.2). By body weight, the males decreased 52%, and the females 47% in 3,600 years. Using ANOVAs to compare analytical units on the islands, measurements for both sexes were statistically significant ($p < 0.001$). SNK tests indicated no consistent pattern of homogeneous analytical

Table 12.2 (*cont.*)

Measurement	Analytical unit	Male Mean	Male SD	Female Mean	Female SD
Lateral depth	OSS	18.90	0.78	18.14	0.62
	STC	20.22	1.19	18.61	0.74
	SCI	20.74	0.90	19.76	1.10
	ILA	23.47	1.23	22.60	0.68
	F ratio	9.68***		50.54***	
	SNK	OSS STC SCI ILA		OSS STC SCI ILA	
Length	OSS	27.72	1.27	26.31	0.79
	STC	29.98	1.38	27.25	1.15
	SCI	30.55	1.29	28.53	2.74
	ILA	34.29	1.50	32.31	1.21
	F ratio	12.18***		24.43***	
	SNK	OSS STC SCI ILA		OSS STC SCI ILA	

Note: Abbreviations for analytical units are listed in Table 12.1. Also shown are F ratios for ANOVAs comparing analytical units for each measurement and sex. Abbreviations for statistical significance are $*p < 0.05$; $**p < 0.01$; and $***$, $p < 0.001$. Homogeneous groups of analytical units at the 0.05 level were determined by the SNK procedure. Sample sizes (σ, $\circ$) are OSS (5, 12); STC (12, 11); SCI (9, 6) and ILA (2,6).

units. Eventually, the island deer became much smaller than the mainland animals. ANOVAs comparing modern analytical units OSS and SRS indicated that all measurements were statistically significant for both sexes at the 0.001 level.

On the mainland, the decline in size was less precipitous (Table 12.3). Nevertheless, the males decreased 23% in body weight, and the females 15%, over the course of 4,400 years. All measurements for both sexes from the analytical units on the mainland were statistically significant at the 0.05 level or less. However, the middle portions of the curves through time were different for the sexes (Table 12.3, Figure 12.3). For males, analytical unit LS is larger than unit ST. SNK tests indicate that for some measurements the two units are in homogeneous groups; for other measurements they are not (Table 12.3). For females, the results are more consistent: SRS, LS, and ST are always in a homogeneous group (Table 12.3). Regardless of the uncertainty about the shapes of the curves, mainland deer clearly decreased in size during the late Holocene.

In Figure 12.4, the size of deer on the upper coastal plain of the Southeast is contrasted with that of white-tails in central Illinois. Be-

Table 12.3. *Corrected descriptive statistics for* O. virginianus *from the upper coastal plain of Georgia and South Carolina*

Measurement	Analytical unit	Male		Female	
		Mean	SD	Mean	SD
Medial depth	SRS	22.04	1.08	20.84	1.05
	LS	23.12	1.05	20.82	1.57
	ST	22.84	1.46	21.22	1.37
	FH	23.51	0.82	22.42	1.47
	F ratio	5.21**		5.83***	
	SNK	SRS ST LS FH		LS SRS ST FH	
Medial length	SRS	36.34	1.37	34.71	1.30
	LS	36.97	1.29	34.51	2.19
	ST	36.60	1.56	34.69	1.56
	FH	38.02	0.75	36.43	1.82
	F ratio	3.16*		4.92**	
	SNK	SRS ST LS FH		LS ST SRS FH	
Lateral length	SRS	39.07	1.47	37.08	1.35
	LS	40.64	1.74	37.15	2.14
	ST	39.42	1.59	37.19	1.50
	FH	41.05	1.28	38.96	1.50
	F ratio	5.11**		5.76***	
	SNK	SRS ST LS FH		SRS LS ST FH	

tween 4.0 and 5.0 ky BP, deer from the two regions were not different in size. From that time forward, southeastern deer decreased in size. Midwestern animals first increased to very large proportions at 2.0–1.0 ky BP, then declined slightly over the past 1,000 years.

Table 12.3 (*cont.*)

Measurement	Analytical unit	Male		Female	
		Mean	SD	Mean	SD
Distal width	SRS	24.20	1.21	23.02	1.06
	LS	26.27	1.06	23.35	1.29
	ST	24.97	0.98	23.56	0.98
	FH	26.85	1.35	24.71	1.26
	F ratio	15.64***		8.46***	
	SNK	SRS ST LS FH		SRS LS ST FH	
Lateral depth	SRS	21.47	0.98	20.42	0.80
	LS	22.57	1.08	20.65	1.90
	ST	21.97	0.95	20.54	0.92
	FH	23.09	0.79	21.60	0.95
	F ratio	7.55***		4.92**	
	SNK	SRS ST LS FH		SRS ST LS FH	
Length	SRS	31.15	1.18	29.51	1.15
	LS	32.43	1.44	30.19	1.43
	ST	31.70	1.45	29.98	1.24
	FH	33.51	1.25	31.77	1.47
	F ratio	7.95***		11.49***	
	SNK	SRS ST LS FH		SRS ST LS FH	

Note: ANOVAs and SNK tests were the same as described in Table 12.2. See Tables 12.1 and 12.2 for abbreviations. Sample sizes (σ,φ) are SRS (42, 51); LS (8, 9); ST (21, 21); FH (7, 13).

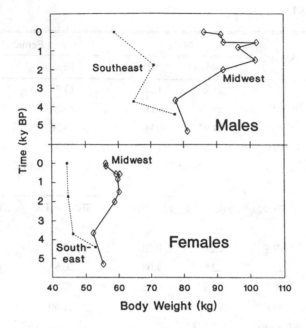

Figure 12.4. Temporal changes in estimated total body weights for *Odocoileus virginianus* in the Southeast and in Illinois. See Table 12.1 for abbreviations of analytical units.

Discussion

Ungulates inhabiting islands often become dwarfed (Case, 1978; Lomolino, 1985). This phenomenon was also common in the past. For example, *Cervus elaphus* during the last interglacial on Jersey, an island off the northern coast of France, displayed a sixfold decrease in size within 6,000 years of the island's separation from the mainland (Lister, 1989). Models explaining the small size of ungulates on islands emphasize that the decrease in size is an adaptation to reduced food resources (Case, 1978; Lomolino, 1985).

Odocoileus virginianus inhabiting Georgia's barrier islands displayed a sharp, unidirectional reduction in size over time. At 3.6 ky BP, the islands were only about 400 years old (DePratter, 1979a; Dolan et al., 1980), and deer size was statistically larger than that for contemporaneous animals on the upper coastal plain. Movement of deer between the mainland and the islands may have occurred occasionally at that time. Likewise, the semiisolated landforms may still have had many ecological characteristics typical of the mainland. With increasing isolation and the passage of time, the island environment gradually developed. Food resources for deer probably followed suit and gradually declined to the modern condition in which forage on the islands (LaGory, 1984) is demonstrably poorer than that on the mainland (Moore, 1967). Although the moderating effects of maritime weather confuse the situation, climate changes in the region, as discussed later, also may have played a role in reducing the islands' food resources.

White-tailed deer have a great capacity for varying in body size (Rees, 1969; Koch, 1986; Geist, 1987; Purdue, 1989). Some of that variability probably is due to nongenetic physiological responses to the quality and duration of summer food (Geist, 1987). But in the case of the deer on the barrier islands, there is evidence that some genetic alterations have occurred. Allele frequency data for deer on the nearby barrier islands of Blackbeard (Hillstead, 1984) and Cumberland (Rowland, 1989) showed a closer genetic relationship among islands than between island and mainland. Rowland (1989) concluded that "these island herds are somewhat [distinctive] in their genetic makeup." Likely, the decrease in body size through time that we have observed for the island deer is due to some combination of genetic and environmental effects, although the relative importance of such effects is unknown. Eventually the island deer became sufficiently small and distinct (Brisbin and Lenarz, 1984) that traditional taxonomists assigned subspecific appellations to the modern populations (Goldman and Kellogg, 1940). In any event, the small-sized modern island deer may be an example of fast evolution in a rapidly changing environment.

The mainland deer also showed a decrease in body size. The lack of concordance between the sexes in the shapes of the size curves through time is troubling. Most likely the difficulty is due to low sample sizes for some of the analytical units. Nevertheless, the statistical analyses verify that deer diminished in size during the late Holocene. As more sites and samples become available, the timing and rate of decline will be clarified.

The decrease in deer size gives support to the hypothesis that climate was the ultimate cause of the environmental changes seen on the upper coastal plain over the past 4,000 years. Simulations of paleoclimate show increased annual dryness due to a reduction in summer monsoon rains in the Southeast over the past 3,000 years (Kutzbach, 1987; Webb et al., 1987). The drier regime would have inhibited the growth of forage for the deer.

The pine forests in the Southeast are maintained in part by periodic fires (Wahlenberg, 1946; Küchler, 1964). No doubt, fires were frequent in the past, and many were set by Native Americans (Hudson, 1976). However, despite all the stimulation that fires provided for plant growth, the negative effect of dry summers on deer forage could not be overcome. We conclude that climate change was the dominate cause of the vegetational shifts in the Southeast.

Over the past 4,400+ years, the sizes of deer in the Southeast and the Midwest have, for the most part, been marked by opposing trends. In the Midwest, the climatic conditions that wrought prairie expansion were ameliorating. The average July temperature, for example, currently is 2°C less than that in the middle Holocene (Bartlein, Webb, and Fleri, 1984; Kutzbach and Guetter, 1986). The conditions that allowed deciduous forest to encroach on grasslands also increased the availability of high-quality food later into the summer (Purdue, 1989, 1991). Consequently, deer grew to greater dimensions. The midwestern animals stand in sharp contrast to their conspecifics in the Southeast that experienced less suitable summer conditions in the late Holocene. The dynamics that we document in deer size reflect the regions' differing environmental histories.

Could factors other that those suggested earlier be responsible for the dynamics in the body sizes of deer? Bergmann's rule, fetal nutrition, seasonality of climate, and anthropogenic influences have been proposed as possible explanations (Purdue, 1989). Although some of these factors can be important in a limited area or for a short time (in the geological sense), climate change and its effect on summer food resources proved the most robust argument for explaining the size shifts in deer (Purdue, 1989). Note, however, the exception offered by the barrier islands. Here, deer size continued to be impacted strongly by limited resources (probably summer forage), but the resources were mainly driven by the developing insularity, not so much by climate.

Genetically, body size in mammals is complex, and its heritability is low, the result of past selective pressure (Falconer, 1989). As summer food resources for deer changed in the past, body size probably shifted rapidly through an ecophenotypic response. However, the changed environment also brought forth a new selection regime, one that would favor an efficient, and probably different, genotype for the new optimum body size. Over the past 4,400 years, genetic shifts most likely occurred in the mainland deer. Unfortunately, the extent of genetic change is not yet known.

Acknowledgments

This study was conducted while the senior author was on sabbatical at The University of Georgia's Savannah River Ecology Laboratory. The research and manuscript preparation were supported in part by contract DE-AC09-SROO-819 from the U.S. Department of Energy. The study was aided by M. J. Brooks, K. E. Sassaman, and staff members at the Savannah River Archaeological Research Program, the South Carolina Institute of Archaeology and Anthropology, and the University of South Carolina. The institute's Columbia staff made available their archaeological collections. For the St. Catherines Island portion of the study, we thank D. H. Thomas of the American Museum of Natural History and R. Hayes, superintendent of St. Catherines Island. Funding for the original faunal identification and analysis was provided through the generosity of the Edward John Noble and St. Catherines Island Foundation. Thanks go also to the staff of the Illinois State Museum. The manuscript was critically read by A. D. Barnosky, M. J. Brooks, A. M. Lister, R. A. Martin, K. E. Sassaman, M. H. Smith, and J. O. Wolff.

References

Bartlein, P. J., T. Webb, III, and E. Fleri (1984). Holocene climatic change in the northern midwest: pollen-derived estimates. *Quaternary Research,* 22:361–74.

Brisbin, I. L., Jr., and M. S. Lenarz (1984). Morphological comparisons of insular and mainland populations of southeastern white-tailed deer. *Journal of Mammalogy,* 65:44–50.

Brooks, M. J., and K. E. Sassaman (1990). Point bar geoarchaeology in the upper coastal plain of the Savannah River valley, South Carolina; a case study. In N. P. Lasca and J. Donahue (eds.), *Archaeological Geology of North America.* Centennial Special Volume 4. Boulder, Colo.: Geological Society of America.

Brooks, M. J., P. A. Stone, D. J. Colquhoun, J. G., Brown, and K. B. Steele. (1986). Geoarchaeological research in the coastal plain portion of the Savannah River valley. *Geoarchaeology,* 1:293–307.

Calder, W. A., III (1984). *Size, Function and Life History.* Cambridge, Mass.: Harvard University Press.

Case, T. J. (1978). A general explanation for insular body size trends in terrestrial vertebrates. *Ecology,* 59:1–18.

Crawford, H. S. (1984). Habitat management. In L. K. Halls (ed.), *White-tailed Deer: Ecology and Management* (pp. 629–46). Harrisburg, Pa.: Stackpole Books.

Delcourt, H. R., and P. A. Delcourt (1985). Quaternary palynology and vegetational history of the southeastern United States. In V. M. Bryant, Jr. and R. G. Holloway (eds.), *Pollen Records of Late-Quaternary North American Sediments* (pp. 1–37). American Association of Stratigraphic Palynologists Foundation.

Delcourt, P. A., and H. R. Delcourt (1987). *Long-Term Forest Dynamics of the Temperate Zone.* New York: Springer-Verlag.

DePratter, C. B. (1979a). Shellmound Archaic on the Georgia coast. *South Carolina Antiquities,* 11:1–69.

(1979b). Ceramics. In D. H. Thomas and C. S. Larsen (eds.), *The Anthropology of St. Catherines Island. 2: The Refuge-Deptford Mortuary Complex* (pp. 109–32). American Museum of Natural History, Anthropological Paper 56.

Dolan, R., B. Hayden, and H. Lins (1980). Barrier islands. *American Scientist,* 68:16–25.

Falconer, D. S. (1989). *Introduction to Quantitative Genetics,* 3rd ed. Essex, U.K.: Longman Scientific.

Fisher, R. A. (1930). *The Genetical Theory of Natural Selection.* Oxford: Clarendon Press.

Geist, V. (1987). On speciation in Ice Age mammals, with special reference to cervids and caprids. *Canadian Journal of Zoology,* 65:1067–84.

Goldman, E. A., and R. Kellogg (1940). Ten new white-tailed deer from North and Middle America. *Proceedings of the Biological Society of Washington,* 53:81–90.

Guthrie, R. D. (1984). Mosaics, allelochemics and nutrients. In P. S. Martin and R. G. Klein (eds.), *Quaternary Extinctions.* Tucson: University of Arizona Press.

Hillstead, H. O. (1984). Stocking and genetic variability of white-tailed deer in the southeastern United States. Ph.D. thesis, University of Georgia, Athens.

Hudson, C. (1976). *The Southeastern Indians.* Knoxville: University of Tennessee Press.

Klein, D. R. (1985). Population biology: the interaction between deer and their food supply. *The Royal Society of New Zealand, Bulletin,* 22:13–22.

Koch, P. L. (1986). Clinal geographic variation in mammals: implications for the study of geoclines. *Paleobiology,* 12:269–81.

Küchler, A. W. (1964). *Potential Natural Vegetation of the Conterminous United States.* American Geographical Society Special Publication 36.

Kutzbach, J. E. (1987). Model simulations of the climatic patterns during the deglaciation of North America. In W. F. Ruddiman and H. E. Wright, Jr.(eds.), *North American and Adjacent Oceans during the Last Deglaciation* (pp. 425–46). (*The Geology of North America, Vol. K-3*). Boulder, Colo.: Geological Society of America.

Kutzbach, J. E., and P. J. Guetter (1986). The influence of changing orbital parameters and surface boundary conditions on climate simulations for the past 18,000 years. *Journal of Atmospheric Sciences,* 43:1726–59.

LaGory, K. E. (1984). Ecological and behavioral correlates of deer sociality: a study of Ossabaw Island white-tailed deer. Ph.D. thesis, Miami University, Oxford, Ohio.

Lister, A. M. (1989). Rapid dwarfing of red deer on Jersey in the last interglacial. *Nature,* 342:539–42.

Lomolino, M. V. (1985). Body size of mammals on islands: the island rule reexamined. *American Naturalist,* 125:310–16.

Moore, W. H. (1967). *Deer Browse Resources of the Atomic Energy Commission's Savannah River Project Area.* U.S. Forest Service Resource Bulletin SE-6.

Peters, R. H. (1983). *The Ecological Implications of Body Size.* Cambridge University Press.

Purdue, J. R. (1983). Methods of determining sex and body size in prehistoric samples of white-tailed deer (*Odocoileus virginianus*). *Transactions of the Illinois State Academy of Science,* 76:351–7.

(1986). The size of white-tailed deer (*Odocoileus virginianus*) during the Archaic period in central Illinois. In S. W. Neusius (ed.), *Foraging, collecting, and harvesting: Archaic period subsistence and settlement in the eastern woodlands* (pp. 65–95). Southern Illinois University at Carbondale, Center for Archaeological Investigations, Occasional Paper 6.

(1987). Estimation of body weight of white-tailed deer (*Odocoileus virginiaus*) from bone size. *Journal of Ethnobiology,* 7:1–12.

(1989). Changes during the Holocene in the size of white-tailed deer (*Odocoileus virginianus*) from central Illinois. *Quaternary Research,* 32:307–16.

(1991). Dynamism in the body size of white-tailed deer (*Odocoileus virginianus*) from southern Illinois. In J. R. Purdue, W. E. Klippel, and B. W. Styles (eds.), *Beamers, Bobwhites, and Blue-points. Tributes to the Career of Paul W. Parmalee* (pp. 277–83). Illinois State Museum, Scientific Papers, Volume 23.

Rees, J. W. (1969). Morphologic variation in the mandible of the white-tailed deer (*Odocoileus virginianus*): a study of populational skeletal variation by principal component and canonical analyses. *Journal of Morphology,* 128:113–30.

Reitz, E. J. (1989). Vertebrate fauna from the St. Catherines Island transect survey. Manuscript on file, Museum of Natural History, University of Georgia, Athens.

Rowland, R. D. (1989). Population genetics of white-tailed deer on Cumberland Island, Georgia. M.S. thesis, University of Georgia, Athens.

Sassaman, K. E. (1991). Economic and social contexts of early ceramic vessel technology in the American southeast. Ph.D. thesis, University of Massachusetts.

Sassaman, K. E., M. J. Brooks, G. T. Hanson, and D. G. Anderson (1989). *Technical Synthesis of Prehistoric Archaeological Investigations on the Savannah River Site, Aiken and Barnwell Counties, South Carolina.* Savannah

River Archaeological Research Program, South Carolina Institute of Archaeology and Anthropology, University of South Carolina.

Wahlenberg, W. G. (1946). *Longleaf Pine*. Washington, D.C.: Charles Lathrop Pack Forestry Foundation, in cooperation with the Forest Service, U.S. Department of Agriculture.

Webb, T., III. (1988). Eastern North America. In B. Huntley and T. Webb, III (eds.), *Vegetation History* (pp. 385–414). Dordrecht: Kluwer Academic.

Webb, T., III, P. J. Bartlein, and J. E. Kutzbach (1987). Climatic change in eastern North America during the past 18,000 years; comparisons of pollen data with model results. In W. F. Ruddiman and H. E. Wright, Jr. (eds.), *North American and Adjacent Oceans During the Last Deglaciation* (pp. 447–62). (*The Geology of North America, Vol. K-3*). Boulder, Colo.: Geological Society of America.

13

Short-term fluctuations in small mammals of the late Pleistocene from eastern Washington

JOHN M. RENSBERGER AND ANTHONY D. BARNOSKY

A sustained local presence of terrestrial vertebrates, followed by burial and preservation, over an interval sufficiently long to record a population's evolution is an uncommon occurrence. In this chapter we describe a deposit that has preserved a relatively continuous faunal record of small land mammals from the late Pleistocene and early Holocene. In 1981 and 1982, field parties from the Burke Museum at the University of Washington collected approximately 15,000 bones and stratigraphic data from a roadside outcrop southwest of Kennewick in eastern Washington (Figure 13.1). Small vertebrates were obtained from the base of the exposed deposits to the uppermost beds. This report describes the depositional setting, the faunal composition and its changes, and the morphological changes in several local populations through the stratigraphic section.

Stratigraphy

General characteristics

The Kennewick Road Cut exposure is located approximately 11 km (7 miles) southwest of Kennewick, on highway 82, in the southwest quarter of Section 4, T. 7 N., R. 29 E., Pasco quadrangle. The exposure extends more than 140 m in a north-south direction along the west side of the highway (Figure 13.2). The locality lies at an elevation of 389 m in the southeast-trending Horse Heaven Hills.

The deposits consist of generally massive, tan-colored, poorly to moderately indurated silt ranging from 21 to 24 m in exposed thickness.

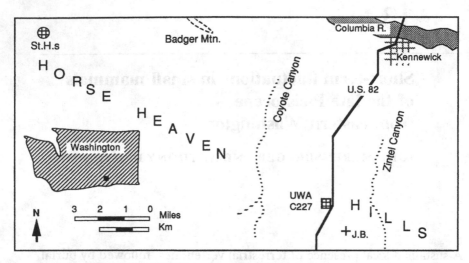

Figure 13.1. Map of the Kennewick Road Cut locality (UWA C227). Abbreviations: J.B., Johnson Butte; St.H.s, exposure of St. Helens S ash. From Waitt (1980).

Variations in thickness result from the north-south rise of the highway bed and the sloping surfaces of the hill. The most prominent sedimentary structures are a series of calcite-cemented layers (calcrete) that are typically less than 2 m in thickness. The calcrete layers provided the primary basis for correlating the fossil samples across most of the outcrop. The massive siltstone units separating the calcrete layers range in thickness from 1 m at the south end of the exposure to as much as 21 m at the northern end, where no calcrete is present. Where calcrete layers are present, the intercalated silt units average 2 m thick. Zones of filamentous calcite, animal burrows, and tephra layers are sometimes present in the otherwise massive silt units. Bedding other than that resulting from the alternating uncemented and cemented units and tephra is rare.

Silt units

The nonbedded units separating the calcrete layers consist of brown to gray-brown silt or sandy silt. When calcite and iron oxides are removed by treatment in HCl, the color becomes light gray.

Sediment grain-size parameters.   Grain-size analyses performed on 21 samples distributed vertically and horizontally through the exposed sec-

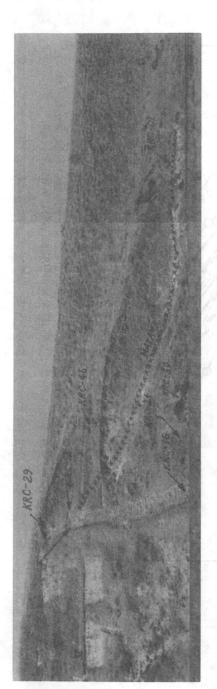

Figure 13.2. Views of the Kennewick Road Cut locality. Northern half of exposure above, and southern below. KRC numbers indicate several collecting sites. X indicates exposed Mazama ash. Dashed lines and Roman numerals show calcrete beds associated with paleosols. GG, trench 1; Y, trench 2; T, trench 3; P, trench 4; J, trench 5.

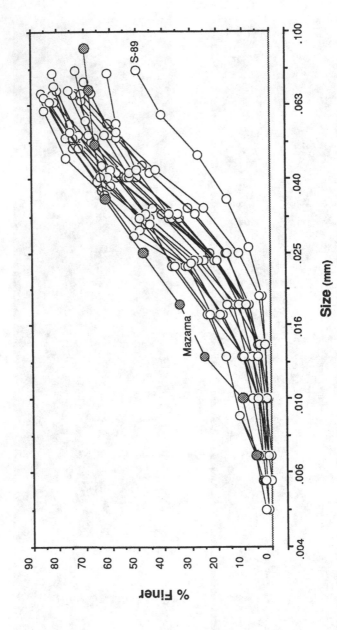

Figure 13.3. Cumulative frequency distributions of grain size in sediment samples S-16, S-25, S-30, S-39, S-42, S-51, S-55, S-59, S-61, S-64 S-84, S-85, S-89, S-91, S-95, S-98, S-106, S-108, S-111, S-Mazama (ash).

Table 13.1. *Dominant minerals of silt units listed in order of decreasing abundance*

Sample	Minerals
Fraction	*>4 Phi (0.067 mm)*
S-98	Quartz, micas, feldspars
S-108	Quartz, feldspars (microcline, albite), ?cordierite
S-51	Quartz, feldspars (microcline, albite), ?cordierite
S-55	Quartz, micas, feldspars (albite, microcline), ?cordierite
S-59	Quartz, feldspars (microcline, albite, orthoclase)
S-61	Quartz
S-64	Quartz, feldspars (microcline, albite)
S-16	Quartz, feldspars (microcline, albite), micas
Fraction	*<4 Phi (0.067 mm)*
S-98	Quartz, micas, feldspars (sanidine, anorthoclase)
S-108	Quartz, microcline, albite, ?cordierite
S-51	Quartz, albite, micas, cordierite, microcline
S-55	Quartz, micas, albite, microcline, ?cordierite
S-59	Quartz, microcline, orthoclase, albite
S-61	Quartz, orthoclase, epidote, ?cordierite, micas
S-64	Quartz, microcline, albite
S-16	Quartz, microcline, albite, micas

tion indicate relative consistency in texture (Figure 13.3). In only one sample did the median size exceed that of silt (0.0037–0.0625 mm), and in none was the median smaller than that of silt (Figure 13.3). The median grain size of the Mazama ash near the top of the section is only slightly smaller than the median of the finest of the other samples. The sand/silt ratio ranges from 0.14 to 0.67. Less than 3% of the sediment consists of clay-sized particles (<0.0039 mm). The particles are consistently well sorted. Trask's sorting coefficient is greater than 2.5 in only one sample (2.85), and less than 1.8 in 17 of 20 samples. Only rarely is there evidence of bimodality in the grain-size distributions (Figure 13.3).

Mineralogy. Microscopic examination and x-ray diffraction analysis of whole samples indicate that quartz is the dominant mineral present, with the remainder comprising micas and microcline, albite, and orthoclase feldspars (Table 13.1). The heavy-mineral percentage for each sample is given in Table 13.2.

Table 13.2. *Heavy-mineral separates*

Sample	Weight of sample	Weight of heavies	Weight of lights	Percentage of heavies
S-16	2.3782	0.1382	2.2400	5.81
S-42	1.7020	0.0765	1.6255	4.49
S-16	2.0503	0.0793	1.9710	3.87
S-51	2.2416	0.0950	2.1466	4.24
S-85	1.6957	0.0940	1.6017	5.54
S-106	2.0645	0.1255	1.9390	6.08
S-91	2.3528	0.1396	2.2132	5.93
S-59	2.5875	0.1106	2.4769	4.27
S-25	2.3350	0.0895	2.2455	3.83
S-98	2.5039	0.0743	2.4296	2.97
Mazama	2.4857	0.0294	2.4562	1.18

Quartz-grain surface textures. Quartz grains sampled vertically and horizontally through the section were examined by scanning electron microscopy for surface features characteristic of specific depositional environments. The samples were cleaned by boiling in concentrated HCl to remove the calcite and iron oxide.

The grains typically are quite angular, but have edges subdued by silica solution and precipitation and have flat precipitation surfaces (Figure 13.4). Cleavage planes and adhesive particles are common. The textures closely resemble those in samples of Pleistocene loess at Kaiserstuhl, West Germany (Krinsley and Doornkamp, 1973), and Wallertheim, West Germany (Pye, 1983). Impact V's characteristic of subaqueous environments are absent from the quartz surfaces. Grains with razor-sharp edges, extreme angularity, and conchoidal fractures characteristic of unmodified grains from glacial environments are rare.

The eolian nature of loess is accepted by most workers (Smalley, 1975). The distributions of the deposits in Germany and Poland indicate derivation from ice-marginal areas and fluvioglacial outwash channels during glacial phases of the late Pleistocene (Fink and Kukla, 1977; Maruszczak, 1980). We conclude that the sediments throughout the section at Kennewick represent airborne dust deposited over vegetated surfaces that held the particles and allowed accumulation of the observed unit thicknesses. The absence of bedding resulted from the relatively homogeneous particle sizes of the sediment and bioturbation by soil-forming processes as discussed later.

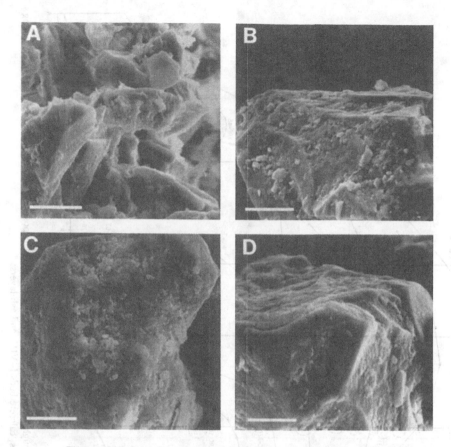

Figure 13.4. Scanning electron micrographs of quartz grains. (A) Compound grain from sediment sample S-31. S-31 is of normal size for the deposit, but is composed of cemented particles smaller than normal, indicating cementation prior to deposition. (B, C, D) Single grains showing silica solution, silica precipitation, and adhering fine particles characteristic of loess: B, sample S-100; C, S-15; D, S-100. Scale = 10 μm.

Pedogenic calcrete zones.   Eight calcrete layers are distinguishable (Figures 13.2 and 13.5). These zones are characterized by a high calcium carbonate content that ranges from coatings on particles to a pervasive medium engulfing the grains to form reddish, weakly cemented nodules 2–10 cm in diameter. The calcrete zones often grade upward into sets of faint carbonate filaments.

The calcrete deposits are similar to *K*-horizon pedogenic carbonate accumulations observed on desert alluvial fans, plains, and grasslands along the Rio Grande of southern New Mexico near Las Cruces (Gile,

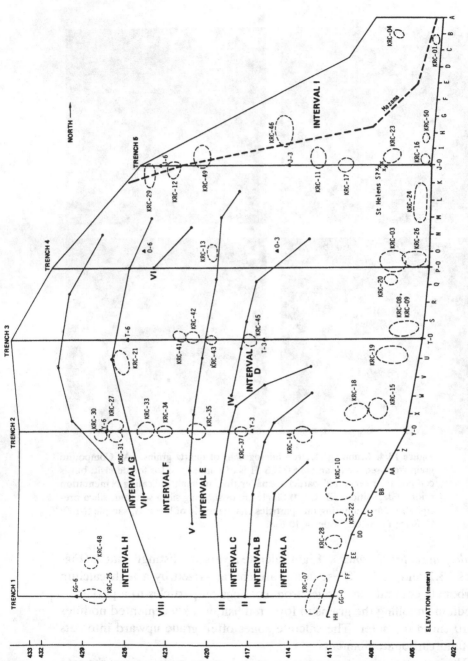

Figure 13.5. Vertical section through Kennewick Road Cut along plane of exposure, showing spatial relations of collecting sites (KRC) to trenches, grid lines, calcrete beds I–VIII, and ash layers.

Peterson, and Grossman, 1965, 1966). Those soils are in thick, well-drained alluvium of either pebbly or fine grained matrix, and the carbonate structures differ in the coarse and fine sediments. Gile et al. (1966) found that in the fine-grained sediments, stage I is characterized by filaments, stage II by rare-to-common nodules, stage III by many nodules and internodule fillings, and stage IV by laminar, heavily impregnated beds. The distinct calcrete zones at the Kennewick Road Cut resemble the fine-sediment stage II or early stage III. The filamentous calcite probably represents stage I.

The youngest geomorphic surfaces on which these stages were observed in New Mexico were (I) from less than 2,600 up to 5,000 years and (II) from less than 5,000 years to latest Pleistocene. Gile et al. (1966) interpreted the carbonate coatings and fillings as having accumulated over long periods of downward percolation of water falling on the surface of porous soils. The water took carbonate into solution from calcareous dust and precipitated the carbonate below the surface. Although eolian dust did not accumulate in large thicknesses in the New Mexico deposits, nor were dunes common, they inferred that the dust deposited by one storm was wetted and then largely removed by the next storm. However, at Kennewick, where coarse clastics are absent, the fineness of the deposit and the apparent relative stability of the structure suggest that the vegetation cover was thick enough much of the time to prevent significant portions of the preceding deposit being blown away during storms.

Even where sediment consists of over 90% calcite, as in coastal dunes, the formation of calcrete is observed to have extended over a lengthy interval of time, commencing soon after stabilization of the dune by vegetation.

Uncemented (up to 6,000 years BP) and cemented, calcreted (up to 80,000 years BP or older) Pleistocene coastal dunes occur on the South Australian coast (Warren, 1983). The stages are analogous in development to those described by Gile et al. (1966) for southern New Mexico: The unconsolidated sand passes in stages from a free-flowing sand to nodular calcrete to massive nodular calcrete to laminar calcrete (Warren, 1983).

The cliff faces of the South Australian Pleistocene dunes show cross-bedded dune intervals separated by calcrete zones. Warren interpreted each sandy unit as a pulse of dune sand deposited during an interglacial and capped by calcrete before the next interglacial. By similar reasoning it is probable that at least some of the loess–calcrete sequences at Kennewick represent glacial–interglacial units.

Seventeen glacial–interglacial episodes are recognized as loess–paleosol units in deposits near Krems, Austria, and Brno, Czechoslovakia (Fink and Kukla, 1977). The sequences consist of windblown loess deposited during glaciation, interlayered with soils deposited during interglacials. Eight major depositional units consisting of loess–soil pairs in the Brunhes and nine within the late and middle Matuyama paleomagnetic epochs were interpreted as glacial–interglacial episodes. Some of the intervening transitions were not recognized as complete glacial–interglacial units, however.

Three soil zones occur beneath the modern soil in 12.5 m of loess at Tyszowce, Poland (Wojtanowicz and Buraczynski, 1978); the lowermost soil is considered to be younger than 115,000 years BP, based on paleomagnetic and thermoluminescence data, and younger dates have been obtained for overlying soils. At Komarow Gorny, Poland, four soils are interbedded in a 9-m section of loess, with a thermoluminescence date of 122,000 years BP from above a thick, lowermost soil that is believed to correlate with the basal soil at Tyszowce (Tucholka, 1977; Pye, 1983). The oldest of a series of thermoluminescence dates from the loess deposit at Wallertheim, West Germany, underlies three weakly developed soils and overlies a fourth.

At the Polish and the German loess deposits, the average span of time represented by each loess–soil unit is as follows: Tyszowce, 38,000 years; Komarow Gorny, 41,000 years; Wallertheim, 14,000 years. The range noted earlier for stage II pedogenic carbonate formation in the southwestern United States is 5,000–10,000 years. At Kennewick, applying the shortest and longest of these reported rates for the formation of sand-calcrete, loess–calcrete, or loess–soil units suggests an interval ranging between 40,000 and 328,000 years for the deposition of the eight loess–calcrete units (Figure 13.5).

Sedimentary structures. Stratification is generally limited to the interfaces between the calcrete, loess, and tephra units. The loess intervals generally lack structure, in part owing to bioturbation, as indicated by outlines and fillings of burrows. The burrows range in diameter from 2 to 5 cm, and therefore all but the smallest could have been made by any of a number of the taxa of rodents found as fossils in the vicinity. Several such burrows occur beneath calcrete bed I in trench 1 (Figure 13.5), and a few similar burrows were found above calcrete bed VIII in the same trench. Several burrows of a different type were seen beneath calcrete bed IV in trench 4 (Figure 13.5). These structures were larger, up to ⅓ m in diameter, and were filled with stratified sediments containing fossil bone material. Except for the presence of laminations, the fillings resemble the overlying sediment. These larger burrows may have been the dens of small carnivores or possibly rabbits. The fine texture of the loess made digging easy, which may account for the abundance of small mammals that build burrows.

Taphonomy. The presence of carnivore dens and of some concentrations of bones that may be attributable to carnivore scat suggest that animals commonly hunted by predators may be overrepresented with respect to their abundance in the living community. However, rodent teeth that under SEM magnifications show no signs of digestion also are present. Because many of the fossil rodents burrowed and the carnivores dug dens, a sampling site should not be regarded as an "instant" in time. Rather, bioturbation may have caused some of the sites to combine mammals that lived up to a few hundred years apart. More pronounced mixing seems to have been unlikely, because living representatives of these animals seldom dig deeper than 2 m (usually less than 1 m), and most sites show no obvious evidence of bioturbation.

Tephra. The stratigraphically highest of two tephra units dips to the northeast in the upper part of trench 5 (Figures 13.2 and 13.5). This tephra is 10 cm thick at the higher end, but thickens to about 30 cm a few meters to the north, suggesting deposition on a slope. The indices of refraction for the majority of the glass shards range between 1.506 and 1.510 and average slightly above 1.508, which is not distinguishable from samples of Mazama ash, but is higher than indices for Glacier Peak pumices (Powers and Wilcox, 1964). The chemical composition of the glass resembles that of Mazama pumice (Powers and Wilcox, 1964) in the percentages of silica, soda, and titanium oxide (Table 13.3). The higher percentages of iron oxides and lime in the Kennewick tephra may be due to contamination from the surrounding loess. These characteristics, together with the high stratigraphic position of this unit above the uppermost calcrete layer, indicate that it represents the Mazama ash (6,600 years BP).

The stratigraphically lower tephra is poorly preserved near the base of trench 5 (Figure 13.5). It was also found between calcrete units V and VI in trench 4, above KRC-13, the *Mammuthus* site. This tephra occurs as an indistinct couplet of units a few centimeters thick and badly contaminated from adjacent sediment. Because of the diffuse nature

Table 13.3. *Compositional comparison of upper tephra, Kennedwick Road Cut*

Compound	Kennewick	Mazama[a]	Glacier Peak[a]
SiO_2	71.95	72.39	73.01
Al_2O_3	14.75	14.79	15.08
FeO's	2.40	1.85	1.79
MgO	0.06	0.52	0.73
CaO	2.13	1.58	2.30
MnO	0.05	0.05	0.05
TiO_2	0.44	0.43	0.30
K_2O	2.75	2.77	2.89
Na_2O	5.27	5.21	3.72
P_2O_5	0.19	0.09	0.13

[a]Data from Powers and Wilcox (1964).

and poor exposure of this unit, determination of the chemistry was not attempted. Waitt (1980, p. 664) found a white silt-tephra couplet of 2.5-cm- and 0.7-cm-thick layers commonly associated with eolian silt in scores of exposures of the Touchet Formation in south-central Washington a short distance (e.g., 6 m) beneath the Mazama ash. Waitt (1980, Table 1) found that samples of this couplet from Badger Coulee, a locality about 24 km (15 miles) northwest of the Kennewick Road Cut, match the St. Helens set S ash in trace-element quantities. This widespread distribution of the St. Helens set S as a couplet suggests that the Kennewick Road Cut couplet may represent that set.

The St. Helens set S is dated at approximately 13,000 years BP, which is younger than expected for this inferred position in interval F, judging from the estimated time for calcrete formation. There are several possible reasons for this discrepancy: (1) The tephra is not the St. Helens set S. (2) The placement of the tephra with respect to the calcrete units is incorrect. (3) The time for formation of the three upper calcrete units averages slightly less than 5,000 years each. This tephra is seen only at the north end of the exposure, where the calcrete units disappear, and one (unit VI) is short and discontinuous.

Discontinuities and bedding structures. The discontinuities and dips of the calcrete zones and tephra (Figure 13.5) suggest that deposition proceeded northward. The dipping northern ends of these layers roughly parallel the present topographic surface, which dips prominently at the northern end of the exposure. Calcrete zones I through V successively extend farther northward. The Mazama ash (Figures 13.2 and 13.5) lies in a plane almost parallel to the northern topographic surface.

The southern end of the section (trench 1) lacks recognizable calcrete layers IV, V, VI, and VII, whereas trench 4 is missing layers I, II, and III. These relationships are consistent with erosion, apparently through deflation, occurring at the southern end of the hill after development of paleosol III and prior to development of paleosol VIII, with deposition progressively filling at the northern margin of the deposit. The fine particle size and the fact that the quartz grains do not have features characteristic of saltation indicate that the deposit did not migrate to the north in the fashion of a sand dune. The foreset effect and the antiquity of the calcrete–silt cycles suggest a consistency in storm wind from a southerly direction over a long period of time.

Relationship to neighboring deposits. Neighboring deposits of silt, sand, and conglomerate of late Pleistocene to Holocene age have been called Touchet beds since the use of that term by Flint (1938). Most of the Touchet Formation along river drainages and in coulees is waterlaid, consisting of bar, channel, and slackwater deposits. Rhythmic slackwater deposits comprising up to 39 rhythmites are exposed in the Walla Walla and Yakima valleys (Waitt, 1980) and in Badger Coulee, a few kilometers to the northwest of the Kennewick Road Cut fossil locality (Waitt, 1980; Bunker, 1982). The individual rhythmite typically consists of a 0.5–1-m-thick unit that grades upward from sand to silt. The silt and calcrete deposits of the Kennewick Road Cut section lack graded bedding, in most cases lack the bimodal grain size characteristic Waitt (1980, p. 664) noted for the massive rhythmite layers, lack SEM-visible quartz-grain textures characteristic of waterlaid deposits, and are located topographically above the boundary of the neighboring waterlaid beds as mapped by Waitt (1980, Figure 2). The Kennewick Road Cut deposit more closely resembles the Palouse Formation, which is characteristically a brownish loess and has at least three moderately cemented calcareous horizons (Newcomb, 1961).

Faunal composition

Faunal list

> Amphibia
> > Order Anura (sp. indet.)

Reptilia
Order Squamata
Suborder Serpentes (sp. indet.)
Suborder Lacertilia (sp. indet.)
Aves (sp. indet.)
Mammalia
Order Chiroptera
Myotis (sp. indet.)
Order Insectivora
Sorex palustris
Sorex sp.
Insectivora (sp. indet)
Order Lagomorpha
Family Leporidae
Sylvilagus idahoensis
Sylvilagus nuttallii
Lepus (sp. indet.)
Order Rodentia
Family Sciuridae
Spermophilus townsendii
Eutamias (sp. indet.)
Family Heteromyidae
Dipodomys microps
Dipodomys ordii
Perognathus parvus
Family Geomyidae
Thomomys talpoides
Family Cricetidae
Neotoma lepida
Neotoma cinerea
Peromyscus maniculatus
Lagurus curtatus
Phenacomys intermedius or *Arborimus longicaudus*
Synaptomys cf. *S. borealis* (or closely related species)
Microtus (*Pitymys*) *meadensis*
Microtus sp.
Order Proboscidea
Family Elephantidae
Mammuthus (sp. indet.)
Order Carnivora

Family Canidae (sp. indet.)
Order Artiodactyla (sp. indet.)

Systematic description

The identifications of most of the mammalian species are based on teeth or jaws from tens to hundreds of specimens. Avian, reptilian, and amphibian fossils are sparse and fragmentary and have not been identified to lower taxonomic levels. Large samples of postcranial elements have not been studied. The following brief descriptions give the critical characteristics of the taxa that are identified to the level of species or genera:

Myotis (sp. indet.): Remains: two jaws, one humerus fragment, three premolars present, anteroposteriorly expanded (esp. P_4); P_4 with incipient talonid.

Sorex palustris (northern water shrew): Upper molars wider than in *S. trowbridgei*, *S. merriami*, *S. cinereus*, or *S. obscurus* (Figure 13.6). P_4 less bladelike than in *S. bendirii* or *S. vagrans*. Smaller than *S. bendirii*. Resembles *S. prebla* (Ingles, 1965, pp. 87–8) in size.

Sorex sp.: Smaller than *S. palustris*.

Sylvilagus idahoensis (pygmy rabbit): Anterior premolars lacking complex anterior borders. Anterior enamel of P^2 with only one inflection, and P_2 with none. Enamel separating anterior and posterior lophs of upper molariform teeth lacking crenulation of *S. idahoensis* (Orr, 1940), but with mild undulation in some individuals. Labial margins of protoconid and hypoconid more angular than in *S. nuttallii*. Talonid of lower molariform teeth one-half to two-thirds width of trigonid.

Sylvilagus nuttallii (mountain cottontail): P^2 with three inflections of anterior enamel, and P_2 with one. Single anterior inflection of P_2 lacking crenulation characteristic of *S. auduboni*. Talonid of lower molariform teeth one-half to two-thirds the width of trigonid (in other *Silvilagus*, except *S. idahoensis*, the talonid is four-fifths the width of the trigonid) (Orr, 1940).

Spermophilus townsendii (Townsend ground squirrel): M_3 longer (Figure 13.7) than in *S. washingtoni*, lacking metaloph. Cheek teeth higher-crowned, with less bulbous cusps than in *S. beecheyi*, *S. saturatus*, or *S. lateralis*. Teeth smaller than in *S. columbianus* and *S. richardsonii*. M_3 and M^3 relatively shorter; P_4 lacking prominent anterior cingulum; upper molars with more rounded lingual margins than in *S. columbianus* and *S. beldingi*.

Eutamias sp. (chipmunk): Within range of variation in dentitions of

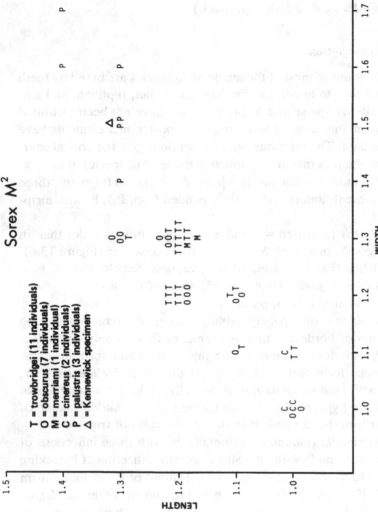

Figure 13.6. Size relationships between fossil *Sorex* from Kennewick Road Cut and several extant species, based on length and width of M². Measurements in millimeters. The Recent specimens are from the Burke Museum, and most of those for which locality data are known are from Washington: *S. cinereus* (Lewis and Pierce counties, WA); *S. bendirii* (Pierce and Whatcom counties, WA); *S. palustris* (Pierce and Ferry counties, WA); *S. obscurus* (Pierce and Whatcom counties, WA); *S. trowbridgei* (King, Pierce, and Whatcom counties, WA); *S. vagrans* (Lincoln, Pierce, King Skagit, Lewis, and Yakima counties, WA, and Glacier Bay, AK).

Spermophilus M₃

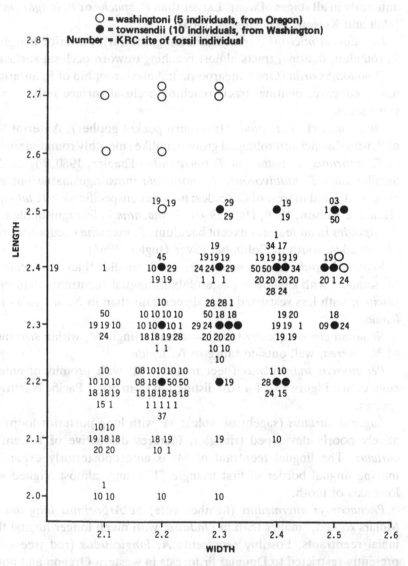

Figure 13.7. Size relationships between fossil *Spermophilus* from Kennewick Road Cut and extant species, based on length and width of M₃. Measurements in millimeters. Measurements were taken to the nearest 0.1 mm; clusters of points all have the same measurement, but are offset for graphic feasibility.

species inhabiting area today (*E. minimus*) and species (*E. amoenus*) ranging nearby (Hall and Kelson, 1959).

Perognathus parvus (Great Basin pocket mouse): Hypolophid bulging anteriorly in all stages of wear. Larger than *P. apache* or *P. longimembris* (Hall and Kelson, 1959; Ingles, 1965).

Dipodomys microps (Great Basin kangaroo rat): Anterior margin of P_4 rounded; dentinal tracts almost reaching unworn occlusal surface.

Dipodomys ordii (Ord kangaroo rat): Anterior lophid of P_4 anteriorly flat to concave; dentinal tracts reaching occlusal surface only in well-worn teeth.

Thomomys cf. *T. talpoides* (northern pocket gopher): Anterior lobe of P_4 with distinct anterolingual groove, unlike smoothly rounded surface in *T. umbrinus, T. bottae,* or *T. townsendii* (Thaeler, 1980, Figure 2b). Smaller than *T. bulbivorous. T. monticola* indistinguishable but generally restricted to west of Cascades; may be conspecific with *T. talpoides* (Hall and Kelson, 1959; Hall, 1980). *T. mazama* indistinguishable from *T. talpoides* in all features except baculum; *T. mazama* occurs only west of Cascades, south of Columbia River (Ingles, 1965).

Neotoma lepida (desert woodrat): Teeth smaller than in *N. cinerea.* M^3 smaller, with shallower posterolabial, lingual reentrants than in *N. fuscipes;* with less restricted lingual reentrant than in *N. albigula* or *N. lepida.*

Neotoma cinerea (bushy-tailed woodrat): Single M_1 within size range of *N. cinerea,* well outside range of *N. lepida.*

Peromyscus maniculatus (deer mouse): M^1 with prominent anteroconule. See Figure 13.8 for size distinction from other Pacific Northwest species.

Lagurus curtatus (sagebrush vole): M^3 with long posterior loop, relatively poorly developed triangles, features distinctive of Recent *L. curtatus.* The lingual reentrant of M^2 is anteroposteriorly expanded, making lingual border of first triangle (T1) long, almost aligned with long axis of tooth.

Phenacomys intermedius (heather vole) or *Arborimus longicaudus:* Molars rooted, smaller than in *Ondatra,* with much longer lingual than labial reentrants. Possibly represents *A. longicaudus* (red tree vole), presently restricted to Douglas fir forests in western Oregon and northwestern California (Johnson, 1973). Subsequent to our analysis, C. A. Repenning borrowed this small collection of isolated molars and is preparing a report in which he discusses the resemblance to *A. longicaudus.*

Synaptomys (*Mictomys*) cf. *S. borealis* (northern bog lemming): Lin-

Peromyscus M¹

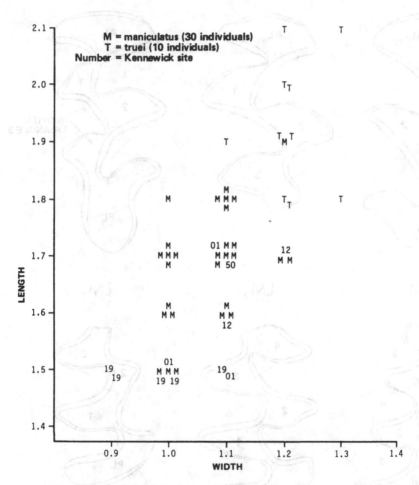

Figure 13.8. Size relationships between fossil *Peromyscus* from Kennewick Road Cut and extant species, based on length and width of M¹. Measurements in millimeters. Plotting of points as in Figure 13.7.

gual reentrants extending to labial edge of the tooth, unlike *S. (Synaptomys) cooperi*. Further study is required to determine whether the specimens represent *S. borealis* or a closely related species.

Microtus (Pitymys) meadensis: Molars with cementum in reentrant angles, lacking roots (Hall and Kelson, 1959; Hall, 1980). M_1 with only three closed triangles (Figure 13.9), with triangles 4 and 5 [primary wings

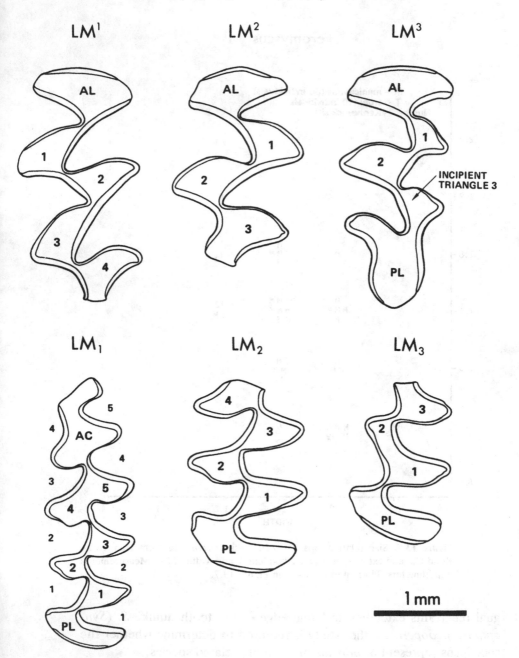

Table 13.4. *Dimensions (mm) of teeth in Kennewick M. (Pitymys) meadensis (KRC-10)*

Parameter	LM/1	WM/1	LM/3	WM/3
Mean	3.14	1.17	1.97	1.02
SD	0.192	0.088	0.134	0.100
OR	2.9-3.5	1.0-1.4	1.7-2.3	0.8-1.2
N	50	53	43	42

of Repenning (1983)] broadly confluent; M^3 with two closed triangles, simple posterior loop, unlike the condition in following subgenera and species of *Microtus*: (*Microtus*) *pennsylvanicus*, (*M.*) *breweri*, (*M.*) *nesophilus*, (*M.*) *montanus*, (*M.*) *californicus*, (*M.*) *townsendii*, (*M.*) *oeconomus*, (*M.*) *longicaudus*, (*M.*) *coronarius*, (*M.*) *mexicanus*, (*M.*) *fulviventer*, (*M.*) *chrotorrhinus*, (*M.*) *xanthognathus*, (*Aulacomys*) *richardsonii*, (*Chilotus*) *oregoni*, (*Stenocranius*) *miurus*, (*S.*) *abbreviatus*. M_1 in (*Herpetomys*) *guatemalensis* somewhat similar, but M^3 with three closed triangles, and M_3 with four triangles.

Prominent closure of anterior two triangles (T3 and T4) of M_2 in *M.* (*Pitymys*) *meadensis*, unlike remaining extant North American species: *M.* (*Pedomys*) *ochrogaster*, *M.* (*Pedomys*) *ludovicianus*, *M.* (*Pedomys*) *parvulus*, *M.* (*Pitymys*) *pinetorum*, *M.* (*Pedomys*) *quasiater*, *M.* (*Orthriomys*) *umbrosus*.

M^2 with three triangles in addition to anterior loop, unlike *M.* (*Pitymys*) *nemoralis* (Repenning, 1983, p. 479).

Prominent constriction separating triangles 4 and 5 from anterior cap in M_1, unlike (*Allophaiomys*) and *Microtus guildayi* (Martin, 1975, van der Meulen, 1978). Lingual reentrant angle 3 almost opposite labial reentrant angle 2, unlike *M. deceitensis* (Guthrie and Matthews, 1971) and *M. paroperarius* (Hibbard, 1944; Paulson, 1961; van der Meulen, 1978). Posterior loop of M^3 less complex than in *M. paroperarius*.

Triangles 3 and 4 on M_2 closed or almost closed, unlike condition in *M. guildayi* (Hibbard, 1955; van der Meulen, 1978), *M.* (*Pitymys*) *mcnowni* (Hibbard, 1937), typical *M.* (*P.*) *aratai* (Martin, 1974, 1987), *M.* (*P.*) *hibbardi* (Holman, 1959), *M.* (*P.*) *cumberlandensis* (van der Meulen, 1978), and *M.* (*P.*) *llanensis* (Hibbard, 1944; Paulson, 1961).

Thinned enamel, as in other samples of *M.* (*Pitymys*) *meadensis*, especially on trailing edges of lower molars. Slightly larger than previously described specimens of *M.* (*Pitymys*) *meadensis* (Hibbard, 1944; Paulson, 1961): mean length of M_1 is 3.1 mm (Table 13.4), compared with 2.9 mm; however, variability of combined samples similar to that in other conspecific samples of arvicolines (Kowalski, 1970, p. 155).

Age

The bones of the Kennewick fauna are low in density and have proved to be collagen-poor; attempts to obtain ^{14}C dates from the two largest fragments available were unsuccessful. Some bone we have seen from neighboring localities in the Touchet Formation resembles the Kennewick material in its low density.

The highest stratigraphic position from which bones were obtained in the Kennewick Road Cut was slightly above the uppermost and more extensive of the tephra layers, the Mazama ash. The age of the youngest of the Kennewick faunas is therefore less than 7,000 years BP.

The age of the oldest of the Kennewick fossils is problematic. Most of the Kennewick taxa have known stratigraphic distributions in North America reaching no further back than the Rancholabrean land-mammal age (Figure 13.10). The Rancholabrean age is characterized by the presence of *Bison* and a high percentage of Recent species (Savage, 1951; Kurtén and Anderson, 1980; Savage and Russell, 1983; Lundelius et al., 1987). The large percentage of Recent species in the Kennewick fauna, even in the lowest stratigraphic levels, is the strongest evidence for a Rancholabrean-to-Recent age range for the deposit. The absence of *Bison* from the Kennewick fauna is of no chronological significance, because large mammals are almost totally lacking, most likely because of taphonomic bias.

Microtus (*Pitymys*) *meadensis* is known elsewhere in the United States only from the late Irvingtonian, and its confinement to the lowest part of the Kennewick section raises the question whether or not the section extends back beyond the Rancholabrean. However, only 41% of the Irvingtonian rodent and lagomorph species survived until the Recent, whereas *M.* (*Pitymys*) *meadensis* is the only extinct species of the eight species present at the base of the Kennewick section. *M.* (*Pitymys*) *meadensis* occurs in a Rancholabrean fauna that also includes *Bison* at El Tajo de Tequixquiac, Mexico (Repenning, 1983). The only other Pacific Coast occurrences of *M.* (*Pitymys*) *meadensis* are Olive Dell Ranch and North Livermore in California, which have been regarded as Irvingtonian (Repenning, 1983). In the North Livermore fauna, *M.* (*Pitymys*) *meadensis* is the only extinct species, and at Olive Dell Ranch the primary evidence for an Irvingtonian age is the stratigraphic position

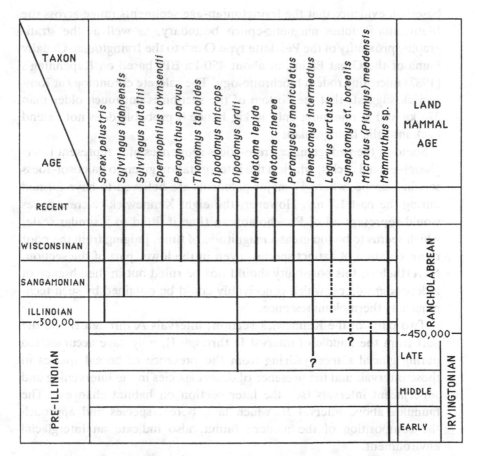

Figure 13.10. North American stratigraphic ranges of Kennewick Road Cut species [dates from Kurtén and Anderson (1980), except that *Lagurus curtatus* is extended as discussed by Barnosky and Rasmussen (1988)].

of the fauna above the late Irvingtonian El Casco fauna. Therefore it is possible that the California faunas containing *M.* (*Pitymys*) *meadensis* correlate with the lower part of the Kennewick section. In the Rocky Mountains of Colorado, *M.* (*Pitymys*) *meadensis* occurs with *Lagurus curtatus,* as it does at Kennewick. Barnosky and Rasmussen (1988) suggested that this association indicates an age of around 400 ka, because at all other localities *M.* (*Pitymys*) *meadensis* disappears by 400 ka, and *L. curtatus* does not appear until after 400 ka.

The age of the Irvingtonian–Rancholabrean boundary has been estimated to be close to 600 ka BP (Lindsay, Johnson, and Opdyke, 1975), based on evidence that the Irvingtonian-age sediments range across the Matuyama-Brunhes magnetic-epoch boundary, as well as the stratigraphic proximity of the Pearlette type O ash to the Irvingtonian Cudahy fauna of the Great Plains, at about 450 ka BP, based on Repenning's (1987) microtine rodent biochronology. The calcrete chronology at Kennewick suggests that the bottom of the section is not much older than 328 ka and at minimum only 40 ka; hence it probably does not extend back into the Irvingtonian.

Some of the pedogenic calcrete–loess sequences may represent interglacial–glacial cycles, which is the interpretation given to paleosol–loess sets in Europe, where 17 major paleosols are believed to have formed during the past 1.7 my. However, the eight Kennewick calcrete zones would represent all of Rancholabrean time if fitted to a similar scale, which seems to be too great a magnitude of time, judging from the poor representation of extinct species, even in the lower part of the section. Nevertheless, this possibility should not be ruled out in the absence of independent dates, which conceivably could be obtained by such techniques as thermoluminescence.

Two parts of the Kennewick section, intervals A through B and the part from the middle of interval D through E, may have accumulated during glacial stages, judging from the presence of boreal species in those intervals and the presence of desert species in the intervening and superjacent intervals (see the later section on habitat changes). The faunules above interval E, which lack boreal species and approach the composition of the modern fauna, also indicate an interglacial environment.

If the two colder intervals indeed represent glacial stages and the maximum rate of calcrete formation applies, then the younger of the two may have been deposited during the Fraser glaciation (25,000–10,000 years BP), and the older during the Salmon Springs glaciation (>36,000 years BP). These Pacific Northwest stages appear to correlate with the Pinedale glaciation and the Bull Lake glaciation, respectively, in the Rocky Mountain region (Richmond et al., 1965). Correlation of the younger cold interval at interval D in the Kennewick section with the youngest glacial stage yields an average of 5,000 years for each of the succeeding five pedogenic calcrete units to form, which is at the low end of the rates observed in other regions. An alternative hypothesis is that the two cold intervals at Kennewick are older and correlate with the Salmon Springs and Stuck glaciations, with the Fraser being unrecognized in the upper part of the section where the fauna is more poorly represented. If the basal part of the Kennewick section is pre-Salmon Springs glaciation, the loess there differs from the pre-Salmon Springs loess as it has been described in other areas of southeastern Washington and Idaho, where it is characterized by mature weathering zones more than 3 m thick and thicker calcified layers (Richmond et al., 1965; Baker, 1978).

At the top of the section, dated at approximately 7,000 years BP because of its association with the Mazama ash, is a faunule with *Dipodomys microps,* a group 4 species (see the later section on habitat changes) that has a more southern distribution today. This occurrence is consistent with conclusions reached by Lyman and Livingston (1983) and C. Barnosky (1984, 1985), based on faunal and floral data, that from about 8,500 to 4,000 years BP a regime of less effective moisture and warmer temperatures prevailed in the Columbia Basin than after 4,000 years BP.

Faunal and phyletic changes

Almost all of the microfauna was collected by passing sediment through 1/16-inch mesh screens. Because the same procedure was used throughout the fieldwork, differences in relative abundances of taxa sampled from site to site and through the stratigraphic section should reflect differences in abundances in the deposit.

Changes in distribution

In order to analyze the vertical ranges, the taxa are grouped according to stratigraphic intervals from A at the base of the section through I at the top, with some intervals subdivided into low and high, or low, middle, and high parts, depending upon our ability to relate sites to key beds (Figure 13.5). The most abundant taxon (Table 13.5) is *Spermophilus townsendii,* which ranges throughout the section (Figure 13.11). Other taxa have more restricted stratal ranges whose endpoints have the potential to allow correlation beyond the limits of the Kennewick locality, or reconstruction of the biogeographic history of the species, or inferral of climatic change.

An early faunal change occurred between intervals A and B, when the extinct vole *Microtus (Pitymys) meadensis* was replaced by an extant but undifferentiable species of *Microtus* (Figure 13.11), and *Sylvilagus nuttallii* appeared for the first time.

Table 13.5. *Taxonomic abundances of Kennewick Road Cut species based on teeth*

STRATIG. INTERVAL	LOCALITY	Myotis	Sorex palustris	Sorex sp.	Sylvilagus sp.	Sylvilagus idahoensis	Sylvilagus nuttallii	Lepus sp.	Spermophilus townsendii	Eutamias	Sciuridae	Perognathus parvus	Thomomys talpoides	Dipodomys sp.	Dipodomys microps	Dipodomys ordii	Neotoma lepida	Neotoma cinerea	Peromyscus maniculatus
I	KRC-29								34			3	7						1
I	KRC-12				3				60			12	2		1				2
I	KRC-46												1						
I	KRC-49								1										
H	KRC-25												1						
G LOW H	KRC-30								2				1						
G LOW H	KRC-27																		
G LOW H	KRC-31								1										
F HIGH	KRC-21								24	1			4						
F HIGH	KRC-33								1										
F LOW	KRC-34								17										
F LOW	KRC-41												2						
F LOW	KRC-42								4										
F LOW	KRC-13								1										
E HIGH	KRC-35																		
E HIGH	KRC-43																		
E HIGH	KRC-17				1				1		1	1							
E HIGH	KRC-01		3		7	3	9	2	226	10		71	68	5	2	1			10
E LOW	KRC-23																		1
E LOW	KRC-50				1	1	2	1	55	3		4	5						1
E LOW	KRC-16								3										
D HIGH	KRC-45								5			1							
D HIGH	KRC-24				2	1	2	2	73	2		12	22	1		1		2	2
D MIDDLE	KRC-03								30			1	4						
D MIDDLE	KRC-26																		
D MIDDLE	KRC-20				10	4	2		53										
D LOW	KRC-09								9				4					1	
D LOW	KRC-08				1	3			7				4						
C	KRC-37								5										
C	KRC-19			2	103	33	32		443	10		70	8	15	1	1	14		9
B	KRC-18				44	8	3		192	2		8	1						
B	KRC-15				1				3										
A	KRC-10			2	69	51			274	11	2	7	15						
A	KRC-22								28				7						
A	KRC-28	3							30		1		8						
A	KRC-07				4				9				4						

A second prominent change occurred at interval C, when *Dipodomys microps, Dipodomys ordii, Neotoma lepida,* and *Peromyscus maniculatus* appeared for the first time.

In interval D, *Neotoma lepida* disappeared, and *Phenacomys* cf. *P.*

Table 13.5 (*cont.*)

Cricetinae	Lagurus curtatus	Phenacomys cf. intermedius	Synaptomys cf. borealis	Microtus (Pitymys) meadensis	Microtus sp.	Microtinae	Rodentia	Artiodactyla	Carnivora	Canidae	Insectivora	Mammuthus	Mammalia	Anura	Squamata	Serpentes	Aves	Vertebrata	Totals	
	27						6						3						81	KRC-29
	42				1		2						6						131	KRC-12
	3						4						2		1				11	KRC-46
	2												1						4	KRC-49
							1												2	KRC-25
	2												2						7	KRC-30
													1						1	KRC-27
													1		1				3	KRC-31
	8												3						40	KRC-21
													1						2	KRC-33
													1						18	KRC-34
													1						3	KRC-41
						1							2						7	KRC-42
												1	1						3	KRC-13
													1						1	KRC-35
							1												1	KRC-43
												2							6	KRC-17
	195	3	9		3	5	23							17		23	2			697	KRC-01
			1																2	KRC-23
	40		3		1								4		5				126	KRC-50
													1						4	KRC-16
													1						7	KRC-45
	15		8		2		2						4		11				164	KRC-24
	16	1	1			1	3						3		1				61	KRC-03
	1																		1	KRC-26
	39						5						1						114	KRC-20
	5						3								1				23	KRC-09
												1				1	1		18	KRC-08
																			5	KRC-37
	138		14		7	2	28	1		1		1	1	4		8	4	2	1	952	KRC-19
1	74		4		1	2	9						1		22			1	373	KRC-18
																			4	KRC-15
	19			629	148	71							6	5	6		2	5	1321	KRC-10
				5	5	15							1		1			1	63	KRC-22
	28		4	3	13	15			1	1					10		1	3	120	KRC-28
	6		2		4	13									28	2			72	KRC-07

intermedius, Neotoma cinerea, and *Lepus* appeared for the first time. After interval E, *Sorex palustris, Sylvilagus idahoensis, D. microps, D. ordii, Phenacomys* cf. *P. intermedius,* and *Synaptomys* cf. *P. borealis* disappeared. However, because the fauna from intervals F, G, and H

FAUNAL CHANGES

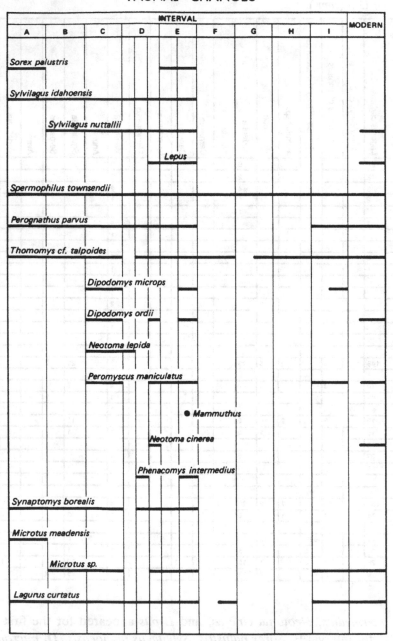

Figure 13.11. Stratigraphic ranges of certain species in Kennewick Road Cut showing changes in faunal composition. Stratigraphic divisions A–I based on positions of collecting sites with respect to key beds (see Figure 13.5).

is sparsely represented (Table 13.5), and only the normally abundant taxa with long ranges were found, it is difficult to determine exactly when the actual disappearances of these taxa occurred. At interval I, *D. microps* makes its last appearance, but only by a single specimen in several large samples.

Habitat changes

Indications of habitat changes are seen when the taxa are grouped according to preferred habitats based on physiographic distributions and ecological studies of modern representatives. See Hall and Kelson (1959), Hall (1946, 1981), Dalquest (1948), Ingles (1965), Baker (1968), Bailey (1936), Beatley (1976), and Finley (1958) for range maps and summaries of habitat preferences.

Group 1.   Taxa whose current distributions are restricted to arid habitats like that of the area today: *Spermophilus townsendii, Perognathus parvus,* and *Lagurus curtatus.* Beginning with interval C, *S. townsendii* and *L. curtatus* are the most abundant species in most samples. Therefore, arid habitats like those of the area today must have predominated in this region through most of the time spanned by the deposit. There are, however, fluctuations in the relative abundances of group 1 species (Figure 13.12). The percentages of *Lagurus* fluctuate extensively, with increases in abundance relative to the abundances of other common taxa at chronologically sequential sites KRC-28, -18, -20, and -50, and decreases at KRC-22, -19, -3, and -24 (probabilities for no change <0.003).

The relative abundance of *Perognathus* also changed in an episodic fashion, with increases at KRC-19, -24, and -1, and decreases at KRC-10, -20, and -21 (p <0.001), with perhaps a decrease at KRC-50 (p = 0.008). The directions of the fluctuations in *Perognathus* are the reverse of those in *Lagurus* at seven positions: KRC-28, -22, -19, -20, -24, -50, and -1.

The nocturnal, seed-eating *P. parvus* prefers habitats with little shrub cover. These pocket mice are restricted to the loose, dry soils of the desert floor (Scheffer, 1983; Kritzman, 1974, p. 176). The diurnal, grass-eating *L. curtatus* prefers habitats within stands of sagebrush, other shrubs, or at least thick grass cover, although it is still limited to arid regions; see Carroll and Genoways (1980) and the references therein. Times of abundant *L. curtatus* and scarce *P. parvus* therefore suggest a relative increase in vegetation, probably shrub cover. Scarce *L. cur-*

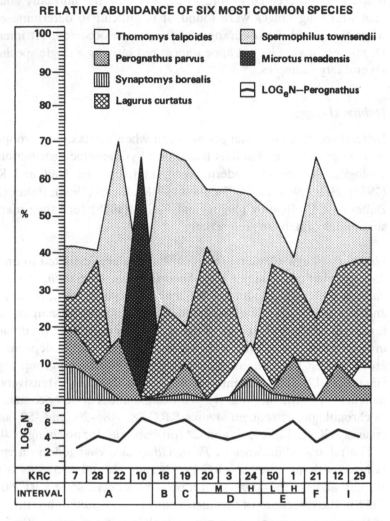

Figure 13.12. Relative abundances of six most common species at 14 best-represented sites in stratigraphic order from left to right. Percentages taken of total numbers of specimens of these six taxa. At bottom is natural logarithm of total number at each site, except *Perognathus parvus*.

tatus and abundant *P. parvus* imply the opposite. Four of the five percentage peaks for *L. curtatus* (at KRC-18, -20, -50, and -29) are stratigraphically just above calcretes, the presence of which also argues for increased vegetation and moisture. Above interval A the fluctuations in the percentage of *P. parvus* seem to correspond with short-term fluctuations in the absolute abundance of the other taxa taken together, as indicated by the graph at the bottom of Figure 13.12.

Group 2. Taxa whose current habitats are diverse and distributions very broad: *Thomomys talpoides, Peromyscus maniculatus,* and *Neotoma cinerea.* Except for *N. cinerea,* these species have long, relatively continuous stratigraphic ranges in the Kennewick section (Figure 13.11). *T.* cf. *T. talpoides* ranges from the base to the top of the section. *P. maniculatus* has a relatively continuous range after its appearance in interval C. *N. cinerea* is found only in the upper part of interval D and in the modern fauna; its poor representation in the stratigraphic section is consistent with its usually sparse representation in the modern fauna (Bailey, 1936, p. 171).

Group 3. Taxa whose current climatic distributions are more boreal or more humid than the Kennewick region today: *Phenacomys* cf. *P. intermedius, Synaptomys borealis,* and *Sorex palustris. P. intermedius* inhabits mountaintops in open grass parks (Dalquest, 1948, p. 339; Guilday and Parmalee, 1972). *P.* cf. *P. intermedius* has also been reported from the Wisconsinan and mid-Holocene of the currently arid Great Basin of Nevada (Grayson, 1981). If this vole represents the related taxon *Arborimus longicaudus,* its preference today is for humid forests (Johnson, 1973). *Synaptomys borealis,* the bog lemming, is confined to alpine environments in Washington today (Dalquest, 1948; Ingles, 1965), and as its name implies, it prefers moist habitats. Presumably, other closely related congeners had similar general preferences. *Sorex palustris* prefers clear, cold streams of alpine cirques and mountainsides (Dalquest, 1948). The restriction of these taxa to the intervals beneath E suggests that the climate during deposition of the lower part of the section was cooler and wetter than at Kennewick today.

Group 4. Taxa whose current distributions are more southern than southeast Washington: *Neotoma lepida, Dipodomys ordii,* and *Dipodomys microps. D. microps* and *N. lepida* today occur only as far north as southeastern Oregon. *D. ordii* occurs in this region today, but this is its northernmost range for Washington, Idaho, and western Montana (it reaches the Canadian border and beyond in eastern Montana). None of these taxa is represented in the large collections from the two lowest intervals A and B. Their simultaneous appearances in the samples of interval C may reflect increasing aridity, because all are desert animals

(Bailey, 1936; Dalquest, 1948; Finley, 1958; Cameron and Rainey, 1972; Beatley, 1976).

Group 5. A taxon (*Sylvilagus idahoensis*) whose current distribution (Bailey, 1936, p. 110; Orr, 1940; Green and Flinders, 1980) is restricted to habitats with denser vegetation than exists in the area today (sage-brush, *Artemisia tridentata,* or rabbitbush, *Chrysothamus*). *S. idahoensis* disappears after interval E, together with group 3 species that prefer cool, wet habitats.

To summarize:

1. The climate in general appears to have been arid in eastern Washington for as long as the section represents – at least 40,000 years and possibly as long as 328,000 years. Within this context there are indications of fluctuations in effective moisture.

2. During intervals A and B, conditions were wetter and cooler than those in that area today.

3. A major decrease in effective moisture occurred between intervals B and C, when desert species appear for the first time.

4. A second decrease in effective moisture and perhaps an increase in temperature occurred between intervals E and I, but the scarcity of fossils above interval E prevents determination of exactly when this event occurred.

5. There are indications of shorter-term fluctuations superimposed on the long-term drying trend. The simultaneous disappearance of *Neotoma lepida* and appearance of *Phenacomys intermedius* (or *Arborimus longicaudus*) and the slightly later appearance of *Neotoma cinerea* may reflect wetter and cooler conditions late in interval D than occurred in interval C. Two taxa with longer records, *Lagurus curtatus* and *Perognathus parvus,* show fluctuations in an inverse abundance relationship that are re-peated in intervals A, C, D, E, and I, implying changes in shrub cover and the moisture that controls it.

6. There was a coexistence in intervals C through E of boreal (group 3) and southern desert (group 4) species that do not live together today.

Sympatry of currently allopatric species is common in late Quater-nary sites of the northeastern and midwestern United States, where the assemblages have been termed "disharmonious" faunas or "no-analogue" faunas and explained by postulating a less strongly seasonal climate than today; see Graham and Lundelius (1984) and Lundelius

et al. (1983) and the references therein. Presumably winters in northeastern and midwestern United States lacked extremely cold temperatures, perhaps because the Laurentide ice sheet restricted Arctic air
to the polar region and split the jet stream (Wright, 1983; Kutzbach
and Wright, 1985; Graham and Mead, 1987), and summers were cooler
than today. Thus, species limited by winter extremes could have expanded their ranges to the north, and those limited by hot summers
could have ranged farther south and to lower elevations. The northern
migration of the group 4 species supports the idea of milder winters
than we have today, because extremely low winter temperatures seem
to limit both *Dipodomys microps* in Nevada (Beatley, 1976) and *Neotoma lepida* in Colorado (Finley, 1958). In addition, and perhaps
alternatively, physical displacement by Cordilleran ice could explain
the presence of boreal species (group 3), whose only option would be
southward migration along suitable microhabitats in the Columbia
River drainage. Although we cannot rule out minor stratigraphic mixing (see the earlier section on taphonomy), the observed amounts of
bioturbation at most of the collecting sites are insufficient to support
a contention that stratigraphic mixing by itself produced the no-
analogue species associations.

Morphological changes within species

Recognition of local evolutionary changes in terrestrial species at the
population level usually is hampered by discontinuities of preservation.
Depositions in lowlands (where most thick sections accumulate) are
intermittently reworked by braiding streams. The Kennewick deposit is unusual in that it was deposited at a high elevation, so that the
stream activity at the lower elevations, even floods, apparently left it
undisturbed.

Changes in the morphology of teeth are recognizable in the Kennewick species that are long-ranging and abundant. Some of the changes
involve size, but others represent modifications of the shapes of occlusal
structures.

Spermophilus townsendii.   The sizes of the M^3 and P_4 in *S. townsendii*
change through the stratigraphic section (Figures 13.13 and 13.14). The
M^3 increases in width at interval B (Fisher's exact test gives a null
probability of 0.037), with probably a further increase at interval C

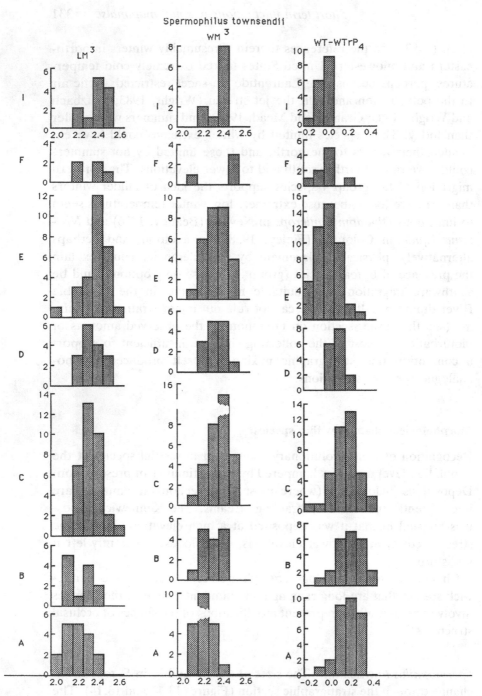

Figure 13.13. Frequency distributions of length and width of M^3 and of difference in width of talonid and trigonid of P_4 (talonid minus trigonid) in *Spermophilus townsendii* at stratigraphic intervals A–I.

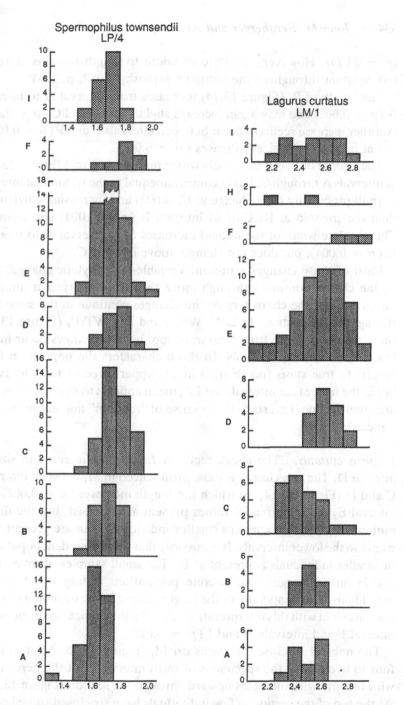

Figure 13.14. Frequency distributions of length of P_4 in *Spermophilus townsendii* (LP$_4$) and length of M_1 in *Lagurus curtatus* (LM$_1$) at stratigraphic intervals A–I.

($p = 0.074$). However, the ratio of width to length remains more or less constant throughout the section (Barnosky, 1987, p. 124).

The length of P_4 (Figure 13.14) increases from interval A to interval B ($p = 0.005$) and may again increase slightly at interval C ($p = 0.14$). Another increase seems to occur between intervals E and F ($p = 0.024$), but at interval I the size decreases ($p = 0.002$).

The talonid of P_4 is transversely wider than the trigonid (Figure 13.13) at intervals A through C, but becomes subequal to the trigonid at interval D (null probability for a change at C: 0.002) and increasingly narrower than the trigonid at KRC-01 of interval E ($p = 0.004$) and upward. The absolute width of the talonid increases from interval A to interval C ($p = 0.004$), but does not change above interval C.

Most of these changes represent variable-rate phyletic change, that is, the change continues through more than one superjacent interval, but in none of the characters do the changes continue in the same rate throughout the section. In LM^3, WM^3, and $WT\text{-}WTrP_4$ (Figure 13.13) there are net changes from bottom to top, but the changes occur in the lower three to four intervals. In three characters the population then reverts to true stasis (no change) in the upper three or four intervals, but in the upper four intervals the LP_4 mean appears to decline, increase, and then decline (i.e., stasis in the sense of "random" fluctuation around a mean).

Lagurus curtatus.   The cheek teeth of *L. curtatus* increase in size at interval D. The size change is most pronounced in M_1 between intervals C and D (Figure 13.14), in which the length increases ($p = 0.002$). At interval E, the same modal values present at D persist, but the distribution is skewed and contains smaller individuals than are present anywhere in the lower intervals. It is possible that a less abundant population of smaller individuals is present at E. The small samples of intervals F and H further suggest two discrete populations of larger and smaller individuals. The bimodality of the larger sample in the overlying interval I is consistent with this interpretation. A reduction in size occurs between interval F and intervals H and I ($p = 0.018$).

The number of closed triangles on M_1 (Figure 13.15A) varies from four to five among the specimens of most intervals, but the percentage with five triangles increases upward through the section (Figure 13.16). At the top of the section, a few individuals have six closed triangles. At intervals A through C, about 35% of the specimens have five closed triangles, whereas more than 75% from interval D have five closed

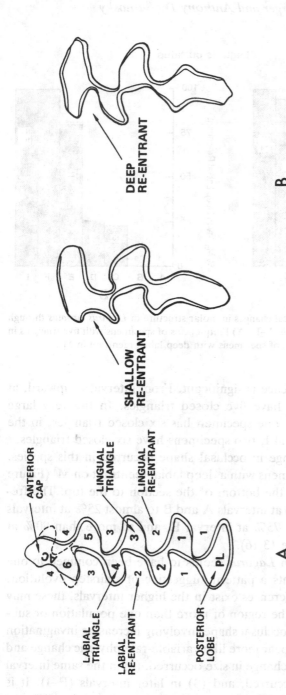

Figure 13.15. Molar structures in *Lagurus curtatus*. (A) Left M₁ with five closed triangles (solid line) and six closed triangles (dashed line). (B) Structure of M³ showing shallow versus deep labial reentrant variants.

Lagurus curtatus

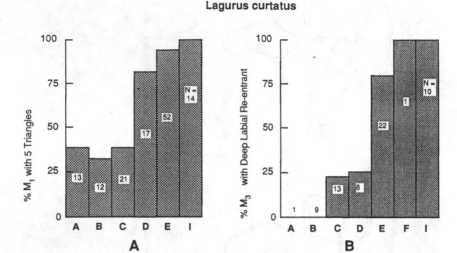

Figure 13.16. Vertical changes in molar structure in *Lagurus curtatus* through stratigraphic intervals A–I. (A) Frequencies of specimens with five triangles in M_1. (B) Frequencies of specimens with deep labial reentrant in M_3.

triangles, and the difference is significant. From interval E upward, at least 90% of the teeth have five closed triangles. In the very large collection of interval E, one specimen has six closed triangles; in the smaller sample of interval I, two specimens have six closed triangles.

Another type of change in occlusal shape occurred in this species. The percentage of specimens with a deep labial reentrant on M^3 (Figure 13.15B) increased from the bottom of the section to the top. The frequency varied from zero at intervals A and B to almost 25% at intervals C and D, to more than 75% at interval E, and to more than 90% at intervals F and I (Figure 13.16).

The increase in size in *Lagurus* seems to have been confined to one interval (D) and presents a pattern suggestive of stairstep evolution. Although some size differences exist in the higher intervals, these may reflect the existence in the region of more than one population or subpopulation. Changes in occlusal shape involving increasing invagination of enamel reentrants appear more like variable-rate phyletic change and occurred (1) before the change in size occurred, (2) in the same interval that the size change occurred, and (3) in later intervals (E–I). It is possible that this pattern of change in occlusal shape (involving different positions changing at different times), like the size changes, is related to the possibility that more than one subpopulation contributed to the fossil sample.

Conclusions

The youngest fauna is from less than 7,000 years BP, based on its position above the St. Helens S ash. The oldest fauna probably is from not less than 40,000 years BP and may be as old as 328 ka, judging from the amount of time attributed to the development of similar stages of calcite cementation in other regions. The faunal content throughout the section is consistent with a Holocene-to-Rancholabrean range. The extinct vole *Microtus* (*Pitymys*) *meadensis,* characteristically found elsewhere in North America in deposits of Irvingtonian age, suggests antiquity for the basal beds, but the predominance of Recent species suggests that even the base of the section is younger than Irvingtonian. *M.* (*Pitymys*) *meadensis* apparently represents a late relict of a once widespread earlier Pleistocene form.

The species turnovers seem to represent environmentally controlled events because of the joint appearances of taxa with present-day ranges that either are more boreal than Kennewick is today or are more southerly. The data suggest a colder period for the base of the section, a warmer interval, a colder interval, and finally a warmer interval probably leading to the present-day fauna. The lower of the colder intervals may correlate with the Salmon Springs glaciation, and the upper with the Fraser glaciation. However, there is a pronounced faunal gap in the Kennewick section beginning immediately above the second cold period, so that one cannot rule out the possibility of another unseen fluctuation there. This leaves the correlation of the cold periods with Pacific Northwest glacial stages uncertain.

The continuous representation of the species *Spermophilus townsendii, Perognathus parvus,* and *Lagurus curtatus,* all of which occur today in the area and have arid distributions elsewhere, indicates that the general conditions throughout at least the past 40,000 years (and possibly as long as 328,000 years) in eastern Washington were grossly similar in effective moisture to current conditions. The fluctuations in relative abundances of these long-ranging taxa, especially the reciprocal relationship of *L. curtatus* and *P. parvus,* are suggestive of shorter-term perturbations in the climate that were superimposed on the longer trends. The short-term fluctuations in *L. curtatus* and *P. parvus* themselves imply changes in effective moisture, because *L. curtatus* prefers substantial stands of low vegetation, whereas *P. parvus* prefers dry desert soils.

Most of the within-species changes in morphology appear to have been independent events in each lineage. The main changes in *S. townsendii* involved the size of the teeth and are explainable as variable-rate phyletic evolution. Changes in *relative* dimensions of both the P_4 and M^3 exhibit stasis. Several of the changes in *L. curtatus* happen to have corresponded to times of immigration of other taxa, but whether these apparent morphological changes reflect immigration of allopatric demes or evolution within one deme cannot be determined without the availability of similar morphological data about *Lagurus* from other regions. The changes in occlusal crests in *L. curtatus* represent long-term evolutionary changes in the population, whether or not the speed of change was influenced by immigrants interbreeding with the previously established deme. These variable-rate phyletic changes include (1) the proportion of specimens with five closed enamel triangles increasing upward through the section, from 50% at the base to over 90% at the top, with a few specimens at the top even having six triangles, and (2) the proportion of specimens with a deep labial enamel reentrant on M^3 increasing from zero at the base of the section to more than 90% at the top.

Acknowledgments

We are indebted to Dr. Harvey S. Rice, who recognized the potential significance of these beds and their fossil content, to the Washington State Department of Transportation for financial support, and to John Alexander, Jacob J. Dickinson, Leland Gibbon, James M. Gibson, Gary D. Johnson, Steve Nelson, Susan L. Richardson, Dr. Patrick Spencer, and Norman H. Wallace for their effort in the field. Dr. Spencer supervised the fieldwork and did the petrologic analyses. Dr. Donald K. Grayson of the Department of Anthropology, University of Washington, and Ellen Kritzman, University of Puget Sound, provided access to their collections of Recent mammals. Dr. Grayson kindly provided measurements of modern species that contributed substantially to this study. Dr. Minze Stuiver and Dr. Pieter Grootes, Department of Geological Sciences, University of Washington, attempted radiocarbon dating of bone from the site. Figures 13.5–13.9, 13.11, 13.12, and 13.15 were made by Mark Orsen, and photographic assistance was provided by Karna Orsen. James and Gary Rensberger wrote software for cataloging the specimens, data management, and sediment grain-size plots. Dr. C. W. Whitlock and Dr. H. E. Wright, Jr., provided helpful suggestions for early versions of the manuscript, and Dr. R. W. Graham and Dr. Robert A. Martin added insights that improved the final version.

References

Bailey, V. (1936). The mammals and life zones of Oregon. *North American Fauna*, 55:1–416.

Baker, R. H. (1968). Habitats and distribution. In J. A. King (ed.), *The Biology of Peromyscus* (pp. 98–126). American Society of Mammalogists.

Baker, V. R. (1978). Quaternary geology of the channeled scabland and adjacent areas. In V. R. Baker and D. Nummedal (eds.), *The Channeled Scabland* (pp. 17–36). Washington, D.C.: Office of Space Science, National Aeronautics and Space Administration.

Barnosky, C. W. (1984). Late Pleistocene and early Holocene environmental history of southwestern Washington State, U.S.A. *Canadian Journal of Earth Sciences*, 21:619–29.

  (1985). Late Quaternary vegetation in the southwestern Columbia Basin, Washington. *Quaternary Research*, 23:109–22.

  (1987). Punctuated equilibrium and phyletic gradualism: some facts from the Quaternary mammalian record. *Current Mammalogy*, 1:109–44.

Barnosky, A. D., and D. L. Rasmussen (1988). Middle Pleistocene arvicoline rodents and environmental change at 2900 meters elevation, Porcupine Cave, South Park, Co. *Annals of Carnegie Museum*, 57:267–92.

Beatley, J. C. (1976). Environments of kangaroo rats (*Dipodomys*) and effects of environmental change on populations in southern Nevada. *Journal of Mammalogy*, 57:67–93.

Bunker, R. C. (1982). Evidence of multiple late-Wisconsinan floods from Glacial Lake Missoula in Badger Coulee, Washington. *Quaternary Research*, 18: 17–31.

Cameron, G. N., and D. G. Rainey (1972). Habitat utilization by *Neotoma lepida* in the Mohave Desert. *Journal of Mammalogy*, 53:251–66.

Carroll, L. E., and H. H. Genoways (1980). *Lagurus curtatus*. *Mammalian Species*, 124:1–6.

Dalquest, W. W. (1948). Mammals of Washington. *University of Kansas Museum of Natural History Publications*, 2:1–444.

Fink, J., and G. J. Kukla (1977). Pleistocene climates in Central Europe: at least 17 interglacials after the Olduvai event. *Quaternary Research*, 7: 363–71.

Finley, R. B., Jr. (1958). The wood rats of Colorado: distribution and ecology. *University of Kansas Museum of Natural History Publications*, 10:213–552.

Flint, R. F. (1938). Origin of the Cheney-Palouse scabland tract, Washington. *Geological Society of America Bulletin*, 49:461–524.

Gile, L. H., F. F. Peterson, and R. B. Grossman (1965). The K horizon: a master soil horizon of carbonate accumulation. *Soil Science*, 99:74–82.

(1966). Morphological and genetic sequences of carbonate accumulations in desert soils. *Soil Science,* 101:347–60.

Graham, R. W., and E. L. Lundelius, Jr. (1984). Coevolutionary disequilibrium and Pleistocene extinctions. In P. S. Martin and R. G. Klein (eds.), *Quaternary Extinctions, a Prehistoric Revolution* (pp. 223–49). Tucson: University of Arizona Press.

Graham, R. W., and J. I. Mead (1987). Environmental fluctuations and evolution of mammalian faunas during the last deglaciation. In W. F. Ruddiman and H. E. Wright, Jr. (eds.). *North American and Adjacent Oceans During the Last Deglaciation* (pp. 371–402). (*The Geology of North America, Vol. K-3*). Boulder, Colo.: Geological Society of America.

Grayson, D. K. (1981). A mid-Holocene record for the heather vole, *Phenacomys* cf. *intermedius,* in the central Great Basin and its biogeographic significance. *Journal of Mammalogy,* 62:115–21.

Green, J. S., and J. T. Flinders (1980). *Brachylagus idahoensis. Mammalian Species,* 125:1–4.

Guilday, J. E., and P. W. Parmalee (1972). Quaternary periglacial records of voles of the genus *Phenacomys* Merriam (Cricetidae; Rodentia). *Quaternary Research,* 2:170–5.

Guthrie, R. D., and J. V. Matthews (1971). The Cape Deceit fauna – early Pleistocene mammalian assemblage from the Alaskan arctic. *Quaternary Research,* 1:474–510.

Hall, E. R. (1946). *Mammals of Nevada.* Berkeley: University of California Press.

(1981). *The Mammals of North America,* 2 vols., 2nd ed. New York: Wiley.

Hall, E. R., and K. R. Kelson (1959). *The Mammals of North America,* 2 vols. New York: Ronald Press.

Hibbard, C. W. (1937). A new *Pitymys* from the Pleistocene of Kansas. *Journal of Mammalogy,* 18:235.

(1944). Stratigraphy and vertebrate paleontology of Pleistocene deposits of southwestern Kansas. *Geological Society of America Bulletin,* 55:718–44.

(1955). Notes on the microtine rodents from Port Kennedy Cave deposit. *Proceedings of the Philadelphia Academy of Natural Science,* 107:87–97.

Holman, J. A. (1959). Birds and mammals from the Pleistocene of Williston, Florida. *Florida State Museum Bulletin,* 5:1–25.

Ingles, L. G. (1965). *Mammals of the Pacific States.* Stanford: Stanford University Press.

Johnson, M. L. (1973). Characters of the heather vole, *Phenacomys,* and the red tree vole, *Arborimus. Journal of Mammalogy,* 54:239–44.

Kowalski, K. (1970). Variation and speciation in fossil voles. *Zoological Society of London, Symposium,* 26:149–61.

Krinsley, D. H., and J. C. Doornkamp (1973). *Atlas of Quartz Sand Surface Textures.* Cambridge University Press.

Kritzman, E. B. (1974). Ecological relationships of *Peromyscus maniculatus* and *Perognathus parvus* in eastern Washington. *Journal of Mammalogy,* 55: 172–88.

Kurtén, B., and E. Anderson (1980). *Pleistocene mammals of North America.* New York: Columbia University Press.

Kutzbach, J. E. (1987). Model simulations of the climatic patterns during the deglaciations of North America. In W. F. Ruddiman and H. E. Wright, Jr. (eds.), *North American and Adjacent Oceans During the Last Deglaciation* (pp. 425–46). (*The Geology of North America, Vol. K-3*). Boulder, Colo.: Geological Society of America.

Kutzbach, J. E., and H. E. Wright, Jr., (1985). Simulation of the climate of 18,000 yr B.P.: results for the North American/ North Atlantic/ European sector. *Quaternary Science Reviews*, 4:147–87.

Lindsay, E. H., N. M. Johnson, and N. D. Opdyke (1975). Preliminary correlation of North American land mammal ages and geomagnetic chronology. In G. R. Smith and N. E. Friedland (eds.), *Studies on Cenozoic Paleontology and Stratigraphy. Claude W. Hibbard Memorial Volume 3* (pp. 111–19). University of Michigan Papers in Paleontology, no. 12.

Lundelius, E. L., Jr., T. Downs, E. H. Lindsay, H.A. Semken, R. A. Zakrzewski, C. S. Churcher, C. R. Harington, G. E. Schultz, and S. D. Webb (1987). In M. O. Woodburne (ed.), *Cenozoic Mammals of North America: Geochronology and Biostratigraphy* (pp. 311–53). Berkeley: University of California Press.

Lundelius, E. L., Jr., R. W. Graham, E. Anderson, J. Guilday, J. A. Holman, D. W. Steadman, and S. D. Webb (1983). Terrestrial vertebrate faunas. In H. E. Wright, Jr. (ed.), *Late-Quaternary Environments of the United States, Vol. 1: The Late Pleistocene* (S. C. Porter, ed.) (pp. 311–53). Minneapolis: University of Minnesota Press.

Lyman, R. L., and S. D. Livingston (1983). Late Quaternary mammalian zoogeography of eastern Washington. *Quaternary Research*, 20:360–73.

Martin, R. A. (1974). Fossil mammals from the Coleman 2A fauna, Sumter County. In S. D. Webb (ed.), *Pleistocene Mammals of Florida* (pp. 35–99). Gainesville: University Presses of Florida.

(1975). *Allophaiomys* Kormos from the Pleistocene of North America. In G. R. Smith and N. E. Friedland (eds.), *Studies on Cenozoic Paleontology and Stratigraphy. Claude W. Hibbard Memorial Volume 3* (pp. 97–100). University of Michigan Papers in Paleontology, no. 12.

(1987). Notes on the classification and evolution of some North American fossil *Microtus* (Mammalia; Rodentia). *Journal of Vertebrate Paleontology*, 7:270–83.

Maruszczak, H. (1980). Stratigraphy and chronology of the Vistulian loesses in Poland. *Quaternary Studies in Poland*, 2:57–76.

Newcomb, R. C. (1961). Age of the Palouse Formation in the Walla Walla and Umatilla River basins, Oregon and Washington. *Northwest Science*, 35: 122–7.

Orr, R. T. (1940). The rabbits of California. *Occasional Papers, California Academy of Science*, 19:1–227.

Paulson, G. R. (1961). The mammals of the Cudahy Fauna. *Papers of the Michigan Academy Science, Arts and Letters*, 46:127–53.

Powers, H. A., and R. E. Wilcox (1964). Volcanic ash from Mount Mazama (Crater Lake) and from Glacier Peak. *Science,* 144:1334–6.

Pye, K. (1983). Grain surface textures and carbonate content of late Pleistocene loess from West Germany and Poland. *Journal of Sedimentary Petrology,* 53:973–80.

Repenning, C. A. (1983). *Pitymys meadensis* Hibbard from the Valley of Mexico and the classification of North American species of *Pitymys* (Rodentia: Cricetidae). *Journal of Vertebrate Paleontology,* 2:471–82.

   (1987). Biochronology of the microtine rodents of the United States. In M. O. Woodburne (ed.), *Cenozoic Mammals of North America: Geochronology and Biostratigraphy* (pp. 236–66). Berkeley: University of California Press.

Richmond, G. M., R. Fryzell, G. E. Neff, and P. L. Weis (1965). The Cordilleran ice sheet of the northern Rocky Mountains and related Quaternary history of the Columbia Plateau. In H. E. Wright and D. G. Frey (eds.), *The Quaternary of the United States* (pp. 231–42). Princeton: Princeton University Press.

Savage, D. E. (1951). Late Cenozoic vertebrates of the San Francisco Bay region. *Bulletin of the Department of Geological Sciences, University of California,* 28:215–314.

Savage, D. E., and D. E. Russell (1983). *Mammalian Paleofaunas of the World.* Reading, Mass.: Addison-Wesley.

Scheffer, T. H. (1983). Pocket mice of Washington and Oregon in relation to agriculture. *U.S. Department of Agriculture Technical Bulletin,* 608:1–15.

Smalley, I. J. (ed.) (1975). *Loess: Lithology and Genesis.* Stroudsburg, Pa.: Dowden, Hutchinson and Ross.

Thaeler, C. S. (1980). Chromosome numbers and systematic relations in the genus *Thomomys* (Rodentia: Geomyidae). *Journal of Mammalogy,* 61:414–22.

Tucholka, P. (1977). Magnetic polarity events in Polish loess profiles. *Biul. Instytutu Geologicznego,* 305:117–23.

van der Meulen, A. J. (1978). *Microtus* and *Pitymys* (Arvicolidae) from Cumberland Cave, Maryland, with a comparison of some New and Old World species. *Annals of the Carnegie Museum,* 47:101–45.

Waitt, R. B., Jr. (1980). About forty last-glacial Lake Missoula jokulhlaups through southern Washington. *Journal of Geology,* 88:653–79.

Warren, J. K. (1983). Pedogenic calcrete as it occurs in Quaternary calcareous dunes in coastal South Australia. *Journal of Sedimentary Petrology,* 53:787–96.

Wojtanowicz, J., and J. Buraczynski (1978). Materials to the absolute chronology of the loesses of Grzeda Sokalska. *Annales Univ. M. Curie-Sklodowska B,* 30/31:37–54.

Wright, H. E., Jr. (1983). Introduction. In H. E. Wright, Jr. (ed.), *Late-Quaternary Environments of the United States, Vol. 2, The Holocene* (pp. xi-xvii). Minneapolis: University of Minnesota Press.

14

Size change in North American Quaternary jaguars

KEVIN SEYMOUR

The jaguar (*Panthera onca*), largest of the New World felids, was formerly widespread in the Neotropical and Nearctic regions. In North America, pre-Wisconsinan jaguars ranged farther north than did those of the Wisconsinan. The present-day range of this species is far to the south of its Wisconsinan range (Figure 14.1). This illustrates a gradual restriction in the range of this species, though this general trend probably was influenced by a sequence of glacial–interglacial shifts in range (Kurtén and Anderson, 1980). The earliest jaguar may have been conspecific with the middle Pleistocene *P. gombaszoegensis* of Eurasia (Hemmer, 1971) that probably dispersed across the Bering land bridge to reach North America at that time (Kurtén, 1973). If that is true, the living species can be considered a relict population of a once more widely distributed Holarctic form (Kurtén and Anderson, 1980).

Concurrent with this restriction in range was a reduction in size. According to Kurtén (1973), size reduction proceeded gradually from the Blancan (Curtis Ranch, now known to be Irvingtonian) (Kurtén and Anderson, 1980; Schultz, Martin, and Schultz, 1985) to the present, although it may have accelerated during the Holocene (Kurtén and Anderson, 1980). Also, there was a gradual shortening of the limbs, especially the metapodials. Hence, the earliest North American jaguar lacked some of the living jaguar's specializations – its limbs were longer and their distal portions were less shortened (Kurtén, 1973). Kurtén and Anderson (1980) estimated that Wisconsinan jaguars exceeded the living species in size by 15% or 20% and that earlier jaguars were even larger. Seymour (1983) calculated that the late Pleistocene jaguar was

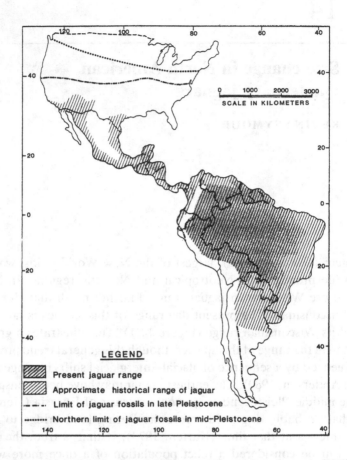

Figure 14.1. Distribution of the jaguar. Adapted from Seymour (1989). North American fossil limits from Kurtén (1973); South American fossil limits from Seymour (1983). Recent range from Swank and Teer (1989).

15% larger than the living species, judging from cranial and dental measurements, or 25% larger, judging from postcranial measurements.

This chapter reexamines the scenario of size change in North American Quaternary jaguars. In particular, data for the Irvingtonian form are compared to those for the Recent jaguar, especially the large southern Brazilian subspecies *P. o. paraguensis*. Body size is estimated by using published regressions, and rates of evolution (in darwins) are calculated. The taxonomic status of the Pleistocene form is reconsidered.

Abbreviations

AMNH, American Museum of Natural History, New York; ANSP, Academy of Natural Sciences, Philadelphia; BMNH, British Museum (Natural History), London; ChM, Charleston Museum, Charleston, South Carolina; CM, Carnegie Museum of Natural History, Pittsburgh; F:AM, Frick Collection, AMNH; FMNH, Field Museum of Natural History, Chicago; IBUNAM, Instituto de Biología, Universidad Nacional Autónomo de México, México City; IGM, Instituto de Geología, México City; KUMNH, University of Kansas Museum of Natural History, Lawrence; LACM, Los Angeles County Museum, Los Angeles; MNHN, Musée National d'Histoire Natural, Paris; NMB, Naturhistorisches Museum Basel, Basel, Switzerland; NMC, National Museum of Canada (now Canadian Museum of Nature), Ottawa; RMM, Red Mountain Museum, Birmingham, Alabama; ROM, Royal Ontario Museum, Toronto; SMU, Southern Methodist University, Dallas; TMM, Texas Memorial Museum, Austin; UCMP, University of California, Museum of Paleontology (now part of the Department of Integrative Biology); UCMVZ, University of California Museum of Vertebrate Zoology, Berkeley; UF, University of Florida, Florida Museum of Natural History, Gainesville; UMZ, Universitets Museum Zoologisches, Copenhagen; UNSM, University of Nebraska State Museum, Lincoln; USNM, United States National Museum, Washington, D.C.; UTEP, University of Texas, El Paso; YPM, Yale Peabody Museum, New Haven, Connecticut.

Localities of fossil jaguars

Schultz et al. (1985) noted that caution should be exercised in the identification of fragmentary felid specimens, as they are notoriously difficult to assign correctly to species. Although Kurtén's (1973) review of the known localities of fossil jaguars is still largely valid today, several localities have been redated, and there are some important new discoveries and reidentifications that should be noted prior to a discussion of the evolution of this form. They are discussed next in order of decreasing age.

Early Irvingtonian jaguars

Curtis Ranch, Cochise Co., Arizona. Although Kurtén (1973) tentatively considered this locality to be Blancan in age, Johnson, Opdyke, and Lindsay (1975) demonstrated that it dates to 1.8–1.9 my BP, or earliest Irvingtonian. Kurtén (1973) mentioned only one specimen from this locality, USNM 12865, a right P_4; this specimen was illustrated by Gazin (1942, p. 505, Figure 45). In addition, I also believe that USNM 12866, a right calcaneum, and USNM 12867, a right metatarsal III, are felid, but I do not think that any of these three specimens represents Felinae, let alone *Panthera*. These specimens need to be reexamined before a positive identification can be made. Nevertheless, reidentifying these specimens as non-*Panthera* removes the only Blancan jaguar from the record.

Inglis IA, Citrus Co., Florida.   Berta (1987) discussed the *Smilodon gracilis* material from this early Irvingtonian locality (correlated with the Curtis Ranch fauna) (Webb and Wilkins, 1984). Certain felid bones that I originally identified as jaguar (Seymour, 1983) were not considered to be *Smilodon* by Berta (1987), but my reidentification indicates that these specimens are small, perhaps female, *Smilodon gracilis*. This material includes the following UF specimens: 45453, right scapholunar; 45348, left femur; 45349, right tibia; 45410, right metatarsal II; 45411, right metatarsal III; 45412, left metatarsal III; 45413, right metatarsal IV. In summary, at present there are no valid Blancan or early Irvingtonian specimens of jaguars.

Middle-to-late Irvingtonian localities

Hamilton Cave, Pendleton Co., West Virginia.   Although its data have not yet been published, a relatively complete skeleton of a jaguar (USNM 299767) appears to date to 820–850 ky BP, judging from biostratigraphic information (Repenning and Grady, 1988). Currently, this is the oldest relatively securely dated jaguar from North America. A skeleton of *Miracinonyx inexpectatus* has also been recovered from this important site (Van Valkenburgh et al., 1990).

McLeod Lime Rock Mine, Levy Co., Florida.   An almost complete skeleton of a jaguar (F:AM 69204–69251) represents the most complete fossil skeleton known to date. Although its data are unpublished, this site is certainly Irvingtonian and dates to 500–700 ky BP (Morgan and Hulbert, in press). For the calculations in this study, I have used 600 ky BP as the age for this site.

Port Kennedy Cave, Montgomery Co., Pennsylvania. There has been much confusion over the identification of the felid material from this locality. Both Simpson (1941a) and Kurtén (1976) mentioned jaguar from Port Kennedy Cave. Certainly some specimens represent *Miracinonyx inexpectatus* (Van Valkenburgh et al., 1990), and others represent *Smilodon gracilis* (Berta, 1987). Upon reexamination of the felid material, I consider three specimens to represent jaguar: ANSP 47, distal part of a left humerus; ANSP 18341, distal part of a right humerus; and ANSP 18343, left metacarpal III. This fauna is considered to be middle to late Irvingtonian in age (Kurtén and Anderson, 1980) and probably is older than the Cumberland Cave fauna (Guilday et al., 1984).

Hanover Quarry, Adams Co., Pennsylvania. This site was briefly reported by Guilday et al. (1984) to be older than Cumberland Cave and perhaps contemporaneous with Port Kennedy Cave. I have identified *Smilodon gracilis, Miracinonyx inexpectatus,* and jaguar from this locality (Seymour, 1983). The jaguar material consists of CM 40434, a right metatarsal V, and probably also a distal metapodial fragment.

Cumberland Cave, Allegany Co., Maryland. The left maxilla (CM 24328) illustrated by Kurtén (1973, Figure 2) represents a jaguar, as does USNM 12840, a left astragalus, calcaneum, and metatarsal III; other postcranial material needs to be reexamined in the light of the presence of *Miracinonyx inexpectatus* (Van Valkenburgh et al., 1990) and also *Smilodon* (e.g., USNM 12840, right scapholunar). This site is considered to be middle Irvingtonian in age (Kurtén and Anderson, 1980).

Conard Fissure, Newton Co., Arkansas. As with Port Kennedy Cave and Cumberland Cave, the presence of *Miracinonyx* (Van Valkenburgh et al., 1990) and possibly also *Smilodon* has caused some confusion in the identification of isolated postcranial bones. All felid bones from this middle Irvingtonian site (approximately 600 ky BP) (Schultz et al., 1985) need to be reexamined.

Loup Fork of Platte River, Nebraska. As discussed by Schultz et al. (1985), this is the locality for the type of *Felis augusta* (Leidy, 1872), the oldest name for a North American fossil jaguar. This specimen (USNM 125, a left maxilla with P^3 and P^4) probably was found near the Mullen locality and is assumed to be middle Irvingtonian in age. There are also two other jaguar specimens from this locality: USNM 147, distal portion of a right humerus, and USNM 5448, a right premaxilla.

Mullen II, Cherry Co., Nebraska.   The most complete fossil jaguar skull known (lacking only the canines) is represented by UNSM 1104. It was illustrated by Kurtén (1973) and Schultz et al. (1985) and is late Irvingtonian in age (Schultz et al., 1985).

Coleman 2A, Sumter Co., Florida.   This latest Irvingtonian or earliest Rancholabrean site, dating between 300 and 400 ky BP (Morgan and Hulbert, in press), contains more individual jaguars than any other locality I know of [I found an MNI = 5, although Kurtén (1973) reported an MNI of 6]. Dental material was studied by Martin (1974); Kurtén (1973) presented data on the postcranial specimens only. Some specimens have since been recatalogued; all specimens were reexamined for this study. For the calculations in this study, I have used 300 ky BP as the age for this site.

Two other localities once thought to be Irvingtonian (Kurtén, 1973) are now considered to be Rancholabrean: Fossil Lake, Lake Co., Oregon (UCMP 2979, distal part of a left radius; UCMP 26914, right metacarpal IV; and UCMP 26966, right metatarsal IV), now reported as Wisconsinan by Kurtén and Anderson (1980), and Edisto Beach, Charleston Co., South Carolina (ChM PV2284, right P^4 not examined by the author), also reported as Wisconsinan (Roth and Laerm, 1980).

Rancholabrean localities

Friesenhahn Cave, Bexar Co., Texas.   The two specimens reported to be jaguar by Kurtén (1973), TMM 933–902, left metacarpal II, and TMM 933–1279, left metacarpal III, actually represent *Smilodon fatalis*.

Devil's Den, Levy Co., Florida.   The following UF bones, originally published by Martin and Webb (1974) as *Canis dirus*, are actually *P. onca*: 7997, right ulna; 7998, left radius; 8000, right femur; 8001, left femur; 8002, right tibia; 8003, left tibia. Also, UF 9023, a right tibia of *Smilodon fatalis*, was reported as jaguar by Kurtén (1973). This site is latest Rancholabrean in age. Some of the smallest jaguar fossils have been found at this locality.

Cutler local fauna, Dade Co., Florida. Additional small fossil jaguar bones have been found at this site, though the data have not yet been published. It is thought to be latest Pleistocene in age (Morgan, 1992).

Materials and methods

A sample comprising 209 Recent jaguar skulls and 17 skeletons was examined; only 145 of these skulls were used in this study, because of the incompleteness of some specimens. Eighty of the 145 skulls had a sex recorded; 65 specimens were sexed using a stepwise discriminant function analysis developed for each of the three subspecies *P. o. hernandesii, P. o. onca,* and *P. o. paraguensis.* This resulted in a data set of 52 female and 93 male skulls. Skulls and skeletons from the following mammalogy collections were measured: AMNH, ANSP, BMNH, FMNH, IBUNAM, KUMNH, MNHN, NMB, NMC, ROM, UCMVZ, UMZ, USNM, YPM. Measurements were taken using dial calipers, as illustrated by von den Driesch (1976): condylobasal length, Figure 17c, measurement 2; zygomatic breadth, Figure 17a, measurement 23; humerus length, Figure 32c, measurement GL; ulna length, Figure 33c, measurement GL; radius length, Figure 33d, measurement GL; femur length, Figure 35b, measurement GL; tibia length, Figure 26b, measurement L; metacarpal III and metatarsal III, Figure 44o, measurement GL. Maximum tooth length was measured at the cingulum and parallel to the long axis of the tooth; for example, for P^4, Figure 14a, measurement L, and for M_1, Figure 23b, measurement L. Two skeletons each of tiger (*P. tigris*), lion (*P. leo*), and leopard (*P. pardus*) from the ROM collections were measured to provide comparative data.

Fossil jaguar material, both cranial and postcranial, was examined from the following collections: AMNH, ANSP, CM, F:AM, FMNH, IGM, LACM, RMM, ROM, SMU, TMM, UF, UCMP, UNSM, USNM, UTEP. Most of this fossil material consisted of disassociated bones of no more than one individual. Of primary importance to this study were specimens of associated skeletons. These included two unpublished Irvingtonian skeletons (USNM 299767, Hamilton Cave, Pendelton Co., West Virginia, and F:AM 69204–69251, McLeod Lime Rock Mine, Levy Co., Florida), several disassociated Irvingtonian skeletons (UF 12128–12165, Coleman 2A, Sumter Co., Florida) (Martin, 1974; Kurtén, 1973), and two published Rancholabrean skeletons: USNM 18262, Salt River Cave, Franklin Co., Tennessee (McCrady, Kirby-Smith, and Templeton, 1951; Guilday and McGinnis, 1972) and TMM

40673, Laubach Cave, Williams Co., Texas (Kurtén, 1965; Slaughter, 1966). Data for *P. leo atrox* from the tar pits of Rancho la Brea were taken from Merriam and Stock (1932).

Log-ratio diagrams

Log-ratio diagrams were constructed using the method described by Simpson (1941a). Because of small sample sizes and incomplete skeletons, three postcranial diagrams were constructed. All postcranial log ratios used the average obtained from measurements of seven *P. o. paraguensis* specimens as the standard. In order to obtain a graphic representation of the average size and limb proportions for the Irvingtonian and Rancholabrean jaguars, the following methods were used:

1. For the Irvingtonian postcranial diagram, specimens from the three major localities were first plotted separately. All are incomplete in some way: The Hamilton specimen lacks a femur, the McLeod specimen lacks an ulna and a metacarpal III, and the Coleman suite lacks a complete humerus and ulna. The McLeod specimen is the smallest, the Coleman suite is the largest, the Hamilton Cave specimen falls near the average, and all are reasonably similar in proportion, except for the length of the radius. To produce a complete plot for the average Irvingtonian specimen, the following method was used. Averages were taken for bones present from all three localities (radius, tibia, metatarsal III), although all the Coleman bones were averaged first so as not to bias the overall average with the greater size of the Coleman jaguar. The intermediate-sized Hamilton specimen was used to represent the "average" humerus, ulna, and metacarpal III, and an average of the Coleman and McLeod (the largest and smallest) was used for the femur.

2. The Rancholabrean postcranial diagram was more problematic. Only the Salt River specimen was reasonably complete; even then the length of metacarpal III was estimated. The Laubach Cave specimen lacks metapodials and appears unlike the Salt River specimen in proportions of the radius and femur. The average of all the Rancholabrean specimens appears more similar to the Salt River specimen than to the Laubach Cave specimen or the Recent jaguar. However, averages of disassociated specimens are easily biased because of the different individuals represented. In this case the sample includes some isolated large bones (which may actually be Irvingtonian in age), with small and unequal sample sizes. It was assumed, therefore, that the Salt River specimen was slightly unusual and that the overall average only coin-

cidentally presents a log-ratio plot similar to that for the Salt River specimen. Therefore the Salt River and the Laubach limb measurements were averaged, while the Rancholabrean overall average for the less variable metapodials was used. The possibility that some specimens represent a different species is discussed later.

3. The Irvingtonian and Rancholabrean "average" plots gained by these manipulations were then plotted with averages for the three subspecies of living jaguar, as well as those for lion, tiger, leopard, and the extinct North American lion (*P. leo atrox*). Because not all the Recent jaguar skeletons were sexed, and because some did not have complete skulls associated with them, the postcranial measurements for the Recent jaguar were plotted by subspecies, not sex. The measurements of one very large Recent male were also plotted.

An additional log-ratio diagram was constructed comparing cranial and dental proportions of fossil and Recent jaguars. The standard was an average of all 93 sexed male skulls; the average of the 52 females was also plotted. Two single individuals were plotted to show the limits of variation in the living species (a large male, BMNH 77.857, and small female, USNM 249826). The Rancholabrean measurements were averaged because the cranial and dental measures were less variable and were based on larger sample sizes than the postcranial plots and were frequently available as complete tooth rows. The McLeod and Hamilton measures were averaged because they were so similar; the Coleman measures were plotted separately.

Body mass estimates

Three published regressions were utilized in order to estimate body mass for fossil and Recent jaguars.

1. Legendre and Roth (1988) published the following regression for the Felidae using the major-axis method:

$$\ln y = 1.54471(\ln x) + 2.91192,$$

where y is body mass, and x is length × width of M_1.

2. Van Valkenburgh (1990) published the following regression for the Felidae using the least-squares method:

$$\log y = 3.05x - 2.15,$$

where y is body mass and x is length of M_1.

These equations were used to estimate body mass for the large, small, and known-mass living jaguars, males and females of each living subspecies, and the average Irvingtonian and Rancholabrean jaguars.

3. Gingerich (1990) published a program, BODYMASS, to predict body mass based on limb lengths and diameters. It used a multiple regression based on the data of Alexander et al. (1979). This program was used to predict body mass for the three subspecies of living jaguar, the specimen with mass recorded (FMNH 70566), and the Rancholabrean and Irvingtonian averages.

Rate of evolution

Rates of evolution were calculated for the jaguar from its "average" Irvingtonian form to its "average" Rancholabrean form, and also from the Rancholabrean "average" to the Recent average. The formula suggested by Haldane (1949) was used:

$$\frac{\ln(x_2) - \ln(x_1)}{t},$$

where x_1 is mean size for the smaller population, x_2 is mean size for the larger population, t is elapsed time, in years. Cranial, dental, and postcranial measures were utilized. The derivation of the "average" postcranial measures was described earlier. Because dental measures were available for all major specimens and were less variable, an overall Irvingtonian (with the Coleman site weighted as 1) plus Rancholabrean average was used. Cranial measures were rare; the McLeod specimen was used to represent the Irvingtonian, and USNM 23733, a skull from Little Airplane Cave, Marion Co., Tennessee (Guilday and McGinnis, 1972), was used to represent the Rancholabrean. The calculation of elapsed time was as follows. Because three Irvingtonian localities of different ages were used to obtain an average measurement, their ages were also averaged. Because most Rancholabrean jaguar localities are not accurately dated, but probably vary between 100,000 and 10,000 years BP, an arbitrary average date of 50,000 years was chosen to represent the population mean. The implications of this will be discussed later.

A simple percentage size change was also calculated using the formula

$$\frac{x_2 - x_1}{x_1} \times 100,$$

where x_1 and x_2 are means of the different samples. For both these

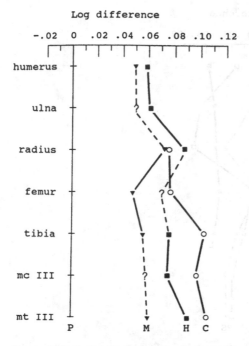

Figure 14.2. Log-ratio diagram for limb measures of Irvingtonian jaguars. The standard is a sample of Recent *P. o. paraguensis*. P, Recent *P. o. paraguensis*; M, McLeod Quarry; H, Hamilton Cave; C, Coleman 2A. Sample size for each locality is 1, except for Coleman 2A, which is as follows: radius 1, femur 3, tibia 2, metacarpal III 3, metatarsal III 3. Specimen catalogue numbers are listed in the Appendix.

formulas, the full suite of cranial, dental, and postcranial measurements was used. As well, an overall average was calculated.

Results

Log-ratio diagrams

The three postcranial log-ratio plots for the Irvingtonian, Rancholabrean, and comparative summary are presented in Figures 14.2, 14.3, and 14.4, respectively. Results for Figures 14.2 and 14.3 are presented here only to demonstrate the derivation of the "average" Irvingtonian and Rancholabrean plot used in Figure 14.4. There are several points of note in Figure 14.4. The postcranial proportions of all three Recent subspecies of jaguar are very similar. The Rancholabrean jaguar is very

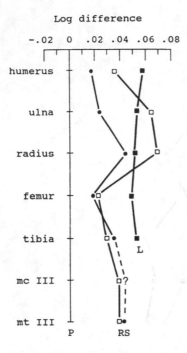

Figure 14.3. Log-ratio diagram for limb measures of Rancholabrean jaguars. The standard is a sample of Recent *P. o. paraguensis*. P, Recent *P. o. para-guensis*; S, Salt River Cave; R, average Rancholabrean; L, Laubach Cave. Sample sizes for the averages are as follows: humerus 2, ulna 1, radius 4, femur 5, tibia 2, metacarpal III 13, metatarsal III 11. Specimen catalogue numbers are listed in the Appendix.

similar in size and proportion to the large Recent individual (ANSP 13949). The Irvingtonian plot shows an animal larger in size and with slightly longer metapodials than the largest living subspecies. The plots for the lion, tiger, leopard, and fossil lion are all more similar to each other than they are to the jaguar. In particular, they all have a relatively longer radius compared with the femur. The leopard has a slightly longer hindfoot compared with the forefoot, whereas the other three large pantherines have a relatively longer forefoot compared with the hindfoot.

The cranial and dental log-ratio plot is given in Figure 14.5. Recent female jaguars have relatively smaller canines than males. All dental measurements of fossil jaguars fall within the range for the living jaguar. The fossil jaguars appear to have slightly shorter P_3 and M_1 than the Recent species. The breadth of the Irvingtonian jaguar skull is slightly

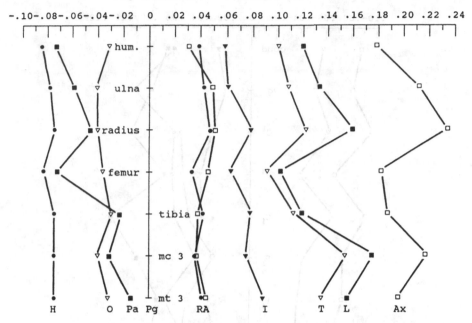

Figure 14.4. Log-ratio diagram for limb measures of fossil and Recent jaguars and other pantherines. The standard is a sample of Recent *P. o. paraguensis*. Sample sizes and abbreviations are as follows: *P. o. hernandesii* 6 (H), *P. o. onca* 4 (O), *P. pardus* 2 (Pa), *P. o. paraguensis* 7 (Pg), Rancholabrean "average" (R): 2 for all limb bones, 13 for metacarpal III, and 11 for metatarsal III; large male *P. o. paraguensis* ANSP 13949 (A), *P. tigris* 2 (T), *P. leo* 2 (L). Sample sizes for the other fossils are listed from humerus to metatarsal III as follows: Irvingtonian "average" (I), 1, 1, 3, 4, 4, 1, 5; *P. leo atrox* (Ax) 9, 7, 10, 9, 10, 2, 2. Specimen catalogue numbers are listed in the Appendix.

larger than that of the living jaguar, although this may be due in part to the fact that the cranial measures were based solely on the McLeod specimen, which was fragmented and reconstructed.

Body mass estimates

Body mass estimates using the equations of Legendre and Roth (1988) and Van Valkenburgh (1990) are very similar (except for the estimate for the large male) and agree with actual published masses; both predict the mass of FMNH 70566 to within 0.6 kg (Table 14.1). The results from the BODYMASS program were not as consistent as the tooth

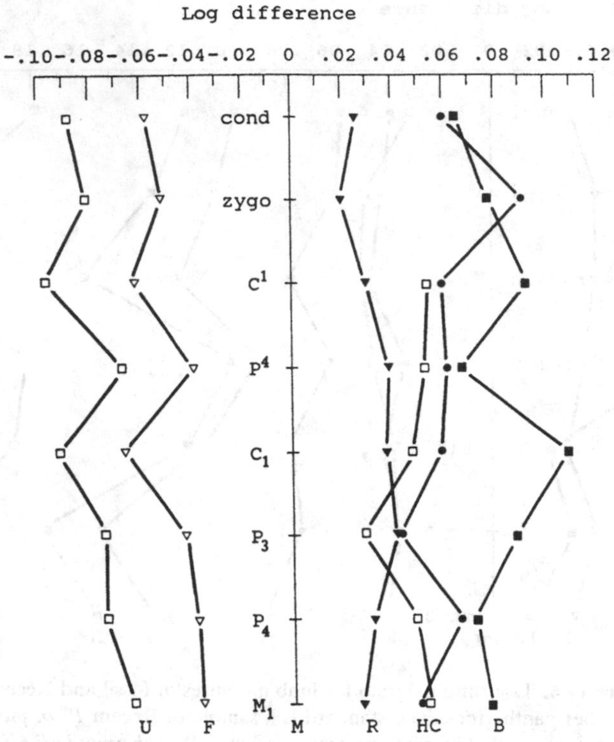

Figure 14.5. Log-ratio diagram for cranial and dental measures of fossil and Recent jaguars. The standard is a sample of Recent male jaguars of all three subspecies; cond, condylobasal length; zygo, zygomatic breadth. Sample sizes and abbreviations are as follows: small Recent female USNM 249826 (U), Recent females 52 (F), Recent males 93 (M), large Recent male BMNH 77.857 (B). Fossil sample sizes are listed from condylobasal length to M_1 length as follows: Rancholabrean average (R) 1, 1, 10, 14, 12, 10, 13, 13; McLeod Quarry and Hamilton Cave averaged (H); Coleman 2A (C) 0, 0, 8, 7, 6, 6, 6, 5. Specimen catalogue numbers are listed in the Appendix.

estimates (Table 14.2). The limb diameter measures usually overestimated, and the limb lengths usually underestimated, the body mass when compared with estimates made on tooth measurements. The overall geometric mean tended to estimate a smaller mass than did the multiple regression. Of the limb estimates, the multiple regression on all species (including Artiodactyla) using all 11 limb and diameter measures made the best prediction for the living subspecies and for FMNH 70566. This regression estimated a larger body mass for the fossil form than did the tooth regression equations.

Table 14.1. *Body mass estimates for the jaguar*

Category	Mass estimate[a] (kg)	Mass estimate[b] (kg)	Published mass[c]	Published sample size[c]
P. o. hernandesii female (USNM 256389)	38.1	36.6	–	_[d]
Avg. *P. o. hernandesii* (females, N=27)	54.9	53.9	–	–
Avg. *P. o. hernandesii* (males, N=37)	60.6	58.5	57.2	6
Avg. *P. o. onca* (females, N=13)	50.1	49.4	56.3	3
Avg. *P. o. onca* (males, N=46)	73.4	68.7	95.0	9
P. o. onca male (FMNH 70566)	75.5	74.2	(75.0)	–
Avg. *P. o. paraguensis* (females, N=12)	71.1	67.1	77.7	3
Avg. *P. o. paraguensis* (males, N=10)	93.0	88.4	94.8	6
P. o. paraguensis male (BMNH 77.857)	140.3	125.2	–	–
Rancholabrean avg. (N=13)	92.9	88.0	–	–
Irvingtonian avg. (N=7)	105.9	104.8	–	–
Largest Irvingtonian (UF 12152)	128.9	128.4	–	–

[a]Mass estimated from the regressions of Legendre and Roth (1988) using lower carnassial area.
[b]Mass estimated from the regressions of Van Valkenburgh (1990) using lower carnassial length.
[c]Published measurements are from the following sources: *P. o. hernandesii* from Rabinowitz (1986); *P. o. onca* from Mondolfi and Hoogesteijn (1986), except FMNH 70566; *P. o. paraguensis* from Schaller and Vasconcelos (1978).
[d]no data.

Rate of evolution

The average percentage size change and the rate of evolution expressed in darwins are presented in Table 14.3 for both transitions (Irvingtonian to Rancholabrean, and Rancholabrean to Recent). For the first transition, the average size change was 6.6%; this was more than doubled for the second transition, to 15.2%. The figure for elapsed time used for the first transition was 520 ky (820 + 600 + 300 / 3 for Hamilton, McLeod, and Coleman respectively); the rates are slow, averaging 0.12 d. The second, larger size change occurred in a much shorter time, here

Table 14.2. *Prediction of jaguar body mass from long-bone lengths and diameters*

Category	Mass estimate, Legendre and Roth (1988) (kg)	Geometric mean (11&D) (kg)	Between maximum minimum & minimum maximum?	Multiple regression estimate	N
Avg. *P. o. hernandesii* (males and females combined)	57.8	37.9	Yes	46.8	6
Avg. *P. o. onca* (males and females combined)	61.8	49.1	No	61.1	4
P. o. onca male FMNH 70566	75.5	53.9	Yes	63.0	1
Avg. *P. o. paraguensis* (males and females combined)	82.1	67.5	No	82.1	7
Rancholabrean average	92.9	89.7	No	117.2	Various
Irvingtonian average	105.9	106.1	Yes	118.7	Various

Note: With sexes combined, the sample sizes are as follows: *P. o. hernandesii* 64, *P. o. onca* 59, *P. o. paraguensis* 22. The first column presents, for comparative purposes, body mass estimates using the regression of Legendre and Roth (1988) using lower carnassial area. The second column shows the geometric mean of 11 long-bone lengths (L) and diameters (D). The third column answers this question: Does the geometric mean fall between the maximum minimum and minimum maximum maximum 95% confidence limits calculated for each of the 11 measures in column two? The fourth column is the body mass estimate based on a multiple regression. The fifth column is the sample size of Recent skeletons used; for specimen numbers of fossils, see Appendix. See text and Gingerich (1990) for details.

Table 14.3. *Percentage size change and change in darwins for 15 cranial, dental, and postcranial measurements of the jaguar*[a]

Measurement	Irvingtonian to Rancholabrean		Rancholabrean to Recent	
	[%]	[d]	[%]	[d]
Condylobasal length	8.3	0.15	10.9	2.1
Zygomatic breadth	17.9	0.32	8.7	1.7
C^1 length	5.4	0.10	13.3	2.5
P^4 length	3.2	0.06	13.7	2.6
C_1 length	1.0	0.02	17.8	3.3
P_3 length	0.0	0.00	12.6	2.4
P_4 length	6.9	0.13	10.7	2.0
M_1 length	4.0	0.08	11.4	2.2
Humerus length	4.5	0.08	18.5	3.4
Ulna length	4.6	0.09	19.1	3.5
Radius length	7.6	0.14	19.9	3.6
Femur length	6.8	0.13	17.3	3.2
Tibia length	8.3	0.15	19.1	3.5
Metacarpal III length	9.2	0.17	17.4	3.2
Metatarsal III length	11.5	0.21	18.2	3.4
Overall average	6.6	0.12	15.2	2.8

[a]Specimen catalogue numbers in Appendix.

roughly estimated at 50 ky (but see the discussion section); the evolutionary rate is about 25 times faster, averaging 2.8 *d*.

Discussion and conclusions

Sample size

To make reasonable judgments concerning relative limb proportions, a statistical sample of associated skeletons is necessary. Presently, not one complete (i.e., skull plus all major limb bones) fossil jaguar skeleton is known. Because the fossil skeletons studied here were incomplete, all conclusions must be considered tentative. The selective averaging process described in the methods section for Figures 14.2 and 14.3 may in the future be viewed as a procedure that obfuscated the real patterns. Clearly, a better sample will be required before we can gain a more complete understanding of these issues.

Sex of fossil animals

Because of a pronounced sexual dimorphism in large felines, Kurtén (1985) recommended that the sex of feline fossils be estimated so that comparisons could be made between samples of the same sex. Although Kurtén (1973) attempted this for some of the fossil jaguar material he studied, he did not use this information in his analyses. I have not attempted to sex the fossil material, because of the small sample available. As there is a sexual dimorphism of about 10% for living subspecies of jaguar in all measurements (Seymour, 1983), sexing of fossil jaguars in the future may lead to a further refinement of this work. A difficulty in correct assignment of sex may lie in the possibility of the jaguar changing size with the glacials and interglacials, as Kurtén (1960) described for several carnivores and rodents. In the words of Stanley (1985), body size is unusually labile in evolution.

How many fossil jaguar species?

Throughout this chapter I have referred to the fossil form of the jaguar as a single entity. Although the taxonomic status of this taxon has been variably treated, as discussed later, all authors have considered the fossil-to-Recent transition to be a single phyletic lineage. An exception was Galusha [unpublished, but mentioned by Neff (1986)], who entertained the idea of two separate species of jaguar that coexisted for some time during the Pleistocene. This is an interesting idea that should be fully explored in the future. It could explain, for instance, the differing limb proportions of the Salt River and Laubach Cave specimens. However, I do not think that there is enough associated material presently available to test this idea. Associated material will be necessary, because good, non-size-related criteria have not yet been fully worked out for the identification of isolated felid bones. As noted by Berta (1987), *Smilodon gracilis* is very similar to the jaguar postcranially in overall size, proportions, and even the shapes of many articular facets. Given that the transition from *S. gracilis* (or *Ischyrosmilus gracilis*) (Churcher, 1984) to *S. fatalis*, with an increase in size, happened at about the same time that the fossil jaguar was decreasing in size, and that *Miracinonyx* frequently co-occurred with these two cats (Van Valkenburgh et al., 1990), there continues to be the possibility of misidentification of isolated remains. Consequently, I shall continue to treat the evolution of the jaguar as a phyletic lineage.

Log-ratio diagrams

McCrady et al. (1951) stated that the usefulness of the log-ratio diagram is completely destroyed when isolated bones from different individuals are mixed together. I suspect that it is for this reason that the graph of the average Rancholabrean postcranial bones presented in Figure 14.3 appears somewhat odd. However, they noted that log-ratio diagrams could work if well-established averages of a single species were used (and I would add even if the bones were not associated). An example of this is the plot for the postcranial bones of *P. leo atrox* in Figure 14.4. The data for this plot were culled from Merriam and Stock (1932), where they presented measurements of a suite of disassociated bones representing the full size range known from Rancho la Brea. Consequently, the average of these measures is probably a reasonable estimate of the average for this species.

Figure 14.4 demonstrates that the Irvingtonian fossil jaguar had limb proportions very similar to those of Recent *P. onca*, with only slightly longer radius and metapodials. This pattern seems to be more typical of the pantherines, as Kurtén (1973) suggested. Allometric scaling probably does not account for these small differences, because all three subspecies and even the largest skeleton measured (ANSP 13949) have virtually identical limb proportions. However, this difference between the fossil and Recent forms is a subtle one, and I would suggest that it is not that different when compared to the limb proportions of the leopard, lion, and tiger plotted in Figure 14.4. In other words, the fossil form probably still looked like a jaguar, except larger. Possible exceptions include the Salt River Cave specimen.

Overall, the Irvingtonian cranial and dental measures average quite large. The only differences compared with the living species appear to be a slightly shorter P_3 and a slightly longer P_4 (Figure 14.5); these are of doubtful significance. It is important to note that all measurements (except skull width of the McLeod specimen, which appears anomalous, possibly because of the vagaries of reconstruction) fall within the range for the living species, albeit at the upper end of the range, and yet the postcranial measures are all clearly larger. Thus, in the evolution of this species from the Irvingtonian to the Recent, the teeth changed proportionately less than did the limbs; in other words, the teeth became relatively larger in the Recent species. This phenomenon has been noted before for fossil herbivores, such as hippopotami (Gould, 1975) and rhinoceroses (Prothero and Sereno, 1982), but apparently not for fossil carnivores. The living species could be considered a phenotypic dwarf of the fossil form (Gould, 1975).

Body mass estimates

Both the tooth equations predicted body mass more consistently than did the postcranial estimates. An animal is more than just teeth, however, and in this case there is evidence of body size and proportions changing more than tooth size. Although the limb estimates were based on equations that included many noncarnivore species, the lion (*P. leo*) was included. Because the fossil jaguar was slightly less jaguar-like and slightly more lion-like in its limb proportions (Figure 14.4), it was felt that the limb regression equations might give a better estimate than the tooth estimates.

In interpreting the BODYMASS prediction results, Gingerich (1990) noted that the maximum of individual minimum limits and the minimum of individual maximum limits can be used to constrain prediction values to a reasonable range. Because three of the six geometric means calculated in Table 14.2 fall just outside of these limits, this program can be judged not to be producing reasonable estimates for this lineage. This may be solely a sample size problem. For instance, the humerus diameter for the Laubach Cave and Salt River Cave specimens is unusually large compared with most of the other Rancholabrean and even Irvingtonian specimens. If that one measurement is changed to include other smaller Rancholabrean specimens, the multiple regression estimate drops to 111.8 kg, and the geometric mean of the 11 limb and diameter measures then falls between the maximum minimum and the minimum maximum. Although more fossils undoubtedly will assist the estimation procedure, other postcranial estimation procedures for the prediction of body mass (e.g., all carnivores or all felids) must be developed before the mass of the fossil jaguar can be estimated with confidence.

Interestingly, the tooth and limb estimates were not that different. We can reasonably conclude, then, that the fossil form probably did not have a mass greater than that of the largest living individuals of the subspecies *P. o. paraguensis,* although it averaged greater than the mean for this subspecies. The Irvingtonian form probably averaged between 100 and 120 kg, and the Rancholabrean form probably averaged between 85 and 100 kg.

Rate of evolution

The formula used was intended originally to measure the rate of evolution of a lineage such that x_1 and x_2 are the population means at the end and beginning of the temporal sequence, respectively, and t is the interval of time (Marshall and Corruccini, 1978). However, the data set of fossil jaguars consisted of isolated bones or individuals; the only populations studied were the Coleman 2A and the Recent samples. Therefore, in order to try to test Kurtén's (1973) assertion that the Pleistocene jaguar became gradually smaller with time, I calculated an average Irvingtonian animal and an average Rancholabrean animal. I also averaged the three Irvingtonian dates to give one endpoint for the calculation of t. The Rancholabrean endpoint was more difficult to calculate, because most Rancholabrean sites are not well dated. Most appear to be Wisconsinan or latest Wisconsinan, but several important specimens are not dated more accurately (e.g., Laubach Cave and Salt River Cave). I arbitrarily chose 50,000 years as the Rancholabrean average age. If I had chosen 10,000 years, the evolutionary rate for the Rancholabrean-to-Recent transition would be five times greater. I therefore consider this to be a minimum value estimate. Whichever way it is calculated, the same trend emerges: The Irvingtonian-to-Rancholabrean transition was slower and of a smaller magnitude than the Rancholabrean-to-Recent transition. The overall pattern, then, could be considered to be a species lineage (Martin, 1984) showing variable-rate phyletic change, as opposed to a punctuated event (Eldredge and Gould, 1972).

Although not included in the calculations, the materials from Devil's Den and the Cutler local fauna give evidence of the smallest Rancholabrean jaguars, approximately the size of the Recent *P. o. paraguensis* or slightly smaller. Both of these sites appear to be very latest Rancholabrean in age. It is probable that these sites give evidence of the last dwarfing event discussed earlier, and without younger Holocene evidence they may indicate the limit of the dwarfing of the eastern North American form, before it went extinct. Indeed, these sites may represent some of the last jaguars in eastern North America, as there is only scant evidence to suggest that the jaguar existed in eastern North America at the time of the arrival of Europeans (Simpson, 1941b). Additional latest Rancholabrean or Holocene sites of fossil jaguars would be of great interest in this respect. Some bones may exist in archaeological sites, but according to Daggett and Henning (1984) large cat bones from North American archaeological sites are routinely identified as *Felis concolor*. This is not surprising, as these two felids can be difficult to separate on the basis of osteological material (Seymour, 1983).

Gingerich (1987) demonstrated that evolutionary rates are not independent of measurement interval ("Gingerich's law"). Generally, the shorter the time interval, the greater the rate. Because the evolutionary rates calculated for the fossil jaguar appear to follow this rule, further specimens and analyses will be necessary before the scenario of variable-rate phyletic change can be corroborated.

Taxonomic considerations

A taxonomic question arises out of the calculation of these rates of evolution. Should the Pleistocene form (Irvingtonian and Rancholabrean) be considered a different species or subspecies from the living form, or should it even be formally recognized at all? Various workers in the past have approached this problem. There is a consensus among those who have studied the fossil jaguar that there are no consistent morphological features on which one may differentiate it from the extant species (Simpson, 1941a; Kurtén, 1965, 1973; Seymour, 1983). Simpson (1941a) and Kurtén (1973) both considered the Pleistocene form to be a different subspecies (*P. o. augusta*), and McCrady et al. (1951) considered it a different species. Marshall and Corruccini (1978) approached a similar problem for dwarfed lineages of marsupials. They suggested that for lineages that experienced about 15% size change in 10,000 years (an evolutionary rate of about 14 *d*), a subspecific rank would be most reasonable. The average rate for the Rancholabrean-to-Recent transition was between 2.8 and 2.8 × 5 = 14 *d*, depending on the timing of the dwarfing event, as discussed earlier. This suggests that the Pleistocene form of the jaguar should be considered at least a subspecies.

I have resisted calling the Pleistocene form a subspecies in the past (Seymour, 1983), because of the frequent confusion of geographic subspecies (lateral variation at one time) and temporal subspecies (vertical variation, i.e., stratigraphic, through time) (Simpson, 1961). In practice, according to Gingerich (1985), fossil subspecies usually are proposed by paleontologists unsure that formal designation of a new taxon is warranted. The North American Pleistocene form probably did not actually change into an average Recent form; rather, it likely became locally extinct in the northern parts of its range. The Rancholabrean-to-Recent rate of change was calculated for an average Recent form; there still exists today a large subspecies of jaguar in southern Brazil (*P. o. paraguensis*) that attains the size of at least the Rancholabrean form. However, given that the northern form probably was locally adapted to prey on some of the large Pleistocene megafauna, and if it were still extant today undoubtedly it would be called a different subspecies, just as is the form in southern Brazil, it would be reasonable to call the Pleistocene form *P. o. augusta*. On the other hand, the Pleistocene form was larger, and it had longer metapodials and relatively smaller teeth compared with skull size. Perhaps it would be reasonable to choose the relatively abrupt change in size that took place at the end of the Pleistocene as the dividing line between the fossil *Panthera augusta* and the Recent *Panthera onca*.

Neither solution is optimal: If these are considered separate species, some will interpret an extinction event at the end of the Pleistocene, and the probable continuity of the lineage will not be understood. If these are considered different subspecies, there will be the potential for confusion of temporal and geographic subspecies; also, size changes and proportional changes, which might be expected to occur at a speciation event, would not be expected to occur between different subspecies. Further confusing the picture, Rose and Bown (1986) note that if successional forms are considered as a single species, the morphological limits of paleospecies become amplified. Either case seems to be an arbitrary decision.

The Pleistocene form in some lineages has been placed in a different species, whereas in other lineages it is kept in the same species as its living descendants (Martin, 1984). Several recent studies have discussed this problem in more detail, in particular Rose and Bown (1986) and Krishtalka and Stucky (1985). For the jaguar, I prefer to use a modified version of the Krishtalka and Stucky (1985) method of recognizing informal lineage segments, as proposed by R. Martin (Chapter 11, this volume). This would result in all the fossil jaguars being recognized as *Panthera onca /augusta*.

This study examined only the North American jaguar and its fossil record. All three living subspecies of jaguar still exist in South America today, and there is a fossil record (albeit meager) for the jaguar from South America. Consequently, future studies of South American fossil material will be crucial in gaining a more complete understanding of the Pleistocene-to-Recent evolution of this animal.

Acknowledgments

I thank C. S. Churcher and M. Engstrom for commenting on an earlier draft of this chapter and R. Martin, T. Barnosky, and E. Anderson for helpful reviews. S. Desser kindly granted travel funds to me from the Department of Zoology, University of Toronto, so that I could attend the symposium on morphological change in Quaternary mammals. F. Grady supplied critical limb measurements for the Hamilton Cave cat, and I. Morrison helped with the figures. The curators and collection managers of all the collections listed here assisted tremendously when I visited their collections or when they lent me material; without their help this study could never have been completed. Although all have been very helpful, I especially would like to thank G. Morgan (UF, Gainesville) and T. Daeschler (ANSP, Philadelphia).

References

Alexander, R. M., A. S. Jayes, G. M. O. Maloiy, and E. M. Wathuta, (1979). Allometry of the limb bones of mammals from shrews (*Sorex*) to elephant (*Loxodonta*). *Journal of Zoology (London,)* 189:305–14.
Berta, A. (1987). The sabercat *Smilodon gracilis* from Florida and a discussion

of its relationships (Mammalia, Felidae, Smilodontini). *Bulletin of the Florida State Museum, Biological Sciences,* 31:1–63.

Churcher, C. S. (1984). The status of *Smilodontopsis* (Brown, 1908) and *Ischyrosmilus* (Merriam, 1918): a taxonomic review of two genera of sabre-tooth cats (Felidae, Machairodontinae). *Royal Ontario Museum, Life Science Contributions,* 140:1–59.

Daggett, P. M., and D. R. Henning (1984). The jaguar in North America. *American Antiquity,* 39:465–9.

Eldredge, N., and S. J. Gould (1972). Punctuated equilibria: an alternative to phyletic gradualism. In T. J. M. Schopf (ed.), *Models in Paleobiology* (pp. 82–115). San Francisco: Freeman Cooper.

Gazin, C. L. (1942). The late Cenozoic vertebrate faunas from the San Pedro Valley, Arizona. *Proceedings of the United States National Museum,* 171: 1–99.

Gingerich, P. D. (1985). Species in the fossil record: concepts, trends and transitions. *Paleobiology,* 11:27–41.

(1987). Evolution and the fossil record: patterns, rates, and processes. *Canadian Journal of Zoology,* 65:1053–60.

(1990). Prediction of body mass in mammalian species from long bone lengths and diameters. *University of Michigan Contributions from the Museum of Paleontology,* 28:79–92.

Gould, S. J. (1975). On the scaling of tooth size in mammals. *American Zoologist,* 15:351–62.

Guilday, J. E., J. F. P. Cotter, D. Cundall, E. B. Evenson, J. B. Gatewood, A. V. Morgan, A. Morgan, A. D. McCrady, D. M. Peteet, R. Stuckenrath, and K. Vanderwal (1984). Paleoecology of an early Pleistocene (Irvingtonian) cenote: preliminary report on the Hanover Quarry No. 1 fissure, Adams Co., Pennsylvania. In W. C. Mahaney (ed.), *Correlation of Quaternary Chronologies* (pp. 119–32). Norwich, U.K.: Geo Books.

Guilday, J. E., and H. McGinnis (1972). Jaguar (*Panthera onca*) remains from Big Bone Cave, Tennessee and east central North America. *National Speleological Society Bulletin,* 34:1–14.

Haldane, J. B. S. (1949). Suggestions as to quantitative measurement of rates of evolution. *Evolution,* 3:51–6.

Hemmer, H. (1971). Zur Charakterisierung und stratigraphischen Bedeutung von *Panthera gombaszoegensis* (Kretzoi, 1938). *Neues Jahrbuch für Geologie und Palaeontologie Monatshefte,* 1971:701–11.

Johnson, N., N. D. Opdyke, and E. Lindsay (1975). Magnetic polarity stratigraphy of Pliocene-Pleistocene terrestrial deposits and vertebrate fauna, San Pedro Valley, Arizona. *Geological Society of America Bulletin,* 86: 5–11.

Krishtalka, L., and R. K. Stucky (1985). Revision of the Wind River faunas, early Eocene of central Wyoming. Part 7. Revision of *Diacodexis* (Mammalia, Artiodactyla). *Annals of the Carnegie Museum of Natural History* 54:413–86.

Kurtén, B. (1960). Chronology and faunal evolution of the earlier European glaciations. *Commentationes Biologicae,* 21:1–62.

(1965). The Pleistocene Felidae of Florida. *Bulletin of the Florida State Museum, Biological Sciences,* 9:215–73.

(1973). Pleistocene jaguars in North America. *Commentationes Biologicae,* 62:1–23.

(1976). Fossil puma (Mammalia, Felidae) in North America. *Netherlands Journal of Zoology,* 26:502–34.

(1985). The Pleistocene lion of Beringia. *Annales Zoologici Fennici,* 22: 117–21.

Kurtén, B., and E. Anderson (1980). *Pleistocene Mammals of North America.* New York: Columbia University Press.

Legendre, S., and C. Roth (1988). Correlation of carnassial tooth size and body weight in Recent carnivores (Mammalia). *Historical Biology,* 1:85–98.

Leidy, J. (1872). Remarks on some extinct vertebrates. *Proceedings, Academy of Natural Sciences, Philadelphia,* 24:38–40.

McCrady, E., H. T. Kirby-Smith, and H. Templeton (1951). New finds of Pleistocene jaguar skeletons from Tennessee caves. *Proceedings of the United States National Museum,* 101:497–511.

Marshall, L. G., and R. S. Corruccini (1978). Variability, evolutionary rates and allometry in dwarfing lineages. *Paleobiology,* 4:101–19.

Martin, L. D. (1984). Phyletic trends and evolutionary rates. In H. H. Genoways and M. R. Dawson (eds.), *Contributions in Quaternary Vertebrate Paleontology: A Volume in Memorial to John E. Guilday* (pp. 526–38). Carnegie Museum of Natural History, Special Publication 8.

Martin, R. A. (1974). Fossil mammals from the Coleman 2A fauna, Sumter County. In S. D. Webb (ed.), *Pleistocene Mammals of Florida* (pp. 35–99). Gainesville: University Presses of Florida.

Martin, R. A., and S. D. Webb (1974). Late Pleistocene mammals from the Devil's Den fauna, Levy County. In S. D. Webb (ed.), *Pleistocene Mammals of Florida* (pp. 114–45). Gainesville: University Presses of Florida.

Merriam, J. C., and C. Stock (1932). *The Felidae of Rancho La Brea.* Carnegie Institute of Washington, Publication 422.

Mondolfi, E., and R. Hoogesteijn (1986). Notes on the biology and status of the jaguar in Venezuela. In S. D. Miller and D. D. Everett (eds.), *Cats of the World: Biology, Conservation, and Management* (pp. 85–123). Washington, D.C.: National Wildlife Federation.

Morgan, G. S. (1992). Neotropical Chiroptera from the Pliocene and Pleistocene of Florida. *Bulletin of the American Museum of Natural History,* 206:176–213.

Morgan, G. S., and R. C. Hulbert, Jr. (in press). Overview of the geology and vertebrate biochronolgy of the Leisey Shell Pit local fauna, Hillsborough County, Florida. In R. C. Hulbert, Jr., G. S. Morgan, and S. D. Webb (eds.), *Geology and Paleontology of the Leisey Shell Pit, Hillsborough County, Florida.* Florida Museum of Natural History, Biological Sciences.

Neff, N. A. (1986). *The Big Cats – The Paintings of Guy Coheleach.* New York: Abradale Press.

Prothero, D. R., and P. C. Sereno (1982). Allometry and paleoecology of medial

Miocene dwarf rhinoceroses from the Texas Gulf Coastal Plain. *Paleobiology*, 8:16–30.

Rabinowitz, A. R. (1986). Jaguar predation on domestic livestock in Belize. *Wildlife Society Bulletin*, 14:170–4.

Repenning, C. A., and F. Grady (1988). The microtine rodents of the Cheetah Room fauna, Hamilton Cave, West Virginia, and the spontaneous origin of *Synaptomys*. *United States Geological Survey Bulletin*, 1853:1–32.

Rose, K. D., and T. M. Bown (1986). Gradual evolution and species discrimination in the fossil record. In K. M. Flanagan and J. A. Lillegraven (eds.), *Vertebrates, phylogeny, and philosophy* (pp. 119–30). Contributions to Geology, University of Wyoming, Special Paper 3.

Roth, J. A., and J. Laerm (1980). A late Pleistocene vertebrate assemblage from Edisto Island, South Carolina. *Brimleyana*, 3:1–29.

Schaller, G. B., and J. M. C. Vasconcelos (1978). Jaguar predation on capybara. *Zeitschrift für Saugetierkunde*, 43:296–301.

Schultz, C. B, L. D. Martin, and M. R. Schultz (1985). A Pleistocene jaguar from north-central Nebraska. *Transactions of the Nebraska Academy of Sciences and Affiliated Societies*, 8:93–8.

Seymour, K. L. (1983). The Felinae (Mammalia: Felidae) from the late Pleistocene tar seeps at Talara, Peru, with a critical examination of the fossil and Recent felines of North and South America. M.Sc. thesis, University of Toronto, Toronto.

(1989). Panthera onca. *Mammalian Species*, 340:1–9.

Simpson, G. G. (1941a). Large Pleistocene felines of North America. *American Museum of Natural History Novitates*, 1136:1–27.

(1941b). Discovery of jaguar bones and footprints in a cave in Tennessee. *American Museum of Natural History Novitates*, 1131:1–12.

(1961). *Principles of Animal Taxonomy*. New York: Columbia University Press.

Slaughter, B. H. (1966). *Platygonus compressus* and associated fauna from the Laubach Cave of Texas. *American Midland Naturalist*, 75:475–94.

Stanley, S. M. (1985). Rates of evolution. *Paleobiology*, 11:13–26.

Swank, W. G., and J. G. Teer (1989). Status of the jaguar – 1987. *Oryx*, 23:14–21.

Van Valkenburgh, B. (1990). Skeletal and dental predictors of body mass in carnivores. In J. Damuth and B. J. MacFadden (eds.), *Body Size in Mammalian Paleobiology* (pp. 181–205). Cambridge University Press.

Van Valkenburgh, B., F. Grady, and B. Kurtén (1990). The Plio-Pleistocene cheetah-like cat *Miracinonyx inexpectatus* of North America. *Journal of Vertebrate Paleontology*, 10:434–54.

von den Driesch, A. (1976). A guide to the measurement of animal bones from archaeological sites. *Peabody Museum of Archaeology and Ethnology, Bulletin*, 1:1–137.

Webb, S. D., and K. T. Wilkins (1984). Historical biogeography of Florida Pleistocene mammals. In H. H. Genoways and M. R. Dawson (eds.), *Contributions in Quaternary Vertebrate Paleontology: A Volume in Memorial to John E. Guilday* (pp. 370–83). Carnegie Museum of Natural History, Special Publication 8.

Appendix

The following are the catalogue numbers of specimens used in the figures and tables.

Figure 14.2

McLeod Quarry: humerus F:AM 69226, ulna 69224, radius 69221, femur 69205, tibia 69205, metacarpal III 69248, metatarsal III 69240. Hamilton Cave: USNM 299767. Coleman 2a: radius UF 12128, femur 12135 (3), tibia 12134 (2), metacarpal III 22757 (2), 22753 (1), metatarsal III 22747, 22748, 22764. Recent *P. o. paraguensis* as in Figure 14.4.

Figure 14.3

Salt River Cave: USNM 18262. Laubach Cave: TMM 40673. Rancholabrean average: humerus UF 3463, CM 24369; ulna UF 3463; radius UF-FGS 6690 (2), UF 3463, 9122; femur LACM HC X-8848, CM 24369, 24699, TMM 1295, 40279–12; tibia LACM HC X-7209, CM 24369; metacarpal III AMNH 92027, LACM 122396, 122397, 122398, 122399, 122400, SMU 60162, UF 10466, 45358, 45365, UTEP 7–1, 27–35, 27–47; metatarsal III UF 45359, 45407, UF-FGA V-4031, LACM 8854, 8858, 122388, 122389, 122390, 122391, 122395, UTEP 27–35. Recent *P. o. paraguensis* as in Figure 14.4.

Figure 14.4

P. o. hernandesii: AMNH 135928, 135929, 139959; BMNH 117J; USNM 155603; YPM 2598. *P. o. onca:* AMNH 209136; FMNH 57199, 70566; UMZ L40. *P. o. paraguensis:* ANSP 13943, 13945, 13947, 13949; FMNH 26552; USNM 12296, 49393. *P. pardus:* ROM 34.2.23.2, 75958. *P. tigris:* ROM 33.8.13.4, 94166. *P. leo:* ROM 33.9.8.1, 32874. *P. leo atrox:* humerus LACM HC 2903-R-1 to R-9 inclusive; ulna LACM HC 2905-R-1 to R-7 inclusive; radius LACM HC 2904-R-1 to R-10 inclusive; femur LACM HC 2907-R-1 to R-8 inclusive, 2907-R-10; tibia LACM HC 2908-R-1 to R-10 inclusive; metacarpal III LACM HC 2913-R-2, 2913-L-2; metatarsal III LACM HC 2918-R2, 2918-L2 [data from Merriam and Stock (1932)]. Rancholabrean "average": limb bones USNM 18262, TMM 40673; metacarpal III and metatarsal III specimen numbers as for Figure 14.3. Irvingtonian "average": specimen numbers as for Figure 14.2.

Figure 14.5

Recent male and Recent female specimen numbers as in Table 14.1. Coleman 2A: C^1: UF 12144, 12154 (4), 12155, 12163, 12164. P^4: UF 12144, 12155, 12156, 12158, 12163, 12161, 12164. C_1: UF 12145, 12154 (5). P_3 and P_4: UF 12145, 12146, 12148, 12150, 12151, 12152. M_1: UF 12145, 12146, 12150, 12151, 12152. McLeod Quarry: F:AM 69204. Hamilton Cave: USNM 299767. Rancholabrean average: skull measures USNM 23733. C^1: AMNH 32635, FMNH P27216–4, LACM HC 1436, TMM 40673–51, UF 2858, 103724, USNM 18262, 23486, 23733, UTEP 27–48. P^4: LACM HC 1436, ROM 12721, RMM 3935, UF 18703, 45394, 45423, 45424. USNM 11411, 18262, 23486, 23733, UTEP 7–1 (2), 27–48. C_1: AMNH 32633, IGM 1442, LACM 1757, 2264, 3025. TMM 40673–49 (2), UF 2858, 14765, USNM 11470, 23486, 23733. P_3: AMNH 32633, FMNH P27216–1, IGM 1442, LACM 2264, 3025, UF 14765, USNM 11470, 23486, 23733, 24580. P_4: AMNH 32633, FMNH P27216–1, IGM 1442, LACM 3025, ROM 38165, TMM 40673–48, 40673–49, UF 14765, 45383, USNM 11470, 23486, 23733, 24580. M_1: AMNH 32633, IGM 1442, LACM 2264, ROM 37165, TMM 40279–13, 40673–50, UF 3078, 14765, 103723, USNM 11470, 23486, 23733, 24580.

Table 14.1

P. o. hernandesii females: AMNH 2306, 25011, 29445, 135928, 135929, ANSP 20070, KUMNH 24556, 32226, 71941, ROM 87968, USNM 8003, 13845, 13846, 231961, 249826. Unknown sex classified as female: AMHH 24689, 38126, 146987, 180275, BMNH 5916, IBUNAM 15983, KUMNH 71940, USNM 8657, 9704, 132519, 155603, 179171. *P. o. hernandesii* males: AMNH 25008, 25009, 25010, 42405, 139959, BMNH 1874.4.8.2, 1934.9.10.64, 1935.3.6.1, 1935.3.6.2, IBUNAM 20106, KUMNH 32227, 93834, 96902, USNM 12176, 14176, 123527, 130362, 225613, 244858, 247337, 249823, 249825, 294541, 268871. Unknown sex classified as male: AMNH 149326, IBUNAM 5424, KUMNH 90987, UCMVZ 35317, USNM 9703, 25118, 25119, 61192, 100541, 131998, 140949, 167894, 179170. *P. o. onca* females: AMNH 98679, BMNH 1845.8.25.21, 1926.12.4.29, 1936.5.26.3, FMNH 86902, 86903, ROM 32079. Unknown sex classified as female: AMNH 16925, FMNH 21392,

53916, UCNVZ 4830, UMZ 4348, USNM 256389. *P. o. onca* males: AMNH 209135, BMNH 1845.8.25.22, 1926.1.12.1, 1928.5.2.139, 1987.236, FMNH 19493, 20020, 39460, 70566, 88477, 98073, ROM 32905, 32944, 33089, UMZ 37, USNM 339678, 362249, 362250, 374849, 395095. Unknown sex classified as male: AMNH 11083, 98671, 98682, 98683, 98684, 98841, 147510, 147511, 147512, 147513, BMNH 1851.8.25.12, FMNH 17768, 55515, 57199; MNHN 1981–1246, NMC 43201, ROM 46328, UMZ 39, USNM 25636, 49393, 100122, 100123, 137039, 256385, 256387, 456774. *P. o. paraguensis* female: AMHH 36949, ANSP 13943, 13946, 13947, 13948, BMNH 1884.2.8.1, 76.671. Unknown sex classified as female: AMNH 331, 37549, ANSP 4719, FMNH 45988, USNM 12296. *P. o. paraguensis* male: AMNH 36950, 37503, ANSP 13944, 13945, 13949, 13950, BMNH 1899.3.2.1. Unknown sex classified as male: AMNH 176373, BMNH 77.857, NMB C2690. Rancholabrean M_1: specimen numbers as in Figure 14.5. Irvingtonian M_1: F:AM 69204, UF 12145, 12146, 12150, 12151, 12152, USNM 299767.

Table 14.2

All specimen numbers as in Figure 14.4.

Table 14.3

Recent cranial/dental specimen as in Table 14.1 (all subspecies grouped). Recent skeleton specimen numbers as in Figure 14.4. Irvingtonian cranial/dental specimens as in Figure 14.5, and postcranial specimens as in Figure 14.2. Rancholabrean cranial/dental specimens as in Figure 14.5, and postcranial specimens as in Figure 14.3.

15

Ontogenetic change of *Ondatra zibethicus* (Arvicolidae, Rodentia) cheek teeth analyzed by digital image processing

LAURENT VIRIOT, JEAN CHALINE, ANDRE SCHAAF, AND ERIC LE BOULENGE

The muskrat (*Ondatra zibethicus* Linnaeus, 1766), a rodent of North American origin, was introduced into Europe in 1905; it spread quickly and is now abundant along coasts, lakes, rivers, and tributaries throughout western Europe.

The stratigraphic implications of *Ondatra* were first published by Semken (1966), who prepared a bivariate graph of M_1 length versus width. The resulting chronocline, which indicated a gradual size increase from *Ondatra idahoensis* through *Ondatra zibethicus*, was correlated with relatively rapid increases in dentine track height (including hypsodonty) and increasing amounts of cement in the reentrant angles. The next year, Hibbard and Zakrzewski (1967) suggested that *Pliopotamys* was probably an ancestor of *Ondatra*. Zakrzewski (1969) graphically added *Pliopotamys minor* and *Pliopotamys meadensis* to the M_1 length/ width bivariate and convincingly demonstrated that *Pliopotamys* was ancestral to *Ondatra*. This evolutionary lineage was reinforced by Schultz, Tanner, and Martin (1972) and Martin (1979, 1984) with additional specimens from critical local faunas of the central Great Plains. Nelson and Semken (1970) added a paleoecological indicator, via modern topocline, to the analysis of North American muskrats. The lineage *Pliopotamys minor–meadensis–Ondatra idahoensis–annectens–nebrascensis–zibethicus* evolved featuring increases in size, hypsodonty, dentine track height (perhaps a function of hypsodonty), and M_1 crown complexity, as well as addition of cement in the reentrant angles. This evolution was parallel to that of the *Mimomys occitanus–Arvicola terrestris* lineage in Eurasia (Chaline, 1987).

The muskrat now living in Europe is a direct descendant of this important North American evolutionary lineage. The aim of this study is to establish clearly the dental ontogeny of *Ondatra* M_1 and M^3. With the advent of computer technology, an improved means of surface quantification through image analysis has been used to describe the morphological changes during dental ontogeny. This technique refines the quantitative value of the muskrat chronocline, better defines the morphology of muskrat teeth, and quantifies the effects of wear on the morphology.

Material and methods

Material

The material consists of 33 individuals captured by Eric Le Boulengé in the Bassin de la Houille, Belgium (Le Boulengé, 1977). Based on comparison with individuals of known age, these specimens were grouped into presumed age categories estimated to range from 38 to 506 days.

Digital image processing

The origin of digital image processing can be traced to the early 1920s, when digitized pictures of world news events were first transmitted between London and New York via the Bartlane submarine transmission system (Gonzalez, Woods, and Swain, 1986). Image analysis is a computerized tool with numerous geological applications (Serra, 1972) in petrography (Haas, Matheron, and Serra, 1967a,b; Vansteelant, 1989), metallography (Chermant et al., 1981), and paleontology (Schmidt-Kittler, 1984, 1986; Schmidt-Kittler and Vianey-Liaud, 1987; Schaaf and Vansteelant, 1988; Viriot, 1989; Barnosky, 1990; Viriot, Chaline, and Schaaf, 1990).

Image analysis makes possible a quantitative description of images in two-dimensional space. Natural objects, though, are never totally planar, and in consequence the captured image is always the projection of a three-dimensional surface of an object onto a plane. The cheek teeth of arvicolids, however, are ideal for image analysis because they have an occlusal (chewing) surface that is almost flat. Projection of the surface onto a two-dimensional plane by drawing it with the aid of a camera lucida introduces only a tiny error. In this study it was necessary to use occlusal surface drawings, rather than the actual teeth, to eliminate problems of topology and image capture encountered with fossil teeth (Viriot, 1989). Fractures, microdissolutions, and differences in color and illumination of enamel parts presented obstacles that prevented automatic extraction of the outline directly from the molar; this remains a problem in capturing images with all but prohibitively expensive hardware and software. In this chapter we portray the occlusal surfaces, corrected for defects only, in the drawings, with enamel in black, dentine in white, and cement in stippled pattern.

Serial-section method

Each molar to be studied was carefully cleaned before being cemented to the chuck of the grinding machine (Figure 15.1). Quick-setting ethyl methacrylate resin fixed the tooth base solidly before the tooth was totally coated with black epoxy resin to enhance the contrast between the various dental components and the resin. A diamond grinding wheel was lowered by a micrometer screw that enabled the user to increment accurately abrasion stages every 0.5 mm. Drawings at each stage were prepared for image analysis (Viriot, 1989; Viriot et al., 1990), and they recorded the morphological changes of the molar occlusal surface during wear (abrasion).

Acquisition, digitalization, and binarization of images

Each drawing was digitalized with a CCD (charge-coupled device) camera; then the natural image was transformed into a numerical and discrete function by the image card (Chermant and Coster, 1984). The resulting image was saved as a table (512 × 512). Each point on such a table is one pixel of the screen, and the pixel is coded on one byte, the value of each pixel ranging from 0 (black) to 255 (white), that is to say, 256 ($= 2^8$) gray levels.

The gray-level histogram display of the image then makes it possible to choose a threshold value X between 0 and 255. The pixels with values lower than or equal to X take the value 0 (black), and those with values higher than X take the value 1 (white). Thus the image is binary, and the pixels are coded on only one bit (black or white: two gray levels).

Mathematical morphology and available parameters

This binary image was analyzed with Visilog © software. The enamel surface can be quantified directly by counting the pixels equivalent to

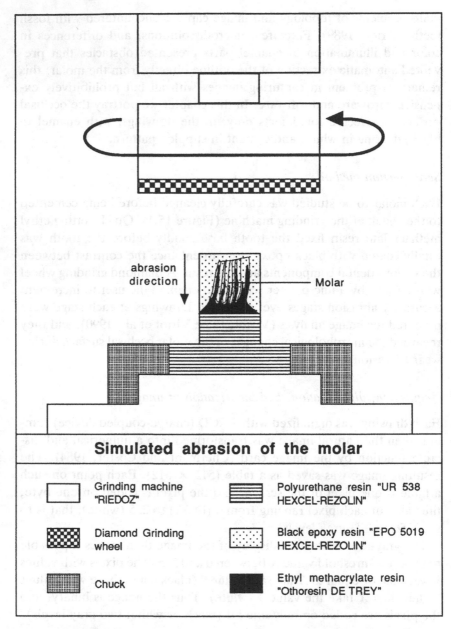

Figure 15.1. Simulated wear of the molar. The molar is cemented to the chuck of the grinding machine. This abrasion provides a simulated wear sequence, and each abrasion stage is drawn every 0.5 mm.

0, but obtaining other parameters, such as the occlusal surface area or the triangle areas, requires the use of mathematical morphology. Mathematical morphology was invented in 1964 by Matheron and Serra to compare an object under analysis with an object of known shape, the structuring element (Serra, 1975; Coster and Chermant, 1985). Visilog mathematical morphology makes use of classical set transformations such as union, intersection, and complementation, but also the all-or-nothing transformation through a hexagonal structuring element. The hexagonal pattern (Golay, 1969) was chosen because it lends itself better to mathematical morphology (Laÿ, 1984). In an all-or-nothing transformation, the origin of the structuring element (center of the hexagon) is moved from pixel to pixel in an image and modifies each pixel or not according to the question asked. The result of each all-or-nothing transformation depends on the size of the structuring element. The choice of structuring element must thus remain rather judicious so as to obtain the most accurate result possible in transformations such as erosion, dilatation, or line-skeleton transformation.

This numerical technique for handling binary images leads to new data for rodent molar morphometry. It affords easy access to parameters such as the outline length, the anterior loop area, or the area of one isolated triangle, and more than 40 surface measurements can be obtained from one molar.

The binary image makes it possible to uncover the morphological differences of the occlusal surface using computed parameters. The pixel, which represents a very small surface of the captured object, is the unit of measurement. To eliminate problems of focal distance with the lens and of distance between the camera and the object, parameters without dimensions provide the best comparative measures.

These include the structural-density parameter and the R parameter. The structural-density (SD) parameter is

$$SD(x) = P(x)^2/4\pi A(x),$$

where P is perimeter and A is area. It compares the degree of distortion of a given surface with that of a circle (Schmidt-Kittler, 1984, 1986). This ratio tends toward 1 when the tooth approaches a circle in shape and takes higher values as the tooth shape becomes more complex.

The R parameter quantitatively describes the extent of narrowing and closing of the triangles and is often qualitatively used in arvicolid systematics. The degree of narrowing of the buccal and lingual tooth reentrants has been difficult to define qualitatively. The R parameter represents a quantitative solution.

R is the ratio between the quadrangular surface of the triangle Q (delimited by the straight lines joining the two reentrants bordering the triangle viewed from the opposite reentrant angle) and T, the triangular surface of the triangle (delimited by the straight line joining the two reentrants bordering the triangle in question).

If R is 1, Q and T have the same area; if $R < 1$, the reentrants are narrow, and if $R > 1$, the reentrants are wide.

Qualitative and quantitative description of molars

The dental formula for *O. zibethicus*, as for all arvicolids, is 1/1 0/0 0/3 3/3, the missing canines and premolars leaving a wide diastema in the mandibles and the maxillae. The first lower molar (M_1) and the third upper molar (M^3) of the arvicolids are the only cheek teeth to have distinctive features at the species level (Chaline, 1972). Thus, to distinguish *O. zibethicus* from other species of *Ondatra*, we shall describe only these two teeth.

Qualitative description of the first lower molar (M_1)

From the very first stage of wear in immature *O. zibethicus* (Galbreath, 1954), the molars immediately develop a planar wearing surface. The occlusal surface wears first to a zigzag pattern and then to a series of triangles or prisms, shown in a section perpendicular to their elongation (Figure 15.2). The occlusal surface is composed, from back to front, of a posterior loop (PL), alternate triangles (T1, T2, T3, . . .), and finally an anterior loop (AL) of varying complexity. Cheek teeth in immature animals are rootless or arhizodont (Chaline, 1972), and the anterior loop has a crenulated outline (Figure 15.3). At the mature or old stage of wear (Galbreath, 1954), immature molars gradually develop roots, and the occlusal surface outlines become less complex, and in old age the triangles appear rounded. The enamel strips, which are narrow in the immature animal, widen significantly with wear as the animal matures, and interruptions appear in the enamel covering at the lingual and buccal ends of the dental triangles. These interruptions in the enamel coating exist in some mammals with rhizodont molars because the roots are never covered with enamel. But among some rodents, the line that separates the dental crown (enamel-coated portion) from the roots (den-

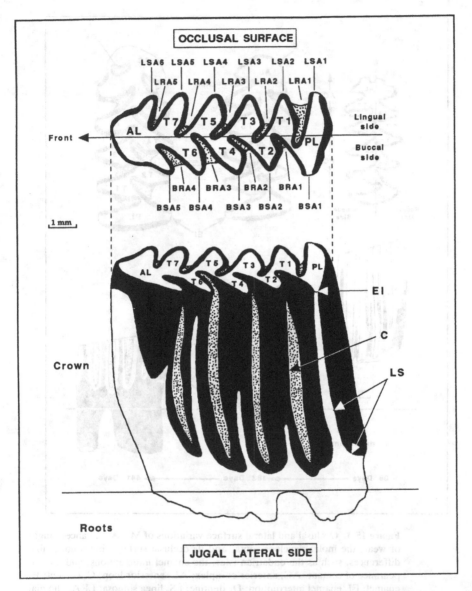

Figure 15.2. Terminology of the first lower molar (M_1). The occlusal surface is composed of a posterior loop, 7 alternate triangles, and an anterior loop. The irregular line that separates enamel from dentine is called the linea sinuosa. Occlusal surface: AL, anterior loop; BSA, buccal salient angle; BRA, buccal reentrant angle; LSA, lingual salient angle; LRA, lingual reentrant angle; PL, posterior loop; T, triangle. Lateral side: C, cement; EI, enamel interruption; LS, linea sinuosa, which borders dentine track.

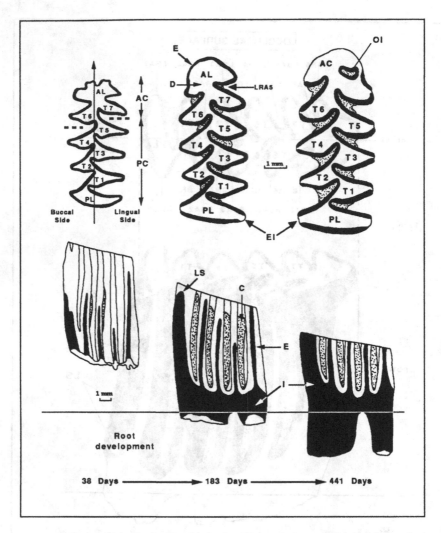

Figure 15.3. Occlusal and lateral surface variations of M_1. At advanced stages of wear, the molar develops roots, and its occlusal surface shows noticeable differences, such as the ondatrian islet, the enamel interruptions, and the appearance of cement. AC, anterior complex; AL, anterior loop, C, cement; E, enamel; EI, enamel interruption; D, dentine; LS, linea sinuosa; LRA5, lingual reentrant angle 5; OI, ondatrian islet; PC, posterior complex; PL, posterior loop; T, triangle.

tine only) is not straight (rectilinear), but tends to rise in places toward the upper part of the crown. This irregular and undulating dividing line is called the linea sinuosa (Rabeder, 1981). In the case of arvicolids, the upturns in the linea sinuosa are not random. They occur at the apex of the dental triangles, on the buccal side of the posterior loop, and in front of the anterior loop.

The biological ages of the individuals from the Bassin de la Houille have been estimated to range from 38 to 506 days. Beginning with the youngest examined specimen, the first lower molar (M_1) of the *O. zibethicus* is always planar. It is composed of a posterior loop with a long lingual dimension and a shorter buccal dimension opening onto seven alternate dental triangles of which the first five are always clearly separated (Figure 15.3). The buccal dental triangles (T2, T4, T6) anterior to the posterior loop are always smaller than the lingual dental triangles (T1, T3, T5, T7). This difference is especially noticeable in immature specimens. These triangles, which are pointed in the immature animals, transform into a more rounded shape in mature animals, but they are never totally closed even if the reentrant angles tend to converge on the anterior edges of the opposite triangle.

The anterior complex is obviously the most variable element in the occlusal surface. Composed of the triangles 6 and 7 and of the anterior loop (Figure 15.3), its morphological change with wear tends to obliterate the distinctiveness between triangles 6 and 7 and the anterior loop. The merging of triangle 7 into the anterior loop, occasionally noticeable in the most advanced stages of wear, produces an intraocclusal isolation of the lingual reentrant angle 5, forming the "ondatrian islet." The isolation process in *Ondatra*, as in *Mimomys*, leads to a reduction in the number of triangles in front of the molar. This number is reduced from 5 to 4 or even to 3 in *Mimomys*, and from 7 to 6 or in extreme cases to 5 in *Ondatra*.

Qualitative description of the third upper molar (M^3)

The M^3 (Figure 15.4) is composed of an anterior loop (AL), three alternate triangles (T1, T2, T3), and one posterior loop (PL). A 38-day-old immature individual shows an M^3 on which the occlusal surface is not yet planar. In fact, at very early stages of wear this surface is convex, and its posterior part (future posterior loop) is separated from the anterior part (future triangles and anterior loop) by a millimetric hole.

The M^3 anterior loop features a sharp, pointed buccal triangle apex and a broad and rounded triangle apex. Triangle 2 is always large in size compared with triangles 1 and 3, and a triangle 3′ included in the posterior loop may be noticeable. The posterior loop, which is morphologically dependent on the state of wear of the molar, is in most

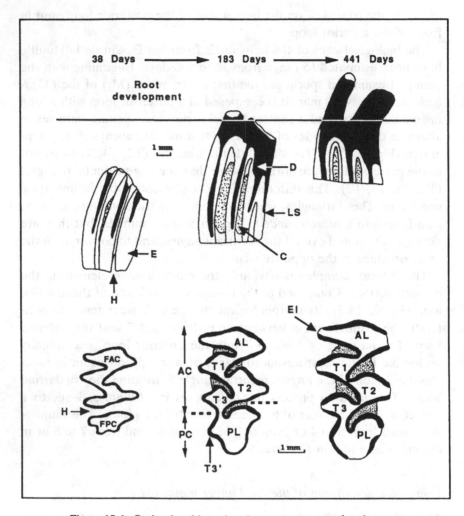

Figure 15.4. Occlusal and lateral surface variations of M^3. M^3 wear shows the same characteristics as for M_1, but at very early stages of wear the anterior part is separated from the posterior part. AC, anterior complex; AL, anterior loop; C, cement; E, enamel; EI, enamel interruption; FAC, future anterior complex; FPC, future posterior complex; H, hole; D, dentine; LS, linea sinuosa; LRA5, lingual reentrant angle 5; OI, ondatrian islet; PC, posterior complex; PL, posterior loop; T, triangle.

cases quite asymmetrical, with a marked lingual lobe in M^3 with three triangles. But some rare types (6% of the M^3 studied) show only two median triangles (T1 and T2) and one enormous anterior "inky cap mushroom-shaped" loop that encloses the third triangle (Figure 15.5).

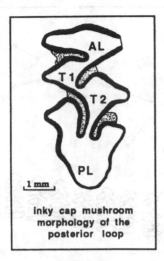

Figure 15.5. Occlusal view of rare type of M^3; 6% of the M^3 studied showed one enormous "inky cap mushroom-shaped" loop that enclosed the third triangle. AL, anterior loop; PL, posterior loop; T, triangle.

Quantitative description of the binary images: simple image parameters

From the immature to the mature stages the M_1 occlusal surface (OA) enlarges from 130% to 160% (Figure 15.6). The enamel area (EA) enlarges even more (160% to 170%) in mature animals than in immature animals, and the ratio EA/DA (enamel area/dentine area) increases as the molar wears, in spite of the appearance of numerous dentine track interruptions of enamel near the crown base that lower the values at the end of the curve. The tooth of an immature individual is bound to wear more quickly than that of a mature or old individual, not only because the amount of enamel is not as large but also because the occlusal surface of the tooth is smaller. Differential wear of the occlusal surface during the growing period must be taken into account in further studies on tooth ontogeny.

Quantitative description of the binary images: structural-density parameter

The structural-density parameter calculated on serial sections of one M_1 of specimen LLN-132 (Figure 15.7) shows a distinct decrease in com-

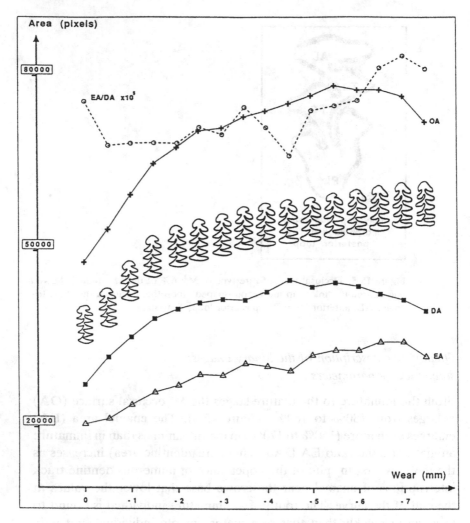

Figure 15.6. Quantitative description of the molar abrasion sequence by simple image parameters. EA, DA, and OA show an increase from immature to mature wear stages, and then a decrease from mature to old wear stages. The ratio EA/DA increases with tooth wear, too, but its increase is irregular. EA, enamel area; DA, dentine area; OA, occlusal area.

plexity of the outline in the most advanced stages of wear. Thus, as in *Mimomys* (Viriot et al., 1990), it would appear that the complex occlusal surfaces of immature animals become simpler with wear. The structural-density parameter of just the anterior loop and triangles 6 and 7 also shows simplification through the abrasion sequence (Figure 15.8). As

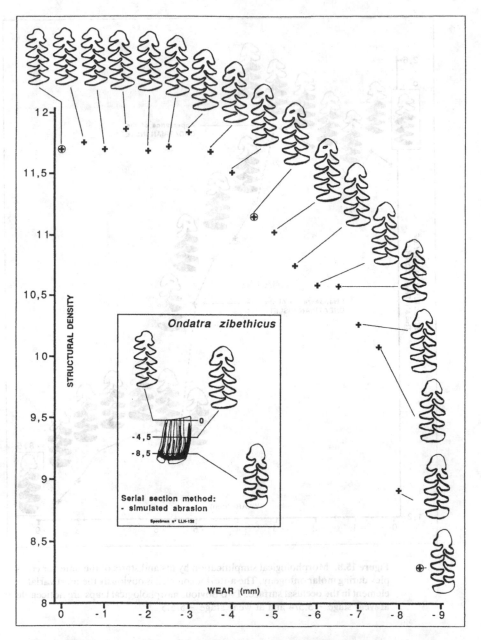

Figure 15.7. Morphological simplification of the occlusal surface with tooth wear. The structural density (SD) compares the degree of distortion of the occlusal surface to the shape of a circle. The structural density of the occlusal surface decreases in the most advanced stages of wear.

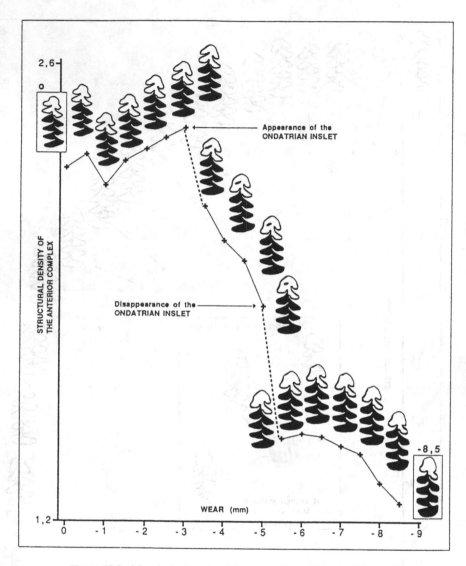

Figure 15.8. Morphological simplification by fits and starts of the anterior complex during molar ontogeny. The anterior complex is obviously the most variable element in the occlusal surface. Two obvious morphological leaps are noticeable at wear stage −3.5/4 and at wear stage −5/5.5.

has been previously observed for the occlusal surface, this simplification is not gradual. Two obvious morphological steps are noticeable: one at wear stage −3.5/4 and another at wear stage −5/5.5. The first step is due to the appearance of the ondatrian islet, and the second step reflects the disappearance of this islet.

Quantitative description of the binary images: the R parameter

The R parameter varies significantly during an abrasion sequence (Figure 15.9). For the R1, R2, R3, and R4 parameters, the reentrants change from a narrow to a wide stage as wear progresses. The R5 and R6 parameters also show increases in the widening of the reentrant angles, but from an already widened stage. R1–6 = $\Sigma Q / \Sigma dT$ demonstrates the general trend toward a widening of the reentrant angles with wear. This widening of the reentrant angles on the molar base may be due to the closeness of roots. But among other species (e.g., *Mimomys*) this ratio may sometimes be stable. Thus it would be interesting to make a detailed study on the R ratio in the ancient species *Pliopotamys minor, P. meadensis, Ondatra idahoensis, O. annectens,* and *O. nebrascensis.*

Discussion

The qualitative examination of the population demonstrates great occlusal and lateral variations with wear in M_1 and M^3 of *O. zibethicus*. These variations show up in (1) the appearance and development of dentine tracks related to root formation and causing enamel interruptions on the lateral faces during dental ontogeny and (2) the appearance of the ondatrian islet on the occlusal surface by isolation of lingual reentrants 5, leading to simplification of the anterior complex and the possible disappearance of triangles 6 and 7, which add their areas to the anterior loop.

The quantitative study of abrasion sequences using image analysis backs up and refines previous qualitative conclusions and brings out other results that cannot be detected by qualitative study alone. These results show that with wear, we see a sizable increase in the chewing surface between immature and old stages (OA), an increase in the enamel area (EA), a decrease in occlusal complexity (SD), steps in the morphological evolution of the anterior complex (SD), and widening of buccal and lingual reentrant angles from the central axis of the molar (R).

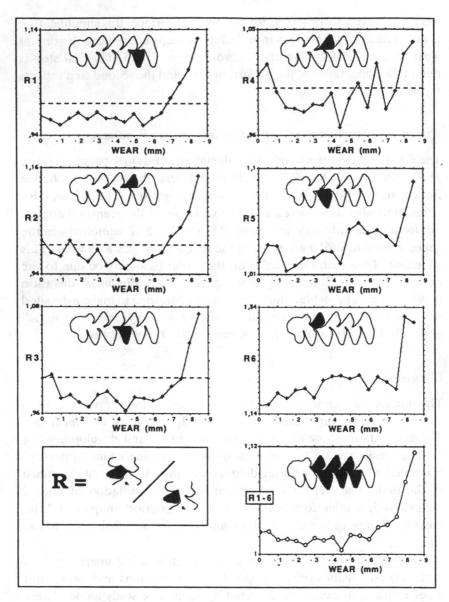

Figure 15.9. Increase of the *R* parameter in proportion with tooth wear. The *R* parameter demonstrates the general trend toward a widening of the reentrant angles with tooth abrasion. The threshold value 1 is figured in dotted line on the graphic.

Image analysis is a high-performance tool for bringing out morphological modifications in planar surfaces, and it makes possible accurate description of morphological transformations in the occlusal surfaces of M_1 and M^3 during *O. zibethicus* dental ontogeny. This preliminary investigation of changing morphology because of wear, conducted on modern specimens, suggests that quantitatively descriptive study of the entire lineage from *Pliopotamys* to *Ondatra* would provide a better explanation for why morphology in this lineage changed as it did.

Acknowledgments

We are indebted to A. D. Barnosky, R. A. Martin, R. Mooi, A. Petitjean, H. A. Semken, and C. Stivin for helpful comments and suggestions, to A. Festeau and A. Godon for technical help, and to V. Parisot for translation. This research was supported by the French C.N.R.S.: programmes "Modalités évolutives: de l'URA 157" and A.S.P. "Evolution," as well as "Approaches interdisciplinaires et développements méthodologiques" de la Direction de la Recherche et des Etudes Doctorales, grant no. EU 1969.

References

Barnosky, A. D. (1990). Evolution of dental traits since latest Pleistocene in meadow voles (*Microtus pennsylvanicus*) from Virginia. *Paleobiology*, 16:370–83.
Chaline, J. (1972). *Les rongeurs du Pléistocène moyen et supérieur de France*. Cahiers de Paléontologie. Paris: CNRS.
  (1987). Arvicolid date (Arvicolidae, Rodentia) and evolutionary concepts. *Evolutionary Biology*, 21:237–310.
Chermant, J.-L., and M. Coster (1984). Application of quantitative metallography to sintering investigations. Part 3: Instrumentation, technics and representativity of measurements in quantitative image analysis. *Prakt. Met.*, 21:472–84.
Chermant, J.-L., M. Coster, J.-P. Pernot, and J.-L. Dupain (1981). Morphological analysis of sintering. *Journal of Microscopy*, 121:89–98.
Coster, M., and J.-L. Chermant (1985). *Précis d'analyse d'images*. Paris: CNRS.
Galbreath, E. C. (1954). Growth and development of teeth in the muskrat. *Transactions of the Kansas Academy of Science*, 57:238–41.
Golay, M. J. (1969). Hexagonal parallel pattern transformations. *IEEE Transactions on Computers*, c18:733–40.
Gonzalez, R. C., R. E. Woods, and W. T. Swain (1986). Digital image processing: an introduction. *Digital Design*, pp.15–20.

Haas, A., G. Matheron, and J. Serra (1967a). Morphologie mathématique et granulométries en place. Part 1. *Annales des Mines*, 11:735–53.

(1967b). Morphologie mathématique et granulométries en place. Part 2. *Annales des Mines*, 12:767–82.

Hibbard, C. W., and R. J. Zakrzewski (1967). Phyletic trends in the late Cenozoic microtine *Ophiomys* gen. nov., from Idaho. *Contributions, Museum of Paleontology, University of Michigan*, 21:1–36.

Laÿ, B. (1984). Description des programmes du logiciel morpholog. In *Rapport du Centre de Géostatistique et de Morphologie Mathématique*. Fontainebleau.

Le Boulengé, E. (1977). Two ageing methods for muskrats: live or dead animals. *Acta Theriologica*, 22:509–20.

Martin, L. D. (1979). The biostratigraphy of arvicoline rodents in North America. *Transactions of the Nebraska Academy of Science*, 7:91–100.

(1984). Phyletic trends and evolutionary rates. In H. Genoways and M. R. Dawson (eds.), *Contributions in Quaternary Vertebrate Paleontology: A Volume in Memorial to John E. Guilday*. Carnegie Museum of Natural History, Special Publication 8.

Nelson, R. S. and H. A. Semken (1970). Paleoecological and stratigraphic significance of the muskrat in Pleistocene deposits. *Geological Society of America Bulletin*, 81:3733–8.

Rabeder, G. (1981). Die Arvicoliden (Rodentia, Mammalia) aus dem Pliozän und dem älteren Pleistozän von Niederösterreich. *Beitrage Paläontologie Osterreich*, 8:1–343.

Schaaf, A., and M. L. Vansteelant (1988). La complexité morphologique linéaire en géologie: comment la caractériser, comment la quantifier? *Bulletin of Sciences Géologiques, Strasbourg*, 41:125–33.

Schmidt-Kittler, N. (1984). Pattern analysis of occlusal surfaces in hyposodont herbivores and its bearing on morphofunctional studies. *Proceedings of the Koninklijke Nederlandse Akademie van Wetenshapen, B*, 87:453–80.

(1986). Evaluation of occlusal patterns of hyposodont rodent dentitions by shape parameters. *Neues Jarhbuch, Paläontologische Abhandlungen*, 173:75–98.

Schmidt-Kittler, N., and M. Vianey-Liaud (1987). Morphometric analysis and evolution of the dental pattern of the genus *Issiodoromys* (Theriodomyidae, Rodentia) of the European Oligocene as a key to its evolution. *Proceedings of the Koninklijke Nederlandse Akademie van Wetenshapen, B*, 90:281–306.

Schultz, C. B., L. G. Tanner, and L. D. Martin (1972). Phyletic trends in certain lineages of Quaternary mammals. *Bulletin of the University of Nebraska State Museum*, 9:183–95.

Semken, H. A. (1966). Stratigraphy and paleontology of the McPherson *Equus* Beds (Sandahl local fauna), McPherson County, Kansas. *Contributions, Museum of Paleontology, University of Michigan*, 20:121–78.

Serra, J. (1972). Morphologie mathématique. In *Traité d'informatique géologique* (pp. 194–238), Paris: Masson et Cie.

(1975). Présentation de la morphologie mathématique. Techniques probabilistes dans l'industrie. *Annales des Mines*, pp. 111–21.

Vansteelant, M.-L. (1989). La vacuité des sédiments biogènes carbonatés: morphologie, granulométrie et relation avec les réflecteurs sismiques. Application à deux forages océaniques (DSDP 586-ODP 709). Thèse de Doctorat de l'Université de Bretagne occidentale. Brest.

Viriot, L. (1989). La lignée évolutive *Mimomys davakosi-Mimomys ostramosensis* (Arvicolidae, Rodentia). Quantification des modifications morphologiques de la surface occlasale à l'aide del'analyse d'images. Mémoire de DEA, Université de Bourgogne, Dijon. (unpublished).

Viriot, L., J. Chaline, and A. Schaaf (1990). Quantification du gradualisme phylétique de *Mimomys occitanus* à *Mimomys ostramosensis* (Arvicolidae, Rodentia) à l'aide de l'analyse d'images. *Comptes Rendus Académie des Sciences, Paris, ser. II,* 310:1755–60.

Zakrzewski, R. J. (1969). The rodents from the Hagerman local fauna, upper Pliocene of Idaho. *Contributions, Museum of Paleontology, University of Michigan,* 20:1–36.

16

Morphological change in woodrat (Rodentia: Cricetidae) molars

RICHARD J. ZAKRZEWSKI

Woodrat (Rodentia: Cricetidae) molars have a number of features that have exhibited morphological change over geological time. These changes have included modification of the occlusal pattern, with a corresponding change in direction of mastication, increasing depth of the reentrant folds, development of dentine tracts, and increasing numbers of anatomical roots. These features and their variations over time can be useful in differentiating between taxa and may have some biostratigraphic and phylogenetic significance as well.

Woodrats have been reported from deposits of Hemphillian age in Texas (Dalquest, 1983) and Kansas (Hibbard, 1967); however, those samples were so small ($N = 2$ in each case) that there was little basis for comparison with younger specimens. Therefore, this study is based on characters and changes observed among Blancan and younger woodrats. Likewise, this study should be considered preliminary, as many more fossil and extant samples need to be examined in detail. However, I hope the information provided herein will be of benefit by delimiting characters and patterns that can be considered by others in subsequent studies.

Appendix A contains a list of fossil woodrats. Included in this list is the name of the fauna from which the fossil was obtained, the location and age of the fauna, the basis for the age assignment, and a primary reference. I have attempted to list all records of extinct species. Representative occurrences of extant species are taken from Harris (1985) and Kurtén and Anderson (1980). Appendix B contains a list of characters and their states discussed herein, when known. Original data were acquired as part of a long-term study of Blancan rodents. Therefore, most of the discussion is based on taxa of that age. I also examined some Irvingtonian samples. I used extant samples of *Neotoma alleni* Merriam and *N. cinerea* (Ord) for comparison. Blancan woodrats from Arizona have been studied recently by Tomida (1987) and Czaplewski (1990). Data for Rancholabrean specimens are primarily from Harris (1984a,b). The stratigraphic placement of the samples primarily follows Lundelius et al. (1987). If the fauna is not listed in the latter publication, reference can be made to the original report.

Occlusal pattern

The occlusal pattern in woodrats is relatively simple, generally consisting of three confluent or offset (= staggered = paratriangular) lophs (lophids). The patterns appear to be divisible into three general types – one represented by Blancan taxa (Figures 16.1C and 16.1D) as the primitive condition, and the other two by the extant *N. alleni* (Figures 16.1E and 16.1F) and *N. cinerea* (Figures 16.1A and 16.1B) as derived conditions. However, the characters are distributed in a mosaic fashion, with each type having some characters in common with each of the other types. Likewise, some characters are variable within populations, and some characters, especially in the occlusal pattern, may be lost or added with wear. A discussion of the variation for each molar in the various groups follows.

The M_1 in woodrats consists of a simple open or confluent posterolophid, a mesolophid that can be confluent or staggered, and an anterolophid that can be confluent or staggered. The M_1 in Blancan taxa tends to have a staggered mesolophid and a confluent anterolophid (Figure 16.1D), whereas the *alleni* (Figure 16.1F) and *cinerea* M_1s (Figure 16.1C) tend to have a confluent mesolophid and staggered anterolophid.

Differences in mesolophids result from the fact that in Blancan taxa the apex of the posterobuccal fold is significantly anterior to the apex of the posterolingual fold; therefore, the buccal end of the mesolophid is well anterior to the lingual. This arrangement imparts a staggered condition to the mesolophid (Figure 16.1D). In *cinerea* these same apices are directly opposite each other (Figure 16.1B), and in *alleni* they are nearly opposite (Figure 16.1F). These latter arrangements impart a confluent condition to the mesolophid. These differences can be quantified by measuring the distance along the midline of the tooth from the approximate midpoint of the anterior face to the midpoint of the apex of each posterior fold (Figure 16.2A–C).

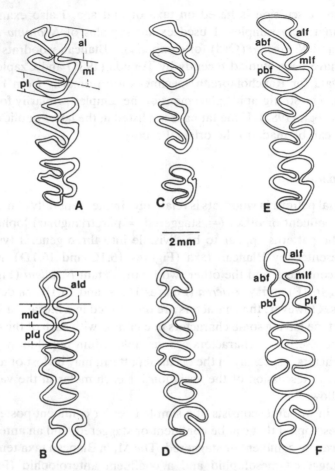

Figure 16.1. Occlusal views of *Neotoma* molars. (A, B) *N. cinerea* (MHP 7037). (A) RM$^{1-3}$ (al, anteroloph; ml, mesoloph; pl, posteroloph). (B) LM$_{1-3}$ (ald, anterolophid; mld, mesolophid; pld, posterolophid). (C, D) *N. leucopetrica* (Blancan, composite). (C) RM^1 (UM 88953), RM^2 (UM 88954), RM^3 (UM 88955). (D) LM$_1$ (UM 88950), LM$_2$ (UM 88951), RM$_3$ (UM 89952, reversed for comparison). (E, F) *N. alleni* (KU 87656). (E) RM$^{1-3}$. (F) LM$_{1-3}$ (abf, anterobuccal fold; alf, anterolingual fold; mlf, mesolingual fold; pbf, poster-obuccal fold).

Differences in anterolophids result from the fact that in Blancan taxa an anterolingual fold is poorly developed or absent (Figure 16.1D); this same fold tends to be better developed in the extant taxa (Figures 16.1B and 16.F). The percentages of occurrence and the mean lengths of the anterolingual folds are listed in Appendix B. The anterolingual fold may

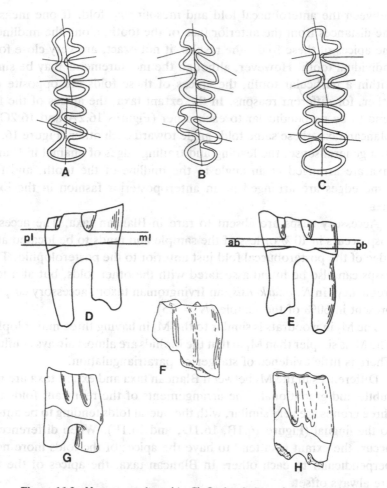

Figure 16.2. *Neotoma* molars. (A–C) Occlusal views showing relationships of fold and lophid development. Center reference axis is drawn from midpoint of anterior edge of tooth to midpoint of posterior edge. (D–H) Lateral views showing relationship of base of fold (pl, posterolingual; ml, mesolingual; ab, anterobuccal; pb, posterobuccal) to base of crown and dentine tract development. Part H modified from Harris (1984).

be lost with wear, and this loss occurs earlier in Blancan taxa. In fact, all folds can be lost if enough wear occurs. The presence of any fold on the occlusal surface is a function of how far that particular fold extends down the side of the crown and how much wear the tooth has undergone. Fold development will be discussed in more detail later.

One variable on the M_1 that is difficult to express is the relationship between the anterobuccal fold and mesolingual fold. If one measures the distances from the anterior face of the tooth, along the midline, to the apices of these folds, the results, if not exact, are very close for an individual tooth. However, although the measurements may be similar within a particular tooth, the apices of these folds are opposite each other, for different reasons. In the extant taxa, the apices of the folds tend to be perpendicular to each other (Figures 16.2A and 16.2C); in Blancan taxa, these same folds angle toward each other (Figure 16.2B). In a general sense, the leading and trailing edges of enamel in Blancan taxa are arranged at an angle to the midline of the tooth, and these same edges are arranged in an anteroposterior fashion in the extant taxa.

Accessory cusps are absent to rare in Blancan taxa. The accessory cusp occurs in 10% or less of the sample and tends to be located at the edge of the posterobuccal fold just anterior to the posterolophid. These cusps can also be found associated with the other folds, but at a lesser frequency. In *N. ozarkensis,* an Irvingtonian taxon, accessory cusps are present in 60% of the sample ($N = 25$).

The M_2 in woodrats is similar to the M_1 in having three major lophids. The M_2 is simpler than M_1 in that the lophids are almost always confluent. There is little evidence of stagger or paratriangulation.

Differences in the M_2 between Blancan taxa and extant taxa are more subtle and gradational. The arrangements of the reentrant folds in all three groups are very similar, with the buccal folds tending to be anterior to the lingual (Figure 16.1B, 16.1D, and 16.1F). When differences do occur, the extant taxa tend to have the apices of the folds more nearly perpendicular to each other. In Blancan taxa, the apices of the folds are always offset.

Another subtle variation in M_2 can be seen in the arrangement of the mesolophid. In *alleni* the apex of the buccal edge of the mesolophid tends to be oriented posteriorly (Figure 16.1F), whereas the same apex in *cinerea* and Blancan taxa is oriented anteriorly (Figure 16.1B and 16.1D). These relationships cause a line that bisects the mesolophid in *alleni* to be nearly perpendicular to a line that bisects the molar in an anteroposterior direction (Figure 16.2C). In *cinerea* and Blancan taxa, a line bisecting the lophids will be at some angle to the anteroposterior bisector (Figures 16.2A and 16.2B). The angle tends to be a little larger in *cinerea* than in Blancan taxa, but not significantly so. These differences are suggestive of a change in masticatory direction and can also be seen in the M_1.

Another variant seen in M_2 in many Blancan taxa is the development of an S-shaped pattern by the time the crown is worn to about 1.00 mm in height. This S pattern is a function of the distances that the bases of the anterobuccal and posterolingual folds are above the base of the crown. The farther the bases of these particular folds are from the base of the crown, the sooner the S pattern develops. Irvingtonian and extant taxa tend to have folds that extend farther down the side of the crown; therefore, these taxa do not generally exhibit an S pattern on M_2. Some Irvingtonian taxa have a shallow anterobuccal fold so that the anterolophid and mesolophid form a C that is attached to the oval posterolophid.

Accessory cusps are slightly more common in the M_2 in Blancan taxa than in M_1, but still occur in less than 25% of the sample. In *ozarkensis*, 86% of the teeth ($N = 15$) have at least one accessory cusp.

The occlusal pattern of M_3 occurs in two morphs. All Blancan taxa, some Irvingtonian, one Rancholabrean, and the extant *Neotoma* (*Hodomys*) *alleni* have an S pattern on M_3 (Figure 16.1D–F), whereas some Irvingtonian and the remaining Rancholabrean and extant taxa have a bilophate pattern (Figure 16.1B). Both of these patterns result from either early loss with wear or absence of anterobuccal and posterolingual folds. The early loss or absence of these folds suggests that reduction of M_3 is occurring. In the S pattern, the apex of the mesolingual fold is directed well anterior of the apex of the posterobuccal fold (Figure 16.1D–F). In the bilophate pattern, the apices of these folds are opposite each other and tend to be perpendicular to an anteroposterior bisector (Figure 16.1D). I have not determined how, or if, the bilophate pattern is derivable from the S pattern.

Accessory cusps are very rare on the M_3. I have seen only 4 teeth out of 125 with such a cusp: one in a Blancan taxon, one each in two different Irvingtonian taxa, and one in an extant taxon.

The M^1 in woodrats consists of a confluent or staggered anteroloph, a confluent or staggered mesoloph, and a simple posteroloph. In Blancan taxa, the anteroloph generally is confluent, although a slightly to poorly developed anterolingual fold may be present (Figure 16.1C). The fold tends to be lost earlier in Blancan taxa. In the extant taxa examined, a well-developed anterolingual fold is present that divides the anteroloph into two subequal parts, the anterior being somewhat lophate, and the posterior paratriangular (Figure 16.1A and 16.1E). Some extant species (*stephansi* and *lepida*) have shallow or no anterolingual folds (Hoffmeister and de la Torre, 1960). The percentage of occurrence and mean length of the anterolingual fold are given for some taxa in Appendix B.

In Blancan taxa and *cinerea*, the anterobuccal fold and mesolingual fold meet at the approximate midline of the tooth, so that the anteroloph appears to be separated from the mesoloph (Figure 16.1A and 16.1C). In *alleni,* the mesolingual fold is posterior to the anterobuccal and extends nearly to the anterior face of the posterobuccal fold (Figure 16.1E). This relationship imparts a triangular appearance to the mesoloph, which with the triangle from the anteroloph imparts a pseudoarvicoline pattern to the M^1 of *alleni.* The posterolophid tends to be simple. Accessory cusps are very rare to absent.

The M^2 is very similar in pattern to M^1, the major difference being that an anterolingual fold is seldom if ever developed, so that the anteroloph of M^2 generally is confluent. Likewise, the mesolingual and anterobuccal folds are arranged so that the anteroloph in Blancan taxa and *cinerea* is set off from the mesoloph (Figures 16.1A and 16.1C), and the mesoloph in *alleni* is divided into an anterior semiloph or pseudotriangle and a posterior triangle (Figure 16.1E).

The M^3 is the only molar that appears to have similar patterns in all three groups, consisting of a confluent anteroloph and mesoloph and a simple posteroloph. The mesolingual and anterobuccal folds are arranged so that the anteroloph is separated from the mesoloph. One exception among extant taxa is the 3-pattern seen in *lepida* (Hoffmeister and de la Torre, 1960), wherein a line through the apex of the mesolingual fold will bisect the mesoloph. Another variation that occurs on an infrequent basis is the development of a posterolingual fold (Figure 16.1A). This fold does not appear restricted to taxon or time, but it does occur more frequently in some samples. For example, Hibbard (1941), in part, based and named the species *quadriplicata* from the Rexroad local fauna on the general occurrence of the posterolingual fold on M^3 in the sample.

Depth of folds

The folds that divide the woodrat tooth into lophs (lophids) extend toward the base of the crown at varying distances (Figure 16.2D–H). In general, the bases of the folds in extant taxa are closer to the base of the crown than they are in extinct taxa (Zakrzewski, 1991). As mentioned earlier, the position of the folds on the side of the crown controls the occlusal pattern. When and if an S pattern will develop, whether an anteroloph (anterolophid) will appear confluent or staggered, and when or if an M^3 will be quadriplicate are functions of fold development. Unfortunately, there seems to be no specific regularity among taxa. For example, Harris (1984a) has shown that the depth of the anterolingual fold on M_1 differs among extant taxa. Likewise, I (Zakrzewski, 1991) have shown that the folds on M_1 of *leucopetrica*, a late Blancan taxa, are significantly longer than those of *quadriplicata*, an early Blancan taxa, but do not differ significantly from the same folds in *alleni;* however, the folds on M_2 in *leucopetrica* are significantly shorter than those of *alleni*, but show no significant difference from those of *quadriplicata*.

Current evidence suggests that an increase in the depth of folds might be a heterochronic phenomenon. Immature woodrat specimens show the folds extending very near or to the base of the crown (Figure 16.2F). As the tooth continues to grow, the bases of the folds form, and then anatomical roots begin to develop. The longer the development of fold bases is delayed before the anatomical roots start to form, the closer to the base of the crown the folds will be. A similar phenomenon has occurred in the evolution of the arvicolines and other rodents with evergrowing cheek teeth. The base and crown in adult arvicolines with evergrowing teeth look like the base and crown in juvenile arvicolines that eventually develop anatomical roots.

Dentine tracts

Dentine tracts are found at the apices of lophs or triangles on the side of the crown in a number of rodent families (Zakrzewski, 1981). As the name implies, these tracts appear to be composed primarily of dentine; however, various workers (Phillips and Oxberry, 1972; van der Meulen, 1973) have shown that in arvicolines the tracts are covered by a thin layer of cementum to which Sharpey's fibers attach. I suspect this same relationship holds for other groups in which tracts have developed. Tract development has been studied most extensively in arvicolines, where it has been used as both a phylogenetic tool and a biostratigraphic tool (Zakrzewski, 1984).

Although not developed to the extent that they are in arvicolines, dentine tracts have been reported in both extant and extinct woodrats by Harris (1984a) and Zakrzewski (1985). In woodrats, the tracts, when present, seem to be best developed on the anterobuccal face of the anterolophid of M_1 (Figure 16.2H). I have seen little (Figure 16.2G) to no (Figure 16.2D) evidence of tract development in this position in Blancan woodrats. Well-developed tracts are found in some Irvingtonian and extant taxa.

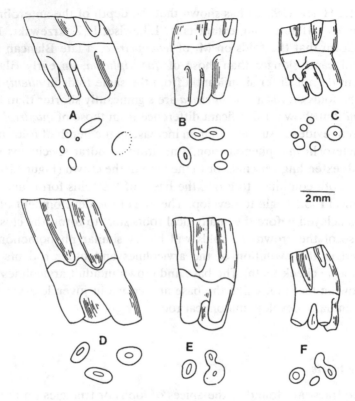

Figure 16.3. Lateral and basal views of *Neotoma* molars showing root development. (A–C) *N. alleni* (KU 103830). (A) M¹. (B) M². (C) M³. (D, E) *N. cinerea* (MHP 7576). (D) M¹. (E) M². (F) M³.

Anatomical roots

The number of anatomical roots in woodrats is another character that appears to be in a state of flux. All Blancan woodrats that I have examined have two major anatomical roots on each of their lower molars, and three on their uppers. A small accessory root may be found, generally equidistant from the three major ones, on the M¹ in Blancan taxa. *N. alleni* is diagnosed, in part, as having four roots on M¹ and M² (Genoways and Birney, 1974). One specimen in my sample (KU 103830) has six roots on M¹ (Figure 16.3A), five roots on M² (Figure 16.3B), and six roots on M³ (Figure 16.3C). Likewise, one *N. cinerea* (MHP 7576) has four roots on M³ (Figure 16.3F). The only M² and M³ of *amplidonta*, an Irvingtonian taxon, have four roots. However, 11 of 17

M²s of *ozarkensis* have the anterior pair of roots starting to fuse, and 4 of the 68 M³s from all the samples exhibit the same condition. Martin (1979) observed that roots are being added in the cotton rat, *Sigmodon*. He attributed the addition to increasing hypsodonty and a shift from browsing to grazing. Perhaps similar causes are at work in the woodrats. Root increase in woodrats is a character that needs further study. Specimens are limited because of breakage of the roots within fossil samples and the difficulty of seeing the roots in extant samples.

Significance of the changes

Biostratigraphic. Analysis of the changes described earlier suggests that they can be used as biostratigraphic indicators only in a very general sense at present. All Blancan woodrats are characterized by the S pattern on M_3. This pattern continues into the Irvingtonian, where it is found on woodrats from Cumberland Cave, Maryland, the Anza-Borrego Section, California, and possibly in the Java local fauna, South Dakota. The Cumberland Cave local fauna is considered to be late Irvingtonian in age, with an approximate date of 700,000 years BP (Lundelius et al., 1987). The other two sites are considered early Irvingtonian in age and may date to as much as 1.8 my BP (Lundelius et al., 1987). The latter two sites also contain woodrats with the bilophate M_3. All Rancholabrean specimens and extant species in the United States have the bilophate M_3. Rancholabrean and extant woodrats with the S pattern on M_3 appear to be restricted to Mexico. In summary, it appears that in the United States during Blancan time only woodrats with the S-shaped M_3 were present; both S-shaped and bilophate forms were present during the Irvingtonian; and only bilophate taxa are found in the Rancholabrean to the present. However, there is one taxon whose status could have a significant bearing on the foregoing: *N. minutus* described by Dalquest (1983) from the Coffee Ranch local fauna (Hemphillian of Texas). I have examined the two specimens, a RM_3 and a LM^3, on which the species is based. The M_3 is bilophate rather than S-shaped. In fact, the M_3 looks very much like an extant *cinerea* M_3. Therefore, the taxon cannot be retained in the subgenus *Paraneotoma*. Likewise, the M^3 was thought to be an M^2 by Dalquest; so the specimen is not minute either. Dalquest (letter of 19 September 1990) now thinks that the material should have been assigned to *Prosigmodon*, a genus described by Jacobs and Lindsay (1981) from the Hemphillian of Arizona and subsequently amplified by them (Lindsay and Jacobs, 1985) on specimens from the Hemphillian and Blancan of Mexico. Although this reassignment remains a possibility, the Coffee Ranch specimens appear to more closely resemble woodrat teeth than they do *Prosigmodon*. Another possibility is that the specimens are intrusive. If additional study demonstrates that the two specimens are those of a Hemphillian woodrat, these questions arise: Where were these animals during Blancan time? What is their relationship to the other taxa?

Another character that may be of general biostratigraphic significance is the relative development of folds along the side of the crown. Folds tend to be farther from the base of the crown in Blancan taxa and become progressively closer to the base in younger specimens. Unfortunately, as discussed earlier, there is a great deal of variability within and between taxa. Data for a particular tooth field are consistent for a taxon, but data for M_1 may not be consistent with data for M_2. Therefore, the distance between the base of the fold and the base of the crown must be used with caution, as the extensions of folds do not appear to be occurring at the same rate in all teeth of a given taxon of woodrat. Mean distances for this character are listed in Appendix B.

Dentine tract development on M_1 and the increase in root number on the upper molars may also be difficult to use as biostratigraphic tools because of variability and irregularity of these characters among the taxa. However, no Blancan taxon that I have observed has well-developed tracts or an excess number of roots. These changes seem to have begun in Irvingtonian time.

Phylogenetic. I think the characters discussed earlier may prove more useful in determining phylogenetic relationships. For example, I suspect that tract development and extension of the folds down the side of the tooth are characters that probably would not be subjected to reversed selection. Therefore, although taxa with tracts and deep folds may be derived from taxa that lack tracts or have shallow folds, the reverse would not be true. Likewise, if roots are being added to the molars of some taxa, these taxa will be derived from taxa with fewer roots; but again the reverse would not be true. For example, Harris (1984a,b) suggested that *N. findleyi*, an extinct taxon from a late Wisconsinan interstadial site in Dry Cave, might be descended from populations of *cinerea*. I would argue against this hypothesis, as the dentine tract on M_1 in *findleyi* is significantly shorter than that in *cinerea*.

The S pattern on the M_3 in Blancan woodrats may have led some workers to believe that they are more closely related to *N. alleni*, the only extant woodrat with the same pattern on M_3. In fact, the Blancan taxa were originally placed in the subgenus *Parahodomys,* a name still applied to the population from Cumberland Cave. I have yet to determine exactly how all these taxa are related to each other, and whether or not *Paraneotoma* is a junior synonym of *Parahodomys.* However, if dental pattern and root number in *N. alleni* are added to the characters discussed by Carleton (1980), a stronger case for raising *Hodomys* to generic status can be made.

There have been few attempts to suggest interrelationships at the specific level for extinct taxa. In addition to his suggestion that *findleyi* descended from *cinerea,* Harris (1984b) suggested that another extinct species, *pygmaea,* from the same site as *findleyi,* might be ancestral to *goldmani.* I (Zakrzewski, 1985) suggested that *amplidonta* could be ancestral to *cinerea.* Although these are viable working hypotheses, I think it is premature to try to establish interrelationships at the specific level, especially for Blancan taxa, until additional data become available.

Functional. Many of the changes seen between the Blancan taxa and extant taxa have been of adaptive significance. Perhaps the most important of these changes has been the extension of the bases of the folds toward the base of the crown. This change, whereby the folds extend farther down the sides of the crown, enables the animal to maintain its occlusal pattern for a longer period of time during its life, thereby allowing a longer span over which the same type and quality of food can be eaten. The increasing development of the anterolingual fold on the first molars increases the enamel shearing surface in a small area where the food first enters the mouth. There appears to be a shift in the direction of mastication from an angular arrangement, as shown by the position of the leading and trailing edges of enamel in Blancan taxa, to an anteroposterior direction, represented by *cinerea.* The adaptive advantage of this shift is not as intuitively obvious.

The addition of anatomical roots on the upper molars would strengthen the ability of the tooth to remain in the jaw during the mastication process. Likewise, the addition of dentine tracts to the anterobuccal face of the M_1 would provide more stability as additional Sharpey's fibers attach in this area. It is of interest that *cinerea* does not appear to be adding anatomical roots, but possesses dentine tracts, in contrast to *alleni,* wherein anatomical roots are being added, but not dentine tracts. Perhaps these are two different approaches to solving the same problem.

The morphological changes discussed earlier suggest a dietary shift among woodrats either to a harsher type of vegetation or to the incorporation of more grit from the soil, or both. Perhaps these morphological changes are responses to the environmental changes brought on by increasing aridity and decreasing equability during the Cenozoic.

Evolution. It is difficult to make any definite statements regarding evolutionary models or patterns. Many of the samples examined have been small, widely distributed in space and time, and not well dated in every case. Therefore, rates of change are difficult to determine, and though change is apparent, it cannot be stated unequivocally whether the pattern of change has been gradual or punctuated or both. Likewise, although some of the changes appear to be directional, they do not seem to be occurring at the same rate in all the teeth of a given taxon. This relationship results in a mosaic pattern, wherein a certain taxon might have an advanced M_1 but a primitive M_2 relative to another taxon, even though the same character is being used for comparison.

Summary

Over time, woodrats have changed in their occlusal patterns and directions of chewing and have extended the bases of the reentrant folds farther down the side of the crown, and some lines appear to be adding dentine tracts or anatomical roots. However, because of the distribution of the samples in space and time and the mosaic nature of the changes, it is difficult to make any statements regarding evolutionary rates and patterns or to use the taxa for biostatigraphic inferences, except in a general sense. The characters under consideration probably have more value in determining phylogenetic relationships and appear to have adaptive significance. These changes suggest that woodrats have shifted their diet to a harsher type of vegetation or have incorporated more grit from the soil, or both. This shift may have been brought on by increasing aridity and decreasing equability during the Cenozoic. Additional work will be necessary to test the propositions and suggestions discussed herein.

Acknowledgments

I thank the curators of the following institutions for access to their collections and for the loan of material: Carnegie Museum of Natural

History, Division of Vertebrate Fossils; Fort Hays State University, Museum of the High Plains (MHP); Natural History Museum of Los Angeles County, Division of Vertebrate Paleontology; South Dakota School of Mines and Technology, Museum of Geology; University of Arizona, Department of Geological Sciences; University of Iowa, Department of Geology; University of Kansas, Museum of Natural History (KU); University of Michigan, Museums of Paleontology and Zoology (UM). I thank John Rensberger for insightful comments on function, and the editors for their constructive comments regarding the manuscript.

References

Alvarez, T. (1966). Roedores fosiles del Pleistoceno de Tequesquinahua, Estado de Mexico, Mexico. *Acta Zoologica Mexicana*, 8:1–16.
Brown, B. (1908). The Conard Fissure, a Pleistocene bone deposit in northern Arkansas: with descriptions of two new genera and twenty new species of mammal. *Memoirs of the American Museum of Natural History*, 9:155–208.
Carleton, M. D. (1980). Phylogenetic relationships in Neotomine-Peromyscine rodents (Muroidea) and a reappraisal of the dichotomy within New World Cricetinae. *Miscellaneous Publications, Museum of Zoology, University of Michigan*, 157:1–146.
Czaplewski, N. J. (1990). The Verde local fauna: small vertebrate fossils from the Verde formation, Arizona. *San Bernardino County Museum Association Quarterly*, 37:1–39.
Dalquest, W. W. (1975). Vertebrate fossils from the Blanco local fauna of Texas. *Occasional Papers, Museum of Texas Technological University*, 30:1–52.
　(1978). Early Blancan mammals of the Beck Ranch local fauna of Texas. *Journal of Mammalogy*, 59:269–98.
　(1983). Mammals of the Coffee Ranch local fauna, Hemphillian of Texas. *Texas Memorial Museum, Pearce-Sellards Series*, 38:1–41.
Genoways, H. H., and E. C. Birney (1974). *Neotoma alleni. Mammalian Species*, 41:1–4.
Gidley, J. W. (1922). Preliminary report on fossil vertebrates of the San Pedro Valley, Arizona, with descriptions of new species of Rodentia and Lagomorpha. *U.S. Geological Survey, Professional Paper*, 131:119–30.
Gidley, J. W., and C. L. Gazin (1938). The Pleistocene vertebrate fauna from Cumberland Cave, Maryland. *United States National Museum Bulletin*, 171:1–99.
Gustafson, E. P. (1978). The vertebrate faunas of the Pliocene Ringold formation, south-central Washington. *University of Oregon, Museum of Natural History Bulletin*, 23:1–62.
Harris, A. H. (1984a). *Neotoma* in the late Pleistocene of New Mexico and Chihuahua. In H. H. Genoways and M. R. Dawson (eds.), *Contributions in Quaternary Vertebrate Paleontology: A Volume in Memorial to John E.*

Guilday (pp. 164–78). Carnegie Museum of Natural History, Special Publication 8.

(1984b). Two new species of late Pleistocene woodrats (Cricetidae: *Neotoma*) from New Mexico. *Journal of Mammalogy*, 65:560–6.

(1985). *Late Pleistocene Vertebrate Paleontology of the West*. Austin: University of Texas Press.

Harrison, J. A. (1978). Mammals of the Wolf Ranch local fauna, Pliocene of the San Pedro Valley, Arizona. *Occasional Papers of the Museum of Natural History, University of Kansas*, 73:1–18.

Hibbard, C. W. (1941). New mammals from the Rexroad fauna, upper Pliocene of Kansas. *American Midland Naturalist*, 26:337–68.

(1950). Mammals of the Rexroad formation from Fox Canyon Kansas. *Contributions, Museum of Paleontology, University of Michigan*, 8:113–92.

(1967). New rodents from the late Cenozoic of Kansas. *Papers of the Michigan Academy of Science, Arts, and Letters*, 52:115–31.

Hoffmeister, D. F., and L. de la Torre (1960). A revision of the wood rat *Neotoma stephansi*. *Journal of Mammalogy*, 41:476–91.

Jacobs, L. L., and E. H. Lindsay (1981). *Prosigmodon oroscoi*, a new sigmodont rodent from the late Tertiary of Mexico. *Journal of Paleontology*, 55: 425–30.

Kurtén, B., and E. Anderson (1980). *Pleistocene Mammals of North America*. New York: Columbia University Press.

Lindsay, E. H., and L. L. Jacobs (1985). Pliocene small mammals from Chihuahua, Mexico. *Universidad Nacional Autonoma de Mexico, Instituto de Geologia, Paleontologia Mexicana*, 51:1–53.

Lundelius, E. L., Jr., C. S. Churcher, T. Downs, C. R. Harington, E. H. Lindsay, G. E. Schultz, H. A. Semken, S. D. Webb, and R. J. Zakrzewski (1987). The North American Quaternary sequence. In M. O. Woodburne (ed.), *Cenozoic Mammals of North America* (pp. 211–35). Berkeley: University of California Press.

Martin, R. A. (1979). Fossil history of the rodent genus *Sigmodon*. *Evolutionary Monographs*, 2:1–36.

Meulen, A. J. van der (1973). Middle Pleistocene smaller mammals from Monte Peglia (Orvieto, Italy) with special reference to the phylogeny of *Microtus* (Arvicolidae, Rodentia). *Quaternaria*, 17:1–144.

Miller, W. E. (1980). The late Pliocene Las Tunas local fauna from southernmost Baja California, Mexico. *Journal of Paleontology*, 54:762–805.

Phillips, C. J., and B. Oxberry (1972). Comparative histology of molar dentitions of *Microtus* and *Clethrionomys*, with comments on dental evolution in microtine rodents. *Journal of Mammalogy*, 53:1–20.

Rogers, K. L., C. A. Repenning, R. M. Forester, E. E. Larson, S. A. Hall, G. R. Smith, E Anderson, and T. J. Brown (1985). Middle Pleistocene (late Irvingtonian: Nebraskan) climatic changes in south-central Colorado. *National Geographic Research*, 1:535–63.

Tomida, Y. (1987). *Small Mammal Fossils and Correlation of Continental Deposits, Safford and Duncan Basins, Arizona, USA*. Tokyo: National Science Museum.

Zakrzewski, R. J. (1969). The rodents from the Hagerman local fauna, upper Pliocene of Idaho. *Contributions, Museum of Paleontology, University of Michigan*, 23:1–36.

(1981). Kangaroo rats from the Borchers local fauna, Blancan, Meade County, Kansas. *Transactions of the Kansas Academy of Sciences*, 84:78–88.

(1984). New arvicolines (Mammalia: Rodentia) from the Blancan of Kansas and Nebraska. In H. H. Genoways and M. R. Dawson (eds.), *Contributions in Quaternary Vertebrate Paleontology: A Volume in Memorial to John E. Guilday* (pp. 200–17). Carnegie Museum of Natural History, Special Publication 8.

(1985). A new species of woodrat (Cricetidae) from the Pleistocene (Irvingtonian) of South Dakota. *Journal of Mammalogy*, 66:770–3.

(1991). New species of Blancan woodrat (Cricetidae) from north-central Kansas. *Journal of Mammalogy*, 72:104–9.

Appendix A. *Locations and ages of fossil* Neotoma

Taxon	Fauna[a]	State	Age[b]	Basis[c]	References
minutus	Coffee Ranch	TX	H	r	Dalquest (1983)
sawrockensis	Saw Rock	KS	H	b	Hibbard (1967)
vaughni	Verde	AZ	B	r/m	Czaplewski (1990)
quadriplicata	Fox Canyon	KS	B	b/m	Hibbard (1950)
quadriplicata	Rexroad	KS	B	b/m	Hibbard (1941), Zakrzewski (1991)
quadripilcata	Country Club	AZ	B	b/m	Tomida (1987)
quadriplicata	Beck Ranch	TX	B	b	Dalquest (1978)
cf. quadriplicata	Hagerman	ID	B	b/r	Zakrzewski (1969)
cf. quadriplicata	White Bluffs	WA	B	b	Gustafson (1978)
cf. quadriplicata	Sand Draw	NE	B	b	Personal observation
cf. quadriplicata	Blanco	TX	B	b	Dalquest (1975)
sp.	Wolf Ranch	AZ	B	b/r	Harrison (1978)
sp.	Las Tunas	MX	B	b	Miller (1980)
sp. a	Taunton Beach	WA	B	b	Personal observation
sp. b	LC/AS	CA	B	b/m	Personal observation
leucopetrica	White Rock	KS	B	b	Zakrzewski (1991)
fossilis	Benson	AZ	B	b/m	Gidley (1922)
fossilis	Duncan	AZ	B	b/m	Tomida (1987)
cf. fossilis	Beck Ranch	TX	B	b	Dalquest (1978), Tomida (1987)
taylori	Borchers	KS	B	b/m/r	Zakrzewski (1991)
taylori	111 Ranch	AZ	B	b/m	Tomida (1987)
sp. c	AS/VC	CA	B/I	b/m	Personal observation
amplidonta	Java	SD	I	b	Zakrzewski (1985)
sp. d	Java	SD	I	b	Zakrzewski (1985)
sp. e	Vallecito	CA	I	b/m	Personal observation
ozarkensis	Conard	AR	I	b	Brown (1908)
ozarkensis	Hansen Bluff	CO	I	r/m	Rogers et al. (1985)
spelaea	Cumberland	MD	I	b	Gidley & Gazin (1938)
floridana	Cumberland	MD	I	b	Gidley & Gazin (1938), Kurtén & Anderson (1980)
floridana	38 sites	V[d]	I/R	b/r	Kurtén & Anderson (1980), Harris (1985)
?stephensi	2 sites	AZ	R	r	Harris (1985)
micropus	10 sites	V[d]	R	b/r	Kurtén & Anderson (1980), Harris (1985)
albigula	20 sites	V[d]	R	b/r	Kurtén & Anderson (1980), Harris (1985)
lepida	12 sites	V[d]	R	b/r	Kurtén & Anderson (1980), Harris (1985)
mexicana	8 sites	V[d]	R	b/r	Kurtén & Anderson (1980), Harris (1985)
fuscipes	6 sites	CA	R	b	Kurtén & Anderson (1980), Harris (1985)
cinerea	>33 sites	V[d]	R	b/r	Kurtén & Anderson (1980), Harris (1985)
findleyi	Lost Valley	NM	R	r	Harris (1984b)
?goldmani	2 sites	NM	R	b/r	Harris (1985)
pygmaea	Lost Valley	NM	R	r	Harris (1984b)
sp.	21 sites	V[d]	R	b/r	Harris (1985)
magnodonta	Tequesquinahua	MX	R	b	Alvarez (1966)

[a]AS, Arroyo Seco; LC, Layer Cake; VC, Vallecito Creek.
[b]H, Hemphillian; B, Blancan, R, Rancholabrean; I, Irvingtonian.
[c]b, biostratigraphy; m, magnetistratigraphy; r, radiometric date.
[d]More than one state.

Appendix B. *Character states in* Neotoma

Taxon	N[a]	DT1[b]	ALFL[c]	%[d]	RFD1[e]	RFD2[e]	RFD3[e]	SM3[f]	ALFU[g]	%[h]	RTSU[i]
minutus	2							b			
vaughni	L	a						s	D	100	N
quadriplicata	30	a	0.20	67	0.93	1.02	0.70	s	0.48	100	A
sp. a	4	a	0.21	75	0.84	1.16		s	0.34	67	
sp. b	46	a	0.24	63	0.86	1.09	0.75	s	0.32	78	N
leucopetrica	8	a	0.06	37	0.65	1.05	0.69	s	0.37	100	N
fossilis	L							s			
taylori	10	a	0.18	70	0.83	1.07	0.69	s	0.27	100	N
sp. c	46	a	0.24	37	0.99	1.31	0.94	s	0.24	84	N
amplidonta	3	0.9	0.48	100	0.53	0.70	0.68	b	>		
sp. d	1	a	0.23	100	0.89						
sp. e	25	a	0.26	87	0.65	0.73	0.60	b	0.31	81	N
ozarkensis	25		0.37	100	0.59	0.69	0.57	b	0.27	100	F
spelaea	9		0.45	77	0.91	0.79		s			
floridana	L	0.02						b			
stephensi	L	1.17						b	S/A		
micropus	L	0.004						b			
albigula	L	0.01						b			
lepida	L	0.28						b	S/A		
mexicana	L	1.32						b	D		
cinerea	8	1.14	0.61					b	0.62	100	N
findleyi	L	0.68						b			
?goldmani	L	0.35						b			
pygmaea	L	0.54						b			
magnodonta	L							s			
alleni	18	a	0.44	100	0.69	0.58	0.49	s	0.58	100	>

[a]N, minimum number of individuals based on count of M_1: L, data taken from literature, see Appendix A or text for references.

[b]DT1, mean height of dentine tract on anterobuccal edge of anterolophid using method of Harris (1984a): a, absent.

[c]ALFL, mean length of anterolingual fold on M_1.

[d]%, percentage of M_1s with anterolingual fold.

[e]RFD, mean of means for all folds of the distance from the base of fold to base of crown; 1, 2, 3, = M_1, M_2, M_3, respectively.

[f]SM3, shape of M_3: s, S-shaped; b, bilophate.

[g]ALFU, mean length of anterolingual fold on M^1: S/A, slight to absent; D, deep.

[h]%, percentage of M^1s with anterolingual fold.

[i]RTSU, number of roots on upper molars: N, normal complement of 3; A, normal complement with high percentage of accessory roots; F, fusion of anterior pair occurs in high percentage; >, more than 3 roots on a regular basis.

Index

adaptive neutrality, 127
adaptive radiation, 270
adaptive zone, 170
Alces
  alces, 179
  gallicus, 179
  latifrons, 179
  antler beam, 179, 183–86
allometric scaling, 361
allometry, 186
allopatric isolate, 102
ameloblasts, 210
anagenesis, 26
anagenetic lineages, 26
antler, 183–85
Anza-Borrego, 401
Appalachian, 27
aquatic mammals, 268
Arborimus longicaudus, 316
artificial selection experiments, 105
Arvicola
  cantiana, 248
  sapidus, 268
  terrestris, 240, 268, 373
Arvicolidae, 373
astragalus, 283

Baker Bluff Cave, 29
barrier islands, 282–83
Bergmann's rule, 71, 294
Bering Strait, 194, 196
Beringia, 188, 196
biochronology, 136
biological species concept, 26
Bison, 320
  antiquus, 100
  bison, 102

Blancan, 393–98
body mass, 90, 244, 351–52, 355–56
body size, 185, 190, 281
BODYMASS program, 355
bootstrapping, 89
bottleneck, 231
Brownian speeds, 265
Bull Lake glaciation, 322

calcrete, 300
canids, 74
Canis
  aureus, 77
  dirus, 78
  latrans, 78
  lupus, 77, 78
carnassial, 73
Castor, 211
Cervalces, 188
Cervus elephus, 100–03
character displacement, 72, 126
Chionomys nivalis, 233
Chlamytherium, 135
chronocline, 167
chronomorph, 234
cladistics, 252
cladogenesis, 25, 233
Clethrionomys glareolus, 230, 238, 242
climate related changes, 32–33, 45–49, 55,
  283–87, 343
cline, 26
coefficient of variation, 112, 143
Coffee Ranch, 401
Coleman 2A, 348
Columbia Basin, 323
Comers Cave, 37
competition, 71, 266

411